Targeting of Drugs 6

Strategies for Stealth Therapeutic Systems

NATO ASI Series

Advanced Science Institutes Series

A series presenting the results of activities sponsored by the NATO Science Committee, which aims at the dissemination of advanced scientific and technological knowledge, with a view to strengthening links between scientific communities.

The series is published by an international board of publishers in conjunction with the NATO Scientific Affairs Division

A	**Life Sciences**	Plenum Publishing Corporation
B	**Physics**	New York and London
C	**Mathematical and Physical Sciences**	Kluwer Academic Publishers Dordrecht, Boston, and London
D	**Behavioral and Social Sciences**	
E	**Applied Sciences**	
F	**Computer and Systems Sciences**	Springer-Verlag
G	**Ecological Sciences**	Berlin, Heidelberg, New York, London,
H	**Cell Biology**	Paris, Tokyo, Hong Kong, and Barcelona
I	**Global Environmental Change**	

PARTNERSHIP SUB-SERIES

1. Disarmament Technologies	Kluwer Academic Publishers
2. Environment	Springer-Verlag
3. High Technology	Kluwer Academic Publishers
4. Science and Technology Policy	Kluwer Academic Publishers
5. Computer Networking	Kluwer Academic Publishers

The Partnership Sub-Series incorporates activities undertaken in collaboration with NATO's Cooperation Partners, the countries of the CIS and Central and Eastern Europe, in Priority Areas of concern to those countries.

Recent Volumes in this Series:

Volume 297 — Acute Respiratory Distress Syndrome: Cellular and Molecular Mechanisms and Clinical Management
edited by Sadis Matalon and Jacob Iasha Sznajder

Volume 298 — Angiogenesis: Models, Modulators, and Clinical Applications
edited by Michael E. Maragoudakis

Volume 299 — Development and Organization of the Retina: From Molecules to Function
edited by Leo M. Chalupa and Barbara L. Finlay

Volume 300 — Targeting of Drugs 6: Strategies for Stealth Therapeutic Systems
edited by Gregory Gregoriadis and Brenda McCormack

Series A: Life Sciences

Targeting of Drugs 6

Strategies for Stealth Therapeutic Systems

Edited by

Gregory Gregoriadis and
Brenda McCormack

School of Pharmacy
University of London
London, England

Plenum Press
New York and London
Published in cooperation with NATO Scientific Affairs Division

Proceedings of a NATO Advanced Study Institute on
Targeting of Drugs: Strategies for Stealth Therapeutic Systems,
held June 24 – July 5, 1997,
at Cape Sounion Beach, Greece

NATO-PCO-DATA BASE

The electronic index to the NATO ASI Series provides full bibliographical references (with keywords and/or abstracts) to about 50,000 contributions from international scientists published in all sections of the NATO ASI Series. Access to the NATO-PCO-DATA BASE is possible via a CD-ROM "NATO Science and Technology Disk" with user-friendly retrieval software in English, French, and German (©WTV GmbH and DATAWARE Technologies, Inc. 1989). The CD-ROM contains the AGARD Aerospace Database.

The CD-ROM can be ordered through any member of the Board of Publishers or through NATO-PCO, Overijse, Belgium.

Library of Congress Cataloging-in-Publication Data

Targeting of drugs 6 : strategies for stealth therapeutic systems / edited by Gregory Gregoriadis and Brenda McCormack.
p. cm. -- (NATO ASI series. Series A, Life sciences ; v. 300)
"Published in cooperation with NATO Scientific AffairsDivision."
"Proceedings of a NATO Advanced Study Institute on Targeting of Drugs: Strategies for Stealth Therapeutic Systems, held June 24-July 5, 1997, at Cape Sounion Beach, Greece"--T.p. verso.
Includes bibliographical references and index.
ISBN 0-306-45937-X
1. Liposomes--Congresses. 2. Drug targeting--Congresses. 3. Drug carriers--Congresses. I. Gregoriadis, Gregory. II. North Atlantic Treaty Organization. Scientific Affairs Division. III. NATO Advanced Study Institute on Targeting of Drugs: Strategies for Stealth Therapeutic Systems (1997 : Akra Sounion, Greece) IV. Series.
[DNLM: 1. Drug Carriers congresses. 2. Liposomes--metabolism congresses. 3. Particle Size. QV 785 T1855 1998]
RS201.L55T37 1998
615'.7--dc21
DNLM/DLC
for Library of Congress 98-29831
CIP

ISBN 0-306-45937-X

A Division of Plenum Publishing Corporation
233 Spring Street, New York, N.Y. 10013

http://www.plenum.com

10 9 8 7 6 5 4 3 2 1

Printed in the United States of America

PREFACE

The NATO Advanced Studies Institute series "Targeting of Drugs" was originated in 1981. It is now a major international forum, held every two years in Cape Sounion, Greece, in which the present and the future of this important area of research in drug delivery is discussed in great depth. Previous ASIs of the series dealt with drug carriers of natural and synthetic origin, their interactions with the biological milieu, ways by which milieu interference is circumvented, approaches to carrier design or modification that contribute to optimal carrier function, and, more recently, with strategies for gene and oligonucleotide delivery in therapy The present book contains the proceedings of the 9th NATO ASI, "Targeting of Drugs: Strategies for Stealth Therapeutic Systems", held in Cape Sounion during 24 June - 5 July 1997. As the title implies, the book deals with a variety of approaches to evade the reticuloendothelial system and thus extend the circulation time of a variety of delivery systems including polymers, biopolymers, liposomes, and other nanoparticles or microparticles.

We express our appreciation to Mrs. Concha Perring for her assistance with the organization of the ASI. The ASI was held under the sponsorship of NATO Scientific Affairs Division and supported by Sequus Pharmaceuticals (Menlo Park, CA, USA), The Liposome Company (Princeton, NJ, USA), NOVO Nordisk (Bagsvaard, Denmark), NeXstar Pharmaceuticals (Boulder, CO, USA), Gene Medicine (Houston, TX, USA), Pfizer Ltd (Sandwich, Kent, UK), Merck (Rahway, NJ, USA), and Biovation Ltd (Aberdeen, UK). The term Stealth™ liposomes used throughout this book is a trade mark of Sequus Pharmaceuticals.

Gregory Gregoriadis
Brenda McCormack

CONTENTS

INTERACTIONS BETWEEN BLOOD COMPONENTS AND ARTIFICIAL SURFACES

Karin D. Caldwell

Center for Biopolymers at Interfaces, Department of Bioengineering
University of Utah, Salt Lake City, UT 84112, USA

INTRODUCTION

Strategies for designing "stealth" therapeutic systems largely coincide with those governing the search for materials suitable to serve as artificial organs and prosthetic devices in contact with blood and other living tissues. In both cases, one strives to design surfaces that will avoid the triggering of inflammatory responses to the foreign material. Many times, there is also a need to link a homing device or bioactive ligand to the surface, in which case the mode of linking must leave the often marginally stable ligand in an active state. Beyond these common concerns, the biomaterials community also focuses on the specific tasks of avoiding a trigger of the clotting cascade and suppressing the potential for bacterial colonization, both issues related to the large size of the typical implant compared to that of a liposome or other circulating drug release vehicle. The following presentation will summarize ongoing efforts to devise some general techniques for preparing surfaces suitable to serve as biomaterials.

PROTEINS AT INTERFACES

It is generally recognized that the fate of a blood-contacting material is regulated by the composition and conformational status of the protein film that typically begins to form within seconds of such contact. It has been pointed out by Vroman and others (Vroman and Adams, 1969; Vroman, 1982; Adams et al., 1984; Andrade and Hlady, 1987; Horbett, 1982; Slack and Horbett, 1995; Wojciechowski and Brash, 1993) that the hundreds of proteins in plasma experience widely differing collision frequencies with respect to an exposed surface due to differences in concentration as well as in diffusivity, i.e. size. Whether or not a collision leads to adsorption depends both on the chemical nature of the surface as well as on the chemistry and the structural stability of the striking protein. Once adsorbed, a protein is more or less readily displaced by others depending on its relative ability to form stable associations with the surface. Thus, on highly hydrophilic glass surfaces one observes the so called Vroman effect, i.e. a time dependent composition of the surface layer, which after short exposure times consists primarily of the most abundant plasma protein, albumin (HSA). As time progresses, the albumin layer is replaced by a sequence of less frequent but more tenaciously binding proteins including immunoglobulin G (IgG), fibrinogen (FIB), fibronectin (FN), high molecular weight kininogen (HWK), among others. This sequence is not as clearly observed on hydrophobic surfaces, as demonstrated by several model studies with mixtures of small numbers of pure plasma proteins (Gölander et al., 1990; Lu and Park, 1991; Malmsten and Lassen, 1994; Elwing et al., 1995; Lassen, 1995). A plausible reason for this is the intricate series of events that takes place as the protein adjusts its conformation to minimize the free energy of the adsorption complex.

According to Lundström, Walton, and others (Lundström and Elwing, 1990; Söderquist and Walton,1980; Sevastianov et al., 1995), protein adsorption to any surface follows a sequence of events which begins with the collision and reversible attachment of the macromolecule, is followed by a train of conformational adjustments to the surface, and eventually terminates in quasi-irreversible fixation. Extensive studies by Norde and coworkers (Norde and Lyklema, 1979; Norde, 1986; Norde, 1992) have demonstrated that the gain in entropy which results from such an unfolding may be the driving force behind adsorption, at times even overwhelming substantial charge repulsions between protein and surface. Therefore, the individual structural stability which is a characteristic of each protein has a strong influence over its adsorption. The nature of the surface is obviously also of importance, with the opportunity for bond formation typically being far greater on a hydrophobic material due to its ability to stabilize the non-polar protein core as it becomes exposed to water during unfolding. Given the great number of variables that govern surface film formation on a blood contacting artificial material, it is not surprising that the biological responses to such materials show wide variations.

CELLULAR RESPONSES TO IMPLANTS

Once deposited, the status of the protein coat which forms on the artificial surface is being explored by circulating blood cells, whose reactions to the surface will reflect the composition of the coat. If, for instance, the surface layer is rich in IgG molecules whose adsorption reactions have prompted the binding of complement factor 3b, generated primarily through the alternative pathway of complement activation (Janatova et al., 1991), then the collision with circulating macrophages will lead to opsonization and an inflammatory response.

One of the most strongly surface active plasma proteins is FIB which, due to its many biological functions, has been the subject of intense research. Of particular interest is the thrombin-induced conversion of fibrinogen to fibrin and the FIB-based activation of platelets which leads to aggregation and clotting. Thus, Saltzmann et al. reported that certain FIB-coated surfaces had the ability to activate platelets whereas others did not (Lindon et al., 1986; Salzman et al., 1987). However, the authors found no direct correlation between platelet activation and the surface concentration of FIB This seemingly puzzling observation was illuminated by the finding that platelet activation paralleled the binding of a polyclonal antibody developed against the intact FIB molecule. Surface adsorbed FIB, that was so distorted that it was impaired in its ability to bind the anti-FIB, had consequently also lost its ability to activate platelets. In a similar vein, Brash and coworkers adsorbed FIB to a polyethylene substrate which was subsequently heated and this heat denaturation resulted in a total loss of platelet adhesion (Woodhouse et al., 1995).

Through a particularly revealing set of experiments, Horbett and coworkers (Chinn et al. 1991; Kiaei et al., 1995) were able to demonstrate several important facts related to the effects of residence time on the structure of adsorbed FIB. From their work it is clear that the surface active FIB molecule adsorbs more readily onto low energy surfaces than to surfaces containing charged groups. The longer the molecule remains in contact with the surface, the more it looses its ability to adhere platelets. This loss of biological function is particularly severe at low surface concentrations of FIB, although by populating empty surface sites with albumin, the authors were able to retain the immunologically recognizable FIB structure for three days. Similarly, long contact times have been shown to reduce the FIB molecule's ability to become displaced from the surface by other surface active species (Elwing et al., 1990).

Although circulating platelets are unaffected by the FIB in plasma, this molecule becomes a powerful activator of the cells once it has undergone a modest surface induced change in conformation. The FIB molecule has been found to possess two separate epitopes that both bind to the platelet specific integrin GPIIb-IIIa. One of these is the ubiquitous RGD peptide of which there are several copies in the FIB molecule, while the other is a 12 amino acid peptide sequence which is buried in the native FIB structure but becomes exposed upon a modest unfolding of the protein (Savage et al., 1995; Kieffer, 1996). A different sequence of similar length has recently been found to bind also to integrins on the surface of neutrophils and other leucocytes. Indeed, Eaton and coworkers (Tang and Eaton, 1993; Tang et al., 1996) have shown that surface adsorbed FIB is responsible for neutrophil attachment and activation by implanted polyurethane devices. These activated neutrophils, in turn, were found to maintain an inflammatory condition on the surface of the device.

In view of the cellular response to artificial materials that is triggered by soluble components adsorbed from plasma, much effort has been directed toward suppressing the surface uptake of plasma proteins, particularly FIB. The strategies selected by the biomaterials community to accomplish this goal have much in common with those used to accomplish drug delivery with "stealth therapeutic systems", as they both attempt to sterically exclude the cellular immune system from recognizing and responding to surfaces foreign to the system.

PROTEIN EXCLUSION FROM SOLUTIONS AND SURFACES

The preferential solvation of proteins in aqueous cosolvents containing polymeric additives was extensively studied in the 1960s for the practical reason that such additives could accomplish the concentration, and ultimately precipitation, of proteins present in dilute solutions. Among a number of water soluble polymers studied for this purpose, the polyethylene glycols (PEG) showed particular promise (Polson et al., 1964; Iverius and Laurent, 1967), due to their high water solubility and weak protein interaction. Their molecular size, and hence excluded volume, increases rapidly with molecular weight in aqueous solutions and, similarly, for a given weight concentration of PEG the efficiency of protein exclusion increases rapidly with increasing molecular weights up to about 6,000 Da, beyond which only marginal exclusion effects are noticed (Ingham, 1978). The efficiency of this exclusion was found to scale with the size of the protein molecule, as expected, so that smaller proteins require higher concentrations of polymer to reach their solubility limits (Atha and Ingham, 1981).

Attempts at applying this exclusion effect to surface protection were first made by Nagaoka et al. (Mori and Nagaoka, 1982; Nagaoka et al., 1984), who modified hydrophobic surfaces by covalent attachment of PEG. The procedure was clearly effective, both in terms of reducing protein adsorption and suppressing platelet adhesion. Since this pioneering work, numerous approaches have been devised to uniformly derivatize various materials with PEG to enhance their biocompatibility. However, it is clear that the presence of a surface will alter the behavior of the PEG chains compared to their behavior in free solution, and in order to most effectively carry out this surface protection one needs to fully understand the influence of both chain length and grafting density on the repulsion of proteins and cells. Unfortunately, the two parameters are typically difficult to vary independently of one another.

SURFACE PROTECTION THROUGH POLYMER ADSORPTION

It was thought that one way around the problems with covalent protection might be instead to let the surface modification proceed through adsorption. Although the PEG (or equivalently PEO, polyethylene oxide) homopolymer is weakly hydrophobic and adsorbs to hydrophobic substrates such as polystyrene latex particles in aqueous environments, the complex is weak and dissociation occurs upon removal of the polymer containing supernatant. An alternative approach is, therefore, to adsorb PEG to the surface in the form of a block co-polymer, containing a hydrophobic block of such length that the generated complex would be stable even in the absence of soluble polymer. Polymeric tri-block surfactants of the poloxamer type (PEO-PPO-PEO, where PPO represents polypropylene oxide) are commercially available in a large number of stoichiometries under the trade name of Pluronic®. Several of these products have been shown to adsorb well to hydrophobic materials with which they form complexes of long term stability in biological environments (Illum et al., 1987; Müller, 1991; Amiji and Park, 1992).

Effects of Layer Thickness and Surface Concentration

Our own efforts to explore the relative effects of PEO chain length and surface density on protein repulsion have involved surface coatings with a series of Pluronic surfactants. The rationale for this choice was that adsorption via a long hydrophobic (PPO) block, with its large number of potential attachment sites, was likely to occur far more rapidly than fixation by covalent linkage, that would involve only 1-2% of the monomer units present in the chain. The latter reaction was thought to be progressively more prone to steric repulsion effects the longer the PEO chain and the higher the surface concentration, while the rapid adsorption reaction would be less subjected to a coupling between surface density and PEO block size.

For this study four Pluronic surfactants were selected which differed primarily in their PEO block lengths (Figure 1), while their PPO blocks were of comparable size. The surfactants

PLURONIC™

(PEO)m(PPO)n(PEO)m
triblock copolymers

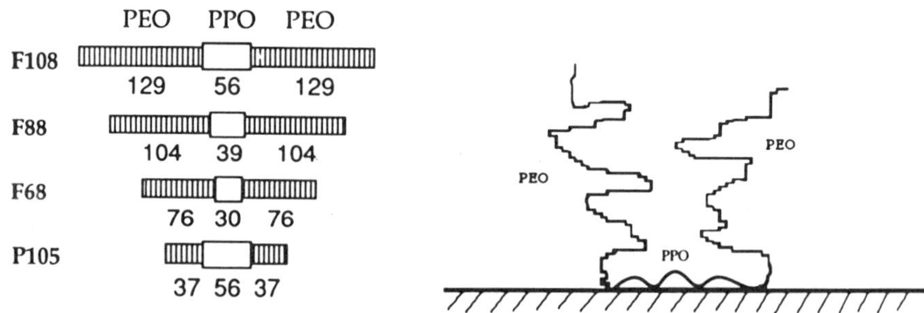

Figure 1. Schematic illustration of the polymeric surfactants comared in terms of their ability to protect hydtophobic surfaces from protein uptake

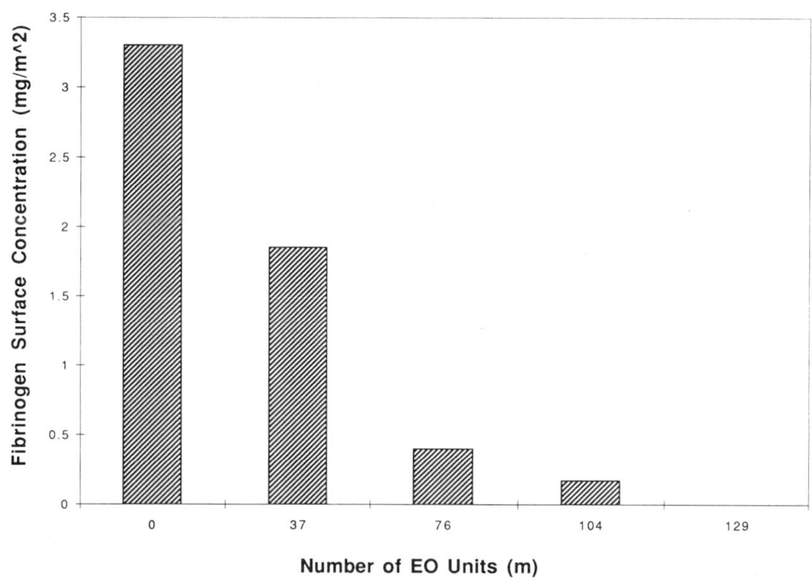

Figure 2. Adsorption of Fibrinogen to polystyrene particles, either bare or coated with Pluronic surfactants with the different PEO block lengths indicated on the abscissa. Prior to exposure to the FIB solution (0.5mg/mL), the 278 nm PS latex particles had been coated with either of the four surfactants (4% w/v) for 24 h. Protein uptakes were determined by depletion analysis

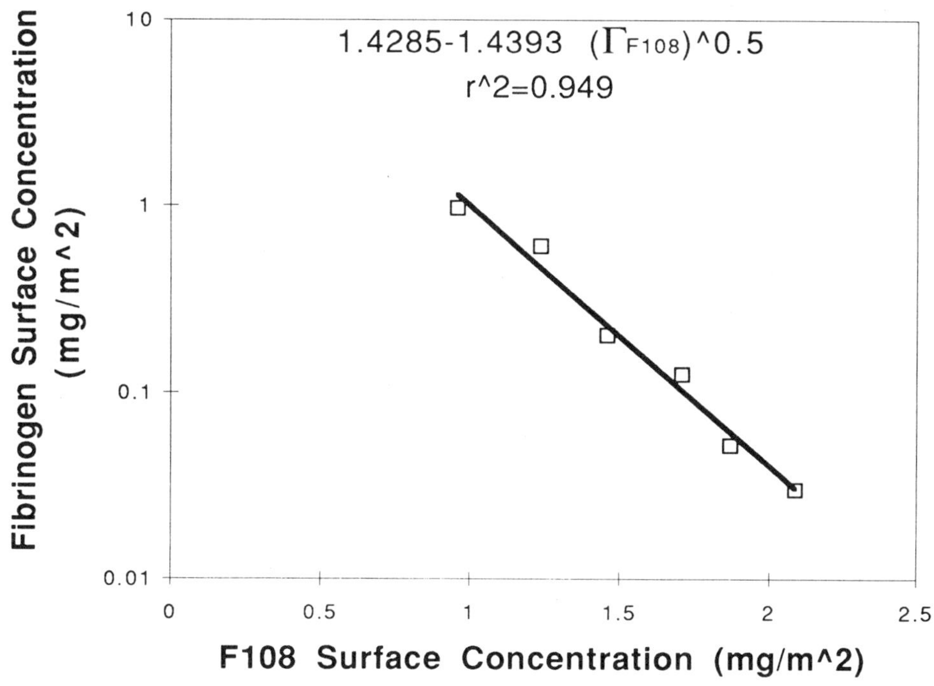

Figure 3. Adsorption of FIB (1 mg/mL) as a function of surface concentration of Pluronic F108 on 252 nm PS particles (Courtesy, Dr M. Bohner)

were adsorbed onto a series of monodisperse model polystyrene (PS) colloids with different diameters, to allow a correction to be made for any curvature effects on the resulting surface concentration . By adsorbing to particles, as opposed to flat surfaces, one had the advantage of increasing the surface area, and hence the amounts adsorbed, without unduly increasing the sample volume. This, in turn, increased the number of analytical techniques that could be used to evaluate the adsorption. In addition to the "indirect methods" that determine depletion of polymer in a solution that is brought into contact with a given amount of particles (i.e. a given surface area), the evaluation also included several "direct methods" which quantified the amount of material adsorbed per particle. The latter strategy was pursued in two ways. First, the surfactant was either labeled with a radioisotope (^{125}I)(Li et al., 1996) or a strong chromophore (benzopyrene or pyridyl disulphide)(Li et al., 1991; Li et al., 1996). Second, the mass increase per particle was determined by sedimentation field-flow fractionation (Li and Caldwell, 1991; Li et al. 1994; Li and Caldwell, 1996) using a procedure that requires no labeling of any of the reactants. The thickness of the ad-layer was determined by photon correlation spectroscopy, as well as by flow field-flow fractionation (Li and Caldwell, 1996). The results of these different analyses were in mutual agreement and showed the following : (a) For any given particle size, adsorption of the four surfactants was comparable. In other words, the surface density was independent of the PEO block length, as we had anticipated; (b) For any given surfactant, the surface density decreased with decreasing particle size. For any given particle size the ad-layer increased with PEO block length, as expected. For highly curved surfaces (particle diameter of 69 nm), the spatial extension of the PEO block was comparable to its dimensions in free solution, while a more extended ("brush-like) conformation was observed on larger particles; (c) the adsorption kinetics for the surfactant of the highest molar mass displayed clear biphasic features, with about 70% of maximum surface concentration reached in the first 30 min. and plateau adsorption reached after 6 hours, under the conditions normally used (Chen and Caldwell, 1997).

These results indicated that if the particle size were held constant, one could use this system to gain insight into the effect of PEO block length on the uptake of protein. Fibrinogen was chosen as the probe molecule owing to its importance in blood compatibility issues mentioned above. The results of its adsorption to particles coated with either of the four surfactants are summarized in Figure 2 and demonstrate that the FIB repulsion is more effective the longer the PEO block, up to a length of about 130 ethylene oxide units. In fact, at maximum close packing, the Pluronic F108-coated surface adsorbs about two orders of magnitude less FIB per unit area than the corresponding bare particles, as seen in Figure 3.

Complex Stability

The adsorbing surfactant forms a complex of significant stability. Indeed, particles coated with isotope-labeled Pluronic F108, when washed and resuspended in a surfactant free buffer, showed no evidence of any leakage of radioactivity even after three days of incubation (Li and Caldwell, 1996). More importantly, F108-coated PS nanoparticles (78 nm in diameter) remained in circulation 24 hours after intravenous injection in rats, with a half-life for clearance of around 13 hours compared to just several minutes for their uncoated counterparts (Tan et al., 1993). These findings parallel those made by Illum and coworkers (Norman et al., 1993; Moghimi et al., 1994). Thus, even in the presence of a highly protein containing medium, displacement of the surfactant does not occur at rates which would prevent their use as protein shields for short term exposures. Our knowledge of their long term stability is limited, but two encouraging experiments are worth reporting: The first derives from a three week-long tissue culture experiment in which a neat PS Petri dish was incubated with an established fibroblast

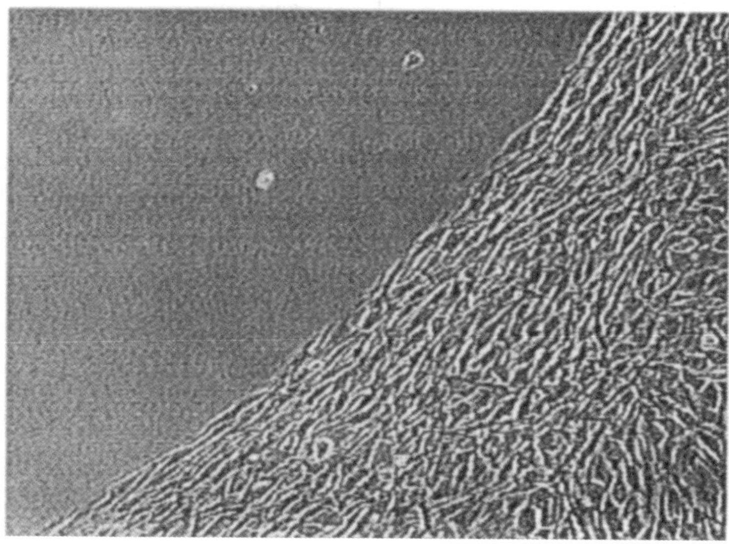

Figure 4. Cell adhesion (NIH 3T3 fibroblasts) to neat polystyrene. The upper left hand corner had been exposed to a drop of 1% Pluronic F108 prior to rinsing and seeding of the cells in serum containing medium (Courtesy, Dr P.A. Tresco)

cell line (NIH 3T3) in 10% serum containing medium. Prior to incubation, one small area of the dish had been exposed to a drop of F108 solution for 30 minutes followed by a rinse. The exposed area remained totally free of cells during the entire observation period, despite the rapid build-up of a confluent layer around its perimeter, as seen in Figure 4 (Neff et al., 1998). The second is a very recent, preliminary *in vivo* study, in which F108-coated polyurethane specimens (Pellethane®) were implanted, together with uncoated material of the same geometry and morphology, in the chest cavity of a sheep. After one month, clear differences could be seen between the tissues surrounding the bare and coated specimens, particularly those with a rough texture. Whereas there was significant tissue ingrowth into the uncoated material, the coated specimen was totally separated from the surrounding tissue, which showed no evidence of any inflammatory response.

Suppressed Bacterial Colonization

The PEO-based barriers to protein adsorption are obviously effectively suppressing the ability of larger structures to approach the surface and attach themselves. Just as the fibroblasts in Fig. 4 shied away from the Pluronic coated surface, so do bacterial cells of different kinds. Work by Bridgett et al (Bridgett et al., 1992) has very clearly demonstrated this effect, as have observations in our own laboratory (C. Durham, ME Thesis in progress). In searching for biological surfactants which present similar barriers to bacterial adhesion we have begun studies of the soluble mucins present in saliva, tear fluid, and gastric mucosa, among other locations. Thus, following a rudimentary purification protocol, a commercially available bovine submaxillary gland mucin was found to adsorb to polymeric surfaces in a manner analogous to the Pluronic F108. However, this mucin coating led to a far greater reduction in contact angle (wettability) than the coating with Pluronic. The efficiency of bacterial repulsion was likewise greater than that produced by Pluronic F108 (Shi, L. et al, manuscript in preparation). Since preliminary results indicate the mucin coatings to be non-inflammatory, these biological surfactants may potentially be of use also in the "stealth coating" of drug delivery vehicles.

Ligand Attachment

Frequently, in preparing surfaces for the attachment of biologically active entities, such as enzymes or cell adhesion molecules, including targeting moieties exemplified by antibodies or lectins, one wishes to control the surface density of the ligand. One very convenient way of accomplishing this is to derivatize the end groups of the Pluronic PEO blocks with a functional group that is resistant to hydrolysis while simultaneously being reactive towards molecules which feature some specific chemical structure. In examining suitable functional groups, we selected the pyridyl disulfide as a nearly ideal candidate. The disulfide bridge is hydrolysis resistant, and a release of the pyridyl thioketone can only take place following disulfide exchange with a thiol-containing moiety on the protein to be attached to the surface, as schematically illustrated in Figure 5 (Li et al., 1996)). The exchange reaction is rapid and insensitive to pH above a value of 5.

By mixing derivatized with underivatized surfactant prior to coating a particular device, one has control over the density of functional groups presented by its surface, since there appears to be no preferential adsorption of either species. The subsequent coupling of a thiol-

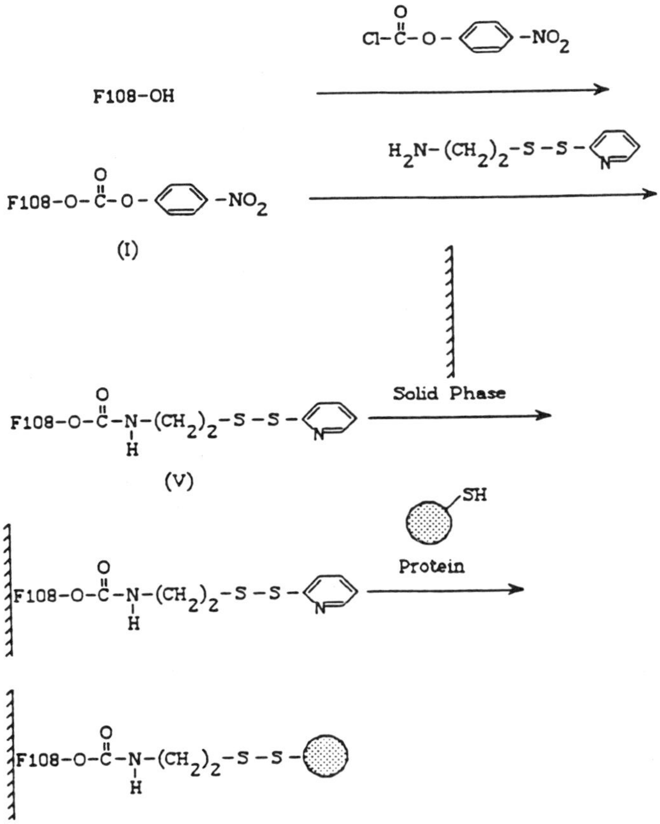

Figure 5. Summary of the steps involved in end-group activation of Pluronic surfactants and the subsequent surface attachment of a thiol containing protein

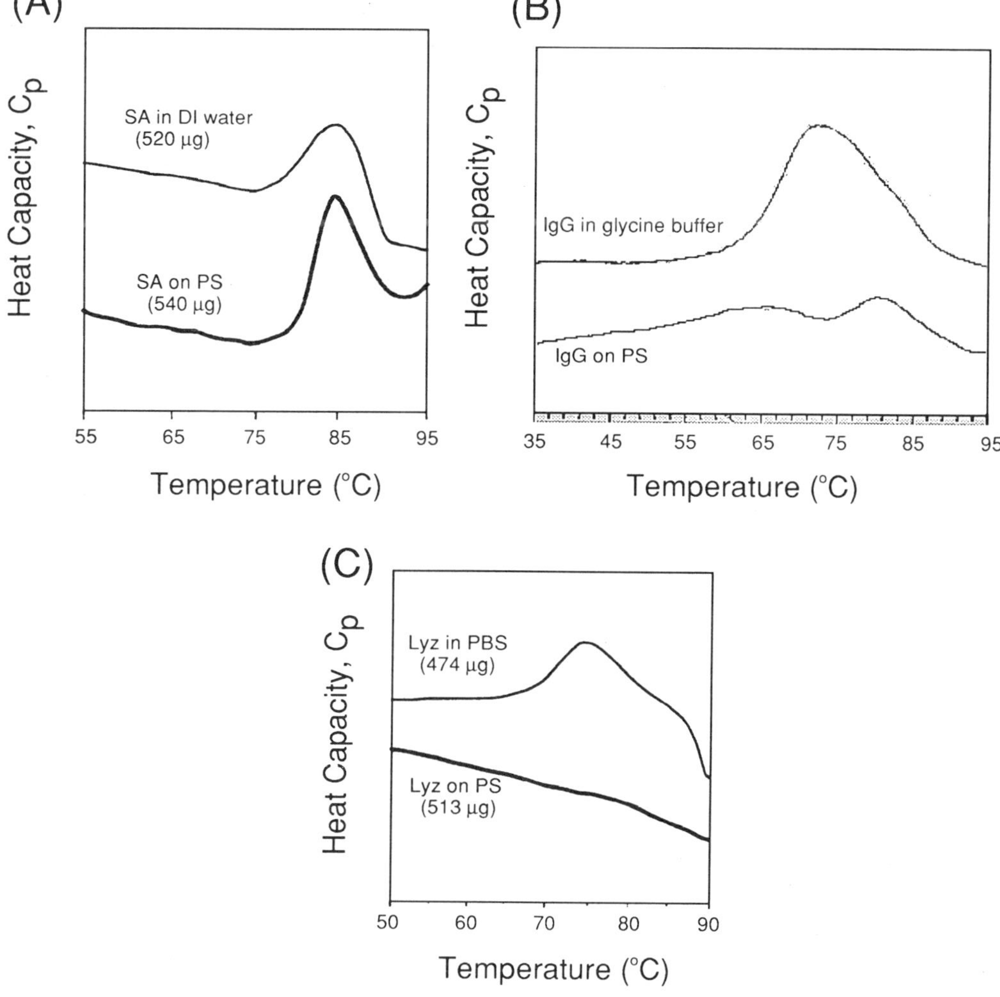

Figure 6. Differencial scanning calorimetric survey of protein conformational stability on hydrophobic surfaces.
A) The very stable Streptavidin in aqueous solution (upper trace) and in adsorption complex with PS latex particles (lower trace); B) IgG is of intermediate stability as seen from the loss of about 60% of the transition enthalpy (area under the curve) upon adsorption (Courtesy, Dr J.-T. Li); C) Lysozyme shows signs of total collapse upon adsorption to PS latex beads, to judge from the featureless lower trace

containing reagent then produces a derivatized surface with a selected density of ligands that are presented for specific biological interaction. Surrounding these ligands is a PEO cushion which protects the surface from non-specific adsorption of inconsequential macromolecules.

The adsorption of proteins to hydrophobic surfaces is often associated with a loss of activity, as discussed above for FIB in terms of its ability to interact with platelets. This activity loss has been found to parallel a structural change that can be monitored e.g. calorimetrically (Haynes and Norde, 1995; Yan et al., 1995). Modern microcalorimeters are sufficiently sensitive to allow thermal analyses to be performed on less than 1 mg of sample, and such quantities of protein are easily accommodated on the surfaces of minute volumes of nanoparticles. Thus, by forming adsorption complexes between PS latex spheres and a variety

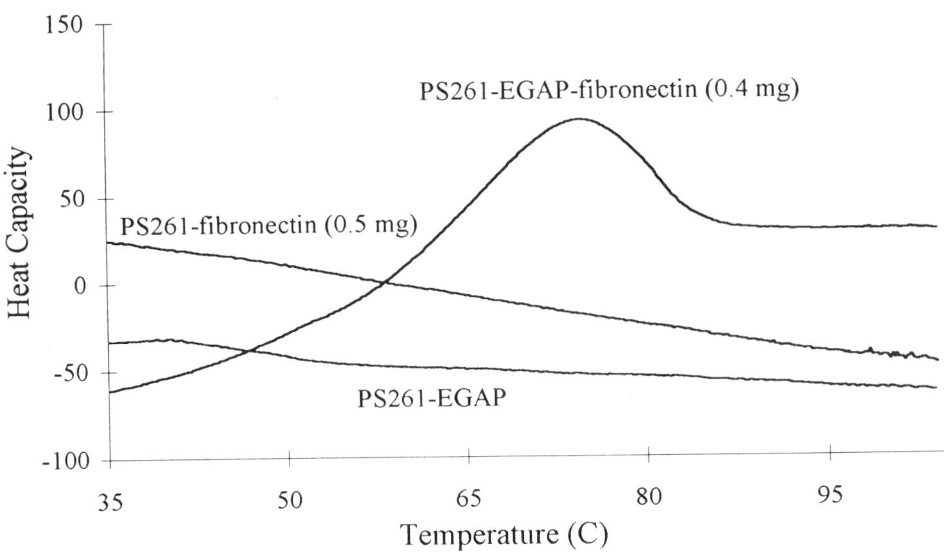

Figure 7. Calorimetric evidence of structural retention upon protein attachment via a PEO tether. The featureless lower traces represent PS particles coated with end-group activated Pluronic F108 Fibronectin, respectively, while the upper trace shows Fibronectin attached to the particle surface via the PEO tether (Courtesy, Dr S.-C. Huang)

of proteins, and comparing the thermal transition profiles for these proteins in solution and in the adsorbed state, respectively, one finds evidence of widely differing surface stabilities. As seen in Figure 6, the effect of the PS substrate on the adsorbed protein ranges from *negligible* for the fully folded streptavidin molecule to *totally destructive* for the thermally featureless lysozyme. An intermediate level of conformational change is seen for polyclonal immunoglobulin G which appears to retain just under half of its calorimetrically detectable folding. This condition has obvious negative implications for the use of IgG or other recognition elements as targeting moieties following adsorption to solid surfaces.

Whenever a surface is to be conferred with a specific activity, e.g. for purposes of antigen recognition, enzymatic catalysis, or cell adhesion and growth, it is essential that attention be paid to the way in which the molecule is linked to the surface. There is evidence to suggest that lifting the molecule off the surface by attaching it through a tether might protect its structure and activity. Indeed, by immobilizing the protein via the end-group activated Pluronic F108 described earlier, we have been able to maintain even fragile structures, such as the cell adhesion protein fibronectin (Figure 7), in their folded state. Tissue culture studies on surfaces modified in this manner, i.e. by attachment of a variety of extracellular matrix proteins, are currently under way with promising results in terms of both cell attachment and spreading.

CONCLUSIONS

The composition of any foreign surface in contact with blood is crucial for the composition of the protein film that rapidly forms on this surface. With time, structural rearrangements occur in these proteins, which result in time-dependent signals to the cellular components of blood, in turn leading to inflammatory and thrombogenic responses to the surface. Protection strategies have successfully involved surface modifications with PEO-based polymers. Since the vast majority of implantable man-made surfaces are hydrophobic in nature, one has the possibility to modify these surfaces, irrespective of size and shape, by simple dip-coating procedures involving aqueous solutions of PEO-containing surfactants. The coatings are stable over several days in protein-containing media, and are even providing surface protection *in vivo*.

Because the introduction of good leaving groups into the ends of the protecting PEO-chains is relatively straightforward, one can easily immobilize more complex molecules, such as affinity ligands and other bio-active entities, to these surfaces. By simply mixing derivatized with underivatized surfactant in desired proportions prior to the adsorption, the surface density of the affinity ligand can be readily controlled. There are two particularly desirable outcomes of this approach: First, it leads to the suppression of non-specific adsorption of biomolecules to the surface and secondly, it protects the often fragile structure of the attached affinity ligand.

This simple but effective immobilization strategy should also prove useful for surface modification of emulsions and other drug delivery vehicles to provide them with targeting moieties such as antibodies, lectins, or other specific recognition elements in a fully active conformation.

REFERENCES

Adams, A.L., Fischer, G.C., Munoz, P.C., and Vroman, L., 1984, Convex-lens-on-slide: human plasma and blood in narrow spaces, *J. Biomed. Materials Res.* 18:643.

Andrade, J.D., and Hlady, V., 1987, Plasma protein adsorption: the big twelve, *Ann. N.Y. Acad. Sci.* 516:158.

Amiji, M., and Park, K., 1992, Prevention of protein adsorption and platelet adhesion on surfaces by PEO/PPO/PEO triblock copolymers, *Biomaterials* 13:682.

Atha, D.H. and Ingham, K.C., 1981, Mechanism of precipitation of proteins by polyethylene glycols, *J. Biol. Chem.* 256:12108.

Bridgett, M.J., Davis, M.C., and Denyer, S.P., 1992, Control of staphylococcal adhesion to polystyrene surfaces by polymer surface modification with surfactants, *Biomaterials* 13:411.

Chen, Q., and Caldwell, K.D., 1997, Field-flow fractionation in the determination of rates of surfactant adsorption to colloidal substrates, *J. Liq. Chrom. Rel. Tech.* 20:2509.

Chinn, J.A., Posso, S.E, Horbett, T.A., and Ratner, B.D., 1991, Postadsorptive transitions in fibrinogen adsorbed to Biomer, *J. Biomed. Materials Res.*, 25:535.

Elwing, H.B., Askendahl, A., and Lundström, I., 1990, Desorption of fibrinogen and gamma-globulin from solid surfaces induced by a non-ionic detergent, *J. Coll. and Interface Sci.*, 128:296.

Elwing, H.B., Li, L., Askendahl, A.R., Nimeri, G.S., and Brash, J.L. 1995, Protein displacement phenomna in blood plasma and serum studied by the wettability gradient method and the lens on surface method, in: *Proteins at Interfaces II: Fundamentals and Applications*, T.A. Horbett and J.L. Brash, eds.,American Chemical Society, Washington.

Gölander, C.-G., Lin, Y.S., Hlady, V., and Andrade, J.D. , 1990, Wetting and plasma-protein adsorption studies using surfaces with a hydrophobicity gradient, *Colloids and Surfaces*, 49:289.

Haynes, C.A., and Norde, W., 1995, Structures and stabilities of adsorbed proteins, *J. Colloid Interface Sci.* 169:313.

Horbett, T.A. 1982, Protein adsorption on biomaterials, in: *Biomaterials: Interfacial Phenomena and Applications*, S.L. Cooper and N.A. Peppas, eds., American Chemical Society, Washington.

Illum, L., Jacobsen, L.O., Müller, R.H., Mak, E. and Davis, S.S., 1987, Surface characteristics and the interaction of colloidal particles with mouse peritoneal macrophages, *Biomaterials* 8:113.

Ingham, K.C., 1978, Precipitation of proteins with polyethylene glycol: Characterization of albumin, *Arch. Biochem. Biophys.* 186:106

Iverius,P.-H. and Laurent, T.C., 1967, Precipitation of some plasma proteins by the addition of dextran or polyethylene glycol, *Biochim. Biophys. Acta* 133:372.

Janatova, J. , Cheung, A.K., and Parker, C.J., 1991, Biomedical polymers differ in their capacity to activate complement, *Complement Inflamm.* 8:61.

Kiaei, D., Hoffman, A.S., Horbett, T.A. and Lew, K.R., 1995, Platelet and monoclonal antibody binding to fibrinogen adsorbed on glow-discharge deposited polymers, *J. Biomed. Mater. Res.* 29:729.

Kieffer, N., 1996, Adhesive platelet glycoproteins and platelet function, in: *Adhesion Receptors as Therapeutic Targets*, M.A. Horton, ed., CRC Press, Boca Raton.

Lassen, B., 1995, Studies of Competitive Protein Adsorption, Doctoral Thesis, University of Gothenburg, Gothenburg, Sweden.

Li, J.-T., Carlsson, J., Lin, J.-N., and Caldwell, K.D., 1996, Chemical modification of surface active poly(ethylene oxide)-poly(propylene oxide) triblock copolymers, *Bioconj. Chem.* 7:592.

Li, J.-T. and Caldwell, K.D., 1991, Sedimentation field-flow fractionation in the determination of surface concentration of adsorbed materials, *Langmuir* 7:2034.

Li, J.-T., Caldwell, K.D. and Tan, J.S., 1991, Size analysis of a block-copolymer coated polystyrene latex, in: *Particle Size Assessment and Characterization*, T. Provder, ed., ACS Symposium Series, vol. 472, American Chemical Society, Washington.

Li, J.-T., Caldwell, K.D. and Rapoport, N., 1994, Surface properties of Pluronic coated polymeric colloids, *Langmuir* 10:4475.

Li, J.-T., and Caldwell, K.D., 1996, Plasma protein interactions with Pluronic™ treated colloids, *Colloids Surfaces B: Biointerfaces* 7:9.

Lindon, J.N., McManama, G., Kushner, L., Merrill, E.W., and Salzman, E.W., 1986, Does the conformation of adsorbed fibrinogen dictate platelet interactions with artificial surfaces?, *Blood* 68:355.

Lu, D.R., and Park, K, 1991, Effect of surface hydrophobicity on the conformational changes in adsorbed fibrinogen, *J. Colloid Interface, Sci.* 144:271.

Lundström, I. and Elwing, H.B., 1990,Simple kinetic models for protein exchange reactions on solid surfaces, *J. Colloid Interface Sci.*136:68.

Malmsten, M., and Lassen, B., 1994, Competitive adsorption at hydrophobic surfaces from binary protein systems, *J. Colloid Interface Sci.* 166:490.

Mori, Y, and Nagaoka,, S., 1982, Polyethylene oxide chains, *Trans. Am. Soc. Art. Int. Org.* 28:459.

Moghimi, S.M., Hawley, A.E., Christy, N.M., Gray, T., Illum, L. and Davis, S.S., 1994, Surface engineered nanospheres with enhanced drainage into lymphatics and uptake by macrophages of the regional lymph nodes. *FEBS Lett.* 344:25.

Müller, R.H., 1991, *Colloidal Carriers for Controlled Drug Delivery and Targeting*, CRC Press, Boca Raton.

Nagaoka, S., Mori, Y., Takiuchi, H., Yokota, K., Tanzawa, H. and Nishiumi, S., 1984, Interaction between blood components and hydrogels with poly(oxyethylene) chains, in: *Bolymers as Biomaterials*, S.W. Shalaby, A.S. Hoffman, B.D. Ratner, and T.A. Horbett, eds., Plenum, New York.

Neff, J.A., Caldwell, K.D., and Tresco, P.A., 1998, A novel method for surface modification to direct cell behavior, *J. Biomed. Mater. Res.*, in press.

Norde, W. and Lyklema, J., 1979, Thermodynamics of Protein Adsorption, *J. Colloid Interface Sci.* 71:350.

Norde, W., 1986, Adsorption of proteins from solution at the solid-liquid interface, *Adv. Colloid Interface Sci.* 24:267.

Norde, W., 1992, The behavior of proteins at interfaces, with special attention to the role of the structure stability of the protein molecule, *Clin. Mater.* 11:85.

Norman, M.E., Williams, P. and Illum, L.,1993, *In vivo* evaluation of protein adsorption to sterically stabilised colloidal particles, *J. Biomed. Mater. Res* 27:861.

Polson, A., Potgieter, G.M., Largier, J.F., Mears, G.E.F., and Joubert, F.J., 1964, The fractionation of protein mixtures by linear polymers of high molecular weight, *Biochim. Biophys. Acta* 82:463.

Salzman, E.W., Lindon, J., McManama,G., and Ware, J.A., 1987, Role of fibrinogen in activation of platelets, *N.Y. Acad. Sci.* 516:184.

Savage, B., Bottini, E., and Ruggieri, Z.M., 1995, Interaction of integrin alpha IIb beta 3 with multiple fibrinogen domains during platelet adhesion, *J. Biol. Chem.* 270:28812.

Sevastianov, V.I., Tremsina, Y.S., Eberhart, R.C.,and Kim, S.W., 1995, Effect of protein competition on surface adsorption-density parameters of polymer-protein interfaces, in: *Proteins at Interfaces II: Fundamentals and Applications*, T.A. Horbett and J.L. Brash, eds., American Chemical Society, Washington.

Slack, S.M. and Horbett, T.A. , 1995, The Vroman effect, in: *Proteins at Interfaces II: Fundamentals and Applications*, T.A. Horbett and J.L. Brash, eds., American Chemical Society, Washington.

Söderquist, M.E., and Walton, A.G., 1980, Structural changes in proteins adsorbed on polymer surfaces, *J. Colloid Interface Sci.* 75:386.

Tan, J.S., Butterfield, D.E., Voycheck, C.L., Caldwell, K.D., and Li, J.-T., 1993, Surface modification of nanoparticles by PEO-PPO block copolymers to minimize interactions with blood components and prolong blood circulation in rats, *Biomaterials* 14:823.

Tang, L., and Eaton, J.W., 1993, Fibrin(ogen) mediates acute inflammatory responses to biomaterials, *J. Exp. Med.* 178:2147.

Tang, L., Ugarova, T.P, Plow, E.F, and Eaton, J.W., 1996, Molecular determinants of acute inflammatory responses to biomaterials, *J. Clin. Invest.* 97:1329.

Vroman, L., and Adams, A.L., 1969, Identification of rapid changes at plasma-solid interfaces, *J. Biomed. Mater. Res.* 3:43.

Vroman, L., 1982, Protein/surface interaction, in:*Biocompatible Polymers*, M. Szycher, ed., Technomic Publ., Lancaster.

Woodhouse, K.A., Skarja, G.A., Bishop, P., and Brash, J.L., 1995, Platelet interactions with cross-linked fibrin and thermally denatured fibrinogen surfaces, *Trans. Soc. Biomat.* 19:33.

Wojciechowski, P.W. and Brash, J.L., 1993, Fibrinogen and albumin adsorption from human blood plasma and from buffer onto chemically functionalized silica substrates, *Colloids and Surfaces B: Biointerfaces* 1:107

Yan, G., Li, J.-T., Huang, S.-C., and Caldwell, K.D., 1995, Calorimetric observations of protein conformation at solid-liquid interfaces, in: *Proteins at Interfaces II: Fundamentals and Applications*, T.A. Horbett and J.L. Brash, eds., American Chemical Society, Washington.

THE MONONUCLEAR PHAGOCYTE SYSTEM:

FEATURES RELEVANT TO INTERACTIONS WITH LIPOSOMES

Siamon Gordon

Sir William Dunn School of Pathology, South Parks Road
Oxford OX1 3RE, UK

INTRODUCTION

Macrophages (MØ) and closely related cells of the mononuclear phagocyte system, previously known as the reticulo-endothelial system (RES), are strategically placed in the body to recognise, remove and respond to particulates, as well as macromolecular ligands (Gordon, 1995). They are highly efficient phagocytes by virtue of their expression of a wide range of plasma membrane receptors for opsonised targets, as well as by direct recognition through so-called pattern recognition receptors (Medzhitov and Janeway, 1997). In addition, the cells are rich in cytoskeletal and other intracellular components which enhance their uptake efficiency. Their interactions with any artificial lipid-enveloped particles (eg. liposomes) will be desirable or a nuisance depending on the viewpoint of the investigator, but certainly inevitable, unless steps are taken to interfere with natural recognition mechanisms. The MØ within different organs in contact with blood, or at portals of entry such as the airway, gut or skin vary considerably in their potential for recognition and clearance of altered host, or foreign components. Their differentiation and activation status is also highly relevant to their capacity for clearance, as are the size, composition and route of entry of liposomes. In order to design better liposome-targeting protocols, it is necessary to learn about the basic cell biology of MØ *in vitro* and their heterogeneous functions *in situ*. Topics relevant to liposome biotechnology and applications will be considered from the viewpoint of the MØ.

EXPERIMENTAL SYSTEMS AVAILABLE

In vitro macrophage systems

A range of culture systems provide MØ in adequate numbers and purity for study of interactions with liposomes *in vitro*. In the mouse, primary MØ, which may express receptors and other properties absent on MØ-like cell lines, are available with a spectrum of physiologic

Targeting of Drugs 6: Strategies for Stealth Therapeutic Systems
Edited by Gregoriadis and McCormack, Plenum Press, New York, 1998

characteristics (Gordon, 1997). Peritoneal MØ are easy to harvest and purify from normal mice ("resident MØ") or can be obtained in increased numbers by prior administration of an inflammatory ("elicited MØ") or infectious stimulus ("immunologically activated MØ"). Thioglycollate-broth yields elicited MØ which adhere and spread rapidly in culture and display high levels of various receptor-mediated uptake activities; biogel polyacrylamide beads, which are too large to ingest, yield elicited MØ which are ideal for endocytic studies since they do not contain intracellular residues (probably agar) derived from the thioglycollate-broth. BCG or *C.parvum* yields activated MØ, which display an altered profile of endocytic and secretory activities, and are cytocidal because of immune priming by γ interferon.

An alternative source, equally convenient, is to generate MØ from bone marrow precursors by cultivation in CSF-1 supplemented medium, a potent and specific growth and differentiation factor. Cells are grown in liquid culture for 7-10 days in bacterial plastic dishes, for easier recovery of the adherent MØ which are essentially pure. Other cytokines such as GM-CSF yield subtly different MØ; a mixture of GM-CSF and IL-4 yields myeloid-type "dendritic cells", which are non-adherent and display enhanced MHC II antigens and a uniquely efficient ability to present foreign peptide antigens to naive T lymphocytes (Inaba et al., 1992). Peritoneal or bone MØ culture-derived MØ can also be modulated by treatment with other cytokines, such as γ interferon, IL-4/13 or IL-10, which alter their endocytic properties selectively.

In man, blood monocytes can be isolated by adherence from the low density mononuclear cell fractions, and cultivated in autologous or pooled human serum, or in defined media (Montaner et al., 1997). After 4-7 days cells differentiate without significant growth, into mature MØ; their phenotype can be manipulated *in vitro* by addition of cytokines as above. In particular, MØ are generated unless cytokine combinations (GM-CSF and IL-4, with additional TNFα or other factors to induce further maturation) are employed to produce dendritic-like cells (Romani et al., 1997).

With both rodent and human, MØ-like cell lines are available for particular applications (Walker, 1997); PMA or other agents, can be used to induce cell maturation. Further details of cell culture, characterisation and functional analysis are available elsewhere (Gordon, 1997a,b).

Transfected cells which stably or transiently express particular MØ receptors in non-MØ cells, eg. CHO cells, are available for molecularly defined receptors.

In vivo macrophage systems

Apart from inbred mouse strains, which might differ in undetermined ways in MØ numbers and functions, there are now several knockout strains available which lack candidate receptors relevant to liposome clearance e.g. for scavenger receptor-A (SR-A) (Suzuki et al., 1997) or type 3 complement receptor (CR3) (Coxon et al., 1996). Clearance assays were used initially to define the mannose receptor (MR), with a labelled glycoprotein (β glucuronidase) and comparing neoglycoproteins as inhibitors (Pontow, 1992). Analysis of tissue distribution of MØ and their phenotype follows standard immunocytochemical methods (Gordon et al., 1992).

SELECTED CHARACTERISTICS OF MØ *IN VITRO*

Plasma membrane receptors for uptake

MØ express a variety of plasma membrane receptors able to bind and endocytose or phagocytose ligands, which can vary in size from particles such as cells, microorganisms or foreign bodies, to macromolecules. These include receptors for opsonins (specific-antibody receptor (FcR), complement (type 3 complement receptor) CR3 as well as non-opsonic receptors, for instance scavenger receptor A(SR-A) or B (SR-B) and mannose receptor (MR). The latter recognise classes of ligand, e.g. carbohydrate or selected polyanions, hence the term pattern recognition receptors. Different scavenger receptors recognise modified lipoproteins and polyanionic lipids (SR-A), or high density lipoproteins (SR-B) (Krieger, 1997). Specific antagonists (eg ab) or class-specific inhibitors (e.g. poly inosinic acid versus poly cytidylate) are available for study.

Whilst several of these receptors interact directly with a surface-ligand, eg. on a bacterium, or after incorporation into the membrane of liposomes, they are also able to bind ligands via an opsonic intermediate. The CR3 provides an example for the binding of a range of ligands (eg I-CAM, iC3b, and others) (Law and Reid, 1988). The interactions of liposomes with the complement cascade can be via the alternate, classical or lectin-pathways. Less-defined opsonins which might act on the target or by activation of MØ receptors such as CR3, include fibronectin (Wright and Detmers, 1988), and the mannose binding protein (MBP), a collectin (Lu, 1997).

Receptors which could be involved in liposome interactions include the GPI-linked CD14 antigen (Haziot et al., 1996), a known receptor for LPS-binding protein (LBP), and MARCO (Elomaa et al., 1995), related to SR-A in its collagenous structure, but a distinct gene product. Receptors for surfactant proteins A and D are not well-characterised.

Receptors can also mediate adhesion to substratum-bound ligands via plasma proteins, such as iC3b, fibronectin, vitronectin and thrombospondin. Several receptors can combine to mediate adhesion and phagocytosis, eg. CR3 and MR (Blackwell et al., 1985), CD36 (a thrombospondin receptor) and vitronectin R (Ren et al., 1995). Other receptors e.g CD44 mediate adhesion, rather than endocytosis, but have also been reported to enhance phagocytosis of apoptotic neutrophils (Hart et al., 1997).

The uptake process

This depends on the size of the ligand; bulk membrane internalisation usually proceeds by a zipper-type mechanism and involves actin-assembly and pseudopod extension, during receptor-mediated phagocytosis (Greenberg and Silverstein, 1993). Phagosomes can be tight or spacious, especially when macropinocytosis and active plasma membrane ruffles fold back to enclose a bound particle. It is not clear whether all phagocytosis is receptor-dependent; in the case of artificial particles such as Latex, the receptors are not known, although SR-A may contribute to the uptake of protein-coated particles. MØ are actively pinocytic, by fluid-phase and receptor-dependent pathways. Recently, caveolin has been detected in MØ (Baorto, 1997), so that specialised domains rich in GPI-linked molecules may also facilitate uptake of selected ligands.

Uptake involves extensive flow of membrane-bound vacuoles and vesicles, selective fusion with endosomes and lysosomes, acidification and ion fluxes. The intracellular machinery includes clathrin, coat (COP) proteins, small GTP-binding (rab) proteins and actin-binding molecules. Signal transduction by specific receptors is mediated by phosphorylation/dephosphorylation, G-protein coupled seven-transmembrane receptors, phosphoinositides etc. For example FcR but not CR3 activate syk selectively (Crowley et al., 1997) and recruit cytoplasmic proteins to the phagocytic MØ (Allen and Aderem, 1996).

Intracellular fate

A wide range of hydrolytic enzymes is delivered to the late endosome/phago-lysosome, capable of degrading all macromolecular substrates at acidic pH. Peptide loading of MHC II molecules protects them from complete degradation and can be retrieved for antigen presentation, especially by dendritic cells. Membrane constituents can be recycled to the cell-surface (the MØ has been calculated to internalise the equivalent of its total surface every 30 minutes). The membrane of the late endosome contains a heavily glycosylated mucin-like molecule characteristic of MØ (CD68 in man, macrosialin in the mouse), which is related to the lysosome-associated membrane protein (LAMP) family, found in many cells (Garni-Wagner and Todd, 1997 and Holnell et al., 1993). Macrosialin is rich in N- and O-linked carbohydrate chains; receptor-mediated phagocytosis induces complex remodelling of glycoforms. These may serve to protect the membrane of late endosomes/phagosomes from an acidic microenvironment. Oxidised LDL has been shown to be a ligand for this MØ-restricted molecule (Ramprasad et al., 1995).

Opsonins such as antibody can deliver a captured ligand to a different compartment, and favour its degradation. Enveloped viruses can penetrate MØ and other cells by acid-catalysed fusion of host and viral membrane, induced by conformational changes of viral glycoproteins, such as haemagglutinins (Helenius et al., 1983). Microorganisms utilise a range of mechanisms to induce or evade uptake by MØ and other cells and can inhibit maturation, fusion or acidification of vacuolar compartments; others survive within phagolysosomes or disrupt the lysosomal membrane to escape into the cytosol (Rabinovitch et al., 1995). Improved understanding of the molecular mechanisms involved should prove of great interest to liposome biologists. Depending on the nature of the ligand and its persistence, the MØ may alter its gene expression pattern considerably, adapt to intracellular requirements and secrete a variety of secretory products (Nathan, 1987). These include low molecular weight oxygen and nitrogen-derived metabolites, chemokines and cytokines, such as IL-1, IL-6, TNFα, which initiate local and systemic inflammatory effects. IL-12 and IL-10 are potent immunoregulators, enhancing and inhibiting T helper lymphocyte activation and γ interferon production, whilst also acting on NK and B lymphocytes, respectively. Other MØ-derived products which can be released by MØ following endocytic uptake and lysosomal storage include neutral proteinases, urokinase, collagenase and elastase.

SELECTED PROPERTIES OF MACROPHAGES *IN VIVO*

The use of the mAb F4/80 directed against a plasma membrane differentiation antigen of unknown function made it possible to localise mature MØ in murine tissues, during development and in the adult (Gordon et al., 1992). The antigen has been shown to have an unusual hybrid structure, consisting of 7 EGF extracellular domains and a seven-transmembrane spanning domain, homologous to a family of G-protein coupled peptide receptors (McKnight and Gordon, 1996). A human homologue, EMR1, has been defined, but no immunocytochemical studies have yet been performed in man. The F4/80 ko mouse has no obvious phenotype. Some MØ eg. in lymphoid organs, lack F4/80, but express another MØ membrane antigen, FA-11 (macrosialin), referred to above.

Detailed analysis with these and other markers, has shown that resident MØ are widely distributed throughout all tissues. Elicited and activated MØ also express these markers, as seen in inflammatory infiltrates, e.g. granulomata. Newly recruited MØ often adopt many of the characteristics of resident cells at particular sites, confirming the importance of the local microenvironment in determining the morphology and phenotype of the MØ.

These studies have brought out several general features: (i) It is possible to construct a putative migration pathway of MØ from the circulation into tissues. The MØ can be "endothelial" in direct contact with blood as in Kupffer cells of the liver, perivascular, interstitial, or associated with epithelia. They are either directly in contact with the external environment (alveolar MØ in lung), or lie close to portals of entry to the body, beneath an epithelium, as in the gut. In the epidermis, specialised Langerhans cells express F4/80 antigen, but lose this upon antigenic stimulation, as the cells migrate into afferent lymph to become potent antigen presenting, myeloid-type dendritic cells. This represents one sublineage of the MPS; another, F4/80 negative, but also FA-11^{+} is the osteoclast of bone. Circulating mononuclear cell precursors differentiate into these specialised bone-resorbing cells under the influence of CSF-1 and other local bone-specific environmental factors. (ii) Resident cells in different tissues display microheterogeneity, differ from one another, and from other MØ in different subcompartments of the same organ. This will be illustrated with regard to spleen, lymph node and the CNS, below. (iii) Resident cells display different properties from newly recruited, elicited and activated MØ. They perform poorly characterised trophic, tissue homeostatic functions whereas recruited cells mediate amplified responses in inflammation and host defence. These MØ differ considerably in their biosynthetic activity and life-span.

In regard to potential liposome interactions and clearance, several points of interest arise from tissue localisation and functional specialisation:

(a) Sinusoidal macrophages correspond to the endothelial component of the "RES" and are prominent in liver, but are also found in bone marrow, spleen and lymph nodes and in endocrine organs such as the pituitary and adrenal glands. They are actively phagocytic and clear particulates and soluble ligands from the circulation. They share properties with sinusoidal true endothelial cells suggesting either a remote lineage relationship or adaptation to a common microenvironment. Clearance of some ligands *in vivo* may be predominantly by endothelial cells rather than MØ, depending on the receptors utilised. Liposomes loaded with toxic drugs such as clodronate have been used by Van Rooijen and his colleagues to ablate MØ by parenteral administration (Van Rooijen, 1997). The route (intravenous, intraperitoneal or intradermal) plays a major role in targeting MØ in different organs. The marginal zone MØ in spleen and the subcapsular sinus MØ in lymph nodes, are most affected because of their direct exposure to blood and lymph, respectively (Kraal 1992). These MØ express specialised receptors for binding and uptake of ligands (particulate or soluble), such as sialoadhesin, a sialic acid-binding lectin of selected MØ, CR3, SR-A and MARCO. Following ablation, it has been possible to observe the repopulation of these specialised MØ and thus to discern their immunological functions.

(b) Bone marrow, foetal liver and spleen red pulp also contain ' stromal' MØ which are part of the haematopoietic micro-environment. These MØ associate with developing blood cells and regulate their differentiation by poorly understood trophic interactions (Crocker et al., 1988). They are actively phagocytic, disposing of erythroid nuclei and naturally dying PMN and B lymphocytes. Clearance of senescent erythrocytes and leukocytes is achieved by these, and sinusoidal MØ, by a variety of receptor-mediated mechanisms not yet fully understood. Receptors that have been implicated in the uptake of apoptotic cells include SR-A (Platt et al., 1996), SR-B and vitronectin receptors (Helenius et al., 1983). Receptors for natural antibody and complement have also been postulated to mediate senescent cell clearance.

(c) Phagocytic MØ (F4/80^{-}, FA-11^{+}) can be found in T lymphocyte-dependent areas, such as in splenic white pulp, germinal centres of lymph node (tangible body MØ) and Peyer's patches. However, the main cell that presents antigens to T lymphocytes is the dendritic cell, a specialised differentiation stage of the MØ lineage mentioned above. A lymphoid type of dendritic cell, which does not share close origin with myeloid cells, has recently been reported (Wu et al., 1996). Dendritic cells express many of the same surface receptors as MØ, are actively endocytic, including macropinocytosis, and poorly phagocytic. They express high levels of MHCII antigens constitutively and also other accessory molecules important for T cell activation.

(d) The nervous system (central and peripheral) provides an interesting special instance where resident MØ become terminally differentiated as microglia (Perry and Gordon, 1991).

Some of these remain in contact with blood, as in the circumventricular organs, whereas most are shielded by the blood brain barrier. Monocytes are among the only haematogenous cells which freely enter the normal neuropil, T lymphocytes do so only when immunologically activated. The cerebral endothelium can transport selected macromolecules, including specific antibodies, across the blood brain barrier, presumably via selected receptors (Reid et al., 1994). Perivascular MØ ("Mato cells") express SR-A and mannose receptor, and can be distinguished from microglia, in which these receptors are down-regulated (Mato et al., 1996). Local stimulation by neurotoxins or LPS, can reactivate microglia to express antigens and receptors (Perry et al., 1993). These cells resemble newly recruited monocytes/MØ.

The functions of microglia are obscure. They are found in the neutrophil throughout life. During development they clear naturally dying neurons, but in the normal adult they associate intimately with neurons and astrocytes. They retain high levels of CR3 on their plasma membrane, but possible ligands for this molecule are unknown. In the pituitary gland, the microglia can engulf parts of neuroendocrine cells, and respond to salt loading. These findings suggest a broad role in tissue homeostasis, as well as responsiveness to injury and cell degeneration.

(e) Activated MØ alter their receptor and secretory profiles, producing a wide range of cytotoxic and pro-inflammatory molecules (Gordon, 1998). These can act locally on endothelium to alter their adhesion and permeability functions, as well as at a distance. TNFα production, for example, is regulated by a complex series of steps, resulting in its surface expression or secretion. TNFα is essential for host defence, as well as mediating tissue injury when produced in excess. Immunologically primed MØ are not only a major source of TNFα production, but also produce inhibitors of TNFα production such as IL-10. The nature of the plasma membrane receptors involved in microbial recognition plays a key role in determining subsequent secretory responses. Whilst FcR and CD14 enhance cytotoxicity, CR3 and SR-A may be silent, or inhibit release of toxic molecules, thus playing an important down-regulatory role (Haworth et al., 1997).

GOALS, OPPORTUNITIES AND LIMITATIONS OF LIPOSOME TARGETING OF MACROPHAGES

It is clear that liposome interactions with MØ can be exploited to direct therapeutic agents to these cells, to enhance or inhibit their functions. Liposomes can be engineered to display specific ligands, or non-ligands and their contents can include enzymes, e.g. for replacement therapy in storage diseases, drugs or DNA. Sophistication in liposome delivery and the achievement of biologic end-results must depend on the detailed understanding of cellular properties as outlined above. A few suggestions follow, as a basis for further investigations: (a) boosting of MØ and/or DC functions would be desirable in vaccine production to polysaccharide antigens which are poorly immunogenic and to enhance effector mechanisms of cell mediated immunity eg HIV infection, or to replace missing functions, as in storage diseases; (b) inhibition of MØ functions could limit autoimmune tissue injury, by neutral proteinase and TNFα overproduction in rheumatoid arthritis and multiple sclerosis, or undesired consequences of chronic inflammation eg. pulmonary fibrosis, or in the vessel wall, in atherosclerosis; (c) enhanced recruitment of MØ to tumours may be achieved by delivery of chemokines, specific for MØ and/or DC; (d) selective ablation of MØ can be used to investigate MØ functions *in vivo*, as noted, and to kill infected MØ serving as reservoirs of tuberculosis or HIV. It may be important to induce apoptosis rather than necrosis, since the fate of the released organism could be different in each circumstance. Apart from therapeutic goals, it should also be possible to use similar strategies for diagnostic purposes, to assess MØ phagocytic and clearance functions *in vivo*.

All of the above pose questions of selectivity of MØ versus non-MØ, and between MØ and DC. Site and subpopulations are also important to consider; joints and airways are readily accessible targets but the CNS is more difficult. Activated cells should be distinguished from key resident MØ performing important trophic functions in bone marrow or brain. In principle, one can target (or avoid) the MØ by non-specific means, using particles, or exploit known receptor-ligand interactions. One can try to control binding, uptake and escape from the MØ vacuolar system, depending on the desired site of action. Survival of DNA is of critical importance after delivery to MØ, in view of their degradative capacity. Apart from a strategy which would deliver a molecule to the cell by specific targeting, it is also possible to enhance MØ-specific gene expression by using appropriate promoter constructs. We have achieved some success in this by utilising a CD68-based DNA construct for high level MØ expression of reporters (Greaves et al., submitted for publication) and a human lysozyme promoter (Clarke et al., 1996) to target immunologically activated and phagocytically active MØ by transgenesis.

Another important goal will be to target MØ functions conditionally e.g. by encoding or delivering prodrugs which, upon further challenge with a specific inducing agent, will generate an active drug.

Finally, it may be desired to avoid MØ uptake, in order to achieve selective uptake by another cell type eg endothelium, hepatocytes or tumour cells. Such cells may express specific receptors that are absent on MØ. Evasion of binding or uptake by MØ could be achieved by interference with known mechanisms of opsonisation eg via complement activation. More could be learnt from micro-organisms which employ capsular polysaccharides to avoid uptake. The recognition of lipids by MØ is also still poorly understood; eg phosphatidylserine has been postulated as a ligand for uptake of apoptotic cells, but the receptor remains uncharacterised. Other lipids are recognised via associated lipoproteins, though scavenger receptors, CD14 or macrosialin. Further work is required in order to learn how to exploit natural systems to the full for future therapeutic benefit.

Acknowledgements

Work in the author's laboratory has been supported by the MRC, ARC, British Heart Foundation, The Wellcome Trust and The Histiocytosis Association of America, Inc. I thank Christine Holt for help in preparing this manuscript.

REFERENCES

Allen, L.A., and Aderem, A., 1996, Molecular definition of distinct cytoskeletal structures involved in complement- and Fc receptor-mediated phagocytosis in macrophages. *J. Exp. Med.* 184:627.

Baorto, D.M., Gao, Z., Malaviya, R., Dustin, M.L., van der Merwe, A., Lublin D.M., and Abraham. S.N., 1997, Survival of FimH-expressing enterobacteria in macrophages relies on glycolipid traffic. *Nature* 389:636.

Blackwell, J.M., Ezekowitz, R.A.B., Roberts, M.B., Channon, J.Y., Sim R.B., and Gordon, S., 1985, Macrophage complement and lectin-like receptors bind *Leishmania* in the absence of serum. *J. Exp. Med.* 162:324.

Clarke, S., Greaves, D.R., Chung, L-P., Tree, P., and Gordon, S., 1996, The human lysozyme promoter directs reporter gene expression to activated myelomonocytic cells in transgenic mice. *Proc.Natl.Acad. Sci. USA* 93:1434.

Coxon, A., Reu, P., Barkalow, F.J., Askari, S., Sharpe, A.H., von Andrian, U.H., Arnaout, M.A., and Mayadas T.N., 1996, A novel role for the β2 integrin CD11b/CD18 in neutrophil apoptosis: A homeostatic mechanism in inflammation. *Immunity* 5:653.

Crocker, P.R., Morris, L., and Gordon, S., 1988, Novel cell surface adhesion receptors involved in interactions between stromal macrophages and haematopoietic cells. *J.Cell Sci. Suppl.* 9:185.

Crowley, M.T., Costello, P.S., Fitzer-Attas, C.J., Turner, M., Meng, F., Lowell, C., Tybulewicz, W.L.J., and DeFranco, A.L., 1997, A critical role for Syk in signal transduction and phagocytosis mediated by Fcγ receptors on macrophages. *J. Exp. Med.* 186:1027.

Elomaa, O., Kangas, M., Sahlberg, C., Tuukkanen, J., Sormunen, R., Liakka, A., Thesleff, I., Kraa, G., and Tryggvason, K., 1995, Cloning of a novel bacteria-binding receptor structurally related to scavenger receptors and expressed in a subset of macrophages. *Cell* 80:603.

Garni-Wagner, B.A., and Todd, R.F., III. 1997, Cluster designation antigens expressed on myeloid lineage cells, in: *Weir's Handbook of Experimental Immunology* 5th ed., vol IV, L.A. Herzenberg, D.M. Weir, L.A. Herzenberg and C. Blackwell, eds., Blackwell Science Inc, Cambridge, MA.

Gordon, S., 1995, Mononuclear phagocyte system and tissue homeostasis, in: *Oxford Textbook of Medicine*, D.J. Weatherall, J.G.G. Ledingham and D.A. Warrell, eds, Oxford University Press, Oxford.

Gordon, S., 1997a, Overview : The myeloid system, in: *Weir's Handbook of Experimental Immunology*, 5th ed., vol IV The integrated immune system. L.A. Herzenberg, D.M. Weir, L.A. Herzenberg and C. Blackwell, eds., Blackwell Science Inc, Cambridge, MA.

Gordon, S., 1997 b, *Weir's Handbook of Experimental Immunology*, 5th ed., L.A. Herzenberg, D.M. Weir, L.A. Herzenberg and C. Blackwell, eds., Blackwell Science Inc, Cambridge, MA.

Gordon, S., 1998, Macrophage Activation in: Encyclopaedia of Immunology 2nd ed., D.J. Weatherall, J.G.G. Ledingham, and D.A. Warrell eds, Academic Press, London. In Press.

Gordon, S., Lawson, L., Rabinowitz, S., Crocker, P. R , Morris L., and Perry, V.H., 1992, Antigen markers of macrophage differentiation in murine tissues, in: *Current Topics in Microbiology and Immunology: Macrophage Biology and Activation*, S. Russell and S. Gordon eds, Springer-Verlag, Berlin.

Greenberg, S., and Silverstein, S.C., 1993, Phagocytosis in: *Fundamental Immunology*. 3rd ed. W. Paul, ed., Raven Press, Philadelphia.

Hart, S.P., Dougherty, G.J., Haslett, C., and Dransfield, I., 1997, CD44 regulates phagocytosis of apoptotic neutrophil granulocytes, but not apoptotic lymphocytes, by human macrophages. *J. Immunol.* 159:919.

Haworth, R., Platt, N., Keshav, S., Hughes, D., Darley, E., Suzuki, H., Kurihara, Y., Kodama, T., and Gordon, S., 1997, The Macrophage Scavenger Receptor Type A (SR-A) is expressed by activated macrophages and protects the host against lethal endotoxic shock. *J. Exp. Med.* 186:1431.

Haziot, A., Ferrero, E., Köntgen, F., Hijiya, N., Yamomoto, S., Silver, J., Stewart, C.L., and Goyert, S.M., 1996, Resistance to endotoxin shock and reduced dissemination of Gram-negative bacteria in CD14-deficient mice. *Immunity* 4:407.

Helenius, A., Mellman, I., Wall, D., and Hubbard, A., 1983, Endosomes. *Trends in Biochemical Sciences* 8:245.

Holness, C.L., da Silva., R.P., Fawcett, J., Gordon, S,. and Simmons, D.L., 1993, Macrosialin, a mouse macrophage restricted glycoprotein, is a member of the lamp/lgp family. *J. Biol.Chem.* 268:9661.

Inaba, K., Inaba, M., Romani, N., Aya, H., Deguchi, M., Ikehara, S., Muramatsu, S., and Steinman, R.M., 1992, Generation of large numbers of dendritic cells from mouse bone marrow culture supplemented with granulocyte/macrophage colony stimulating factor. *J. Exp. Med.* 176:1693.

Kraal, G., 1992, Cells in the marginal zone of the spleen. *Int. Rev. Cytol.* 132:31.

Krieger, M., 1997, The other side of scavenger receptors: pattern recognition for host defense. *Curr. Opinion in Lipidology* 8:275.

Law, S.K.A., and Reid, K.B.M., 1988, *Complement,* 1st ed., In Focus Series, D. Male, ed., IRL Press, Oxford.

Lu, W., 1997, Collectins: collectors of micro-oganisms for the innate immune system. *Bioessays* 19:509.

Mato, M., Ookawara, S., Sakamoto, A., Aikawa, E., Ogawa,T., Mitsuhashi, U., Masuzawa, T., Suzuki, H., Honda, M.,Yazaki, Y., Watanabe, E., Luoma, J., Yla-Herttuala, S., Fraser, I., Gordon, S., and Kodama, T., 1996, Involvement of specific macrophage-lineage cells surrounding arterioles in barrier and scavenger function in brain cortex. *Proc.Natl.Acad.Sci. USA* 93:3269.

McKnight, A.J., and Gordon, S., 1996, EGF-TM7: a novel subfamily of seven-transmembrane-region leukocyte cell-surface molecules. *Immunology Today* 17:283.

Medzhitov, R., and Janeway, C.A., Jr., 1997, Innate Immunity: impact on the adaptive immune response. *Curr. Opin. in Immunol.* 9:4.

Montaner, L.J., Collin, M., and Herbein, G., 1997, Human monocytes: isolation, cultivation and applications, in: *Weir's Handbook of Experimental Immunology*, 5th ed., vol IV, L.A. Herzenberg, D.M. Weir, L.A. Herzenberg and C. Blackwell, eds., Blackwell Science Inc, Cambridge, MA.

Nathan, C., 1987, Secretory products of macrophages. *J. Clin. Invest.* 79:319.

Perry, V.H., Andersson, P-B., and Gordon, S., 1993, Macrophages and inflammation in the central nervous system. *Trends in Neurosciences* 16:268.

Perry, V.H., and Gordon, S., 1991, Macrophages and the nervous system. *Int Rev Cytol* 125:203.

Platt, N., Suzuki, H., Kurihara, Y., Kodama, T., and Gordon, S., 1996, Role for the Class A macrophage scavenger receptor in the phagocytosis of apoptotic thymocytes. *Proc.Natl.Acad.Sci* USA. 93:12456.

Pontow, S.E., Kery, V., and Stahl, P.D., 1992, Mannose receptor. *Int. Rev. Cytol.* 137:221.

Rabinovitch, M., 1995, Phagocytosis. *Trends in Cell Biology*. 15:85.

Ramprasad, M.P., Fischer, W., Witztum, J.L., Sambrano, G.R., Quehenberger, O., Steinberg, D., 1995, The 94- to 97-kDa mouse macrophage membrane protein that recognises oxidized low density lipoprotein and phosphatidyl serine-rich liposomes is identical to macrosialin, the mouse homologue of human CD68. *Proc.Natl.Acad.Sci. USA* 92:9580.

Reid, D.M., Perry, V. H., Andersson, P-B., and Gordon, S., 1994, Mitosis and apoptosis of microglia *in vivo* induced by an anti-CR3 monoclonal antibody which crossed the blood-brain barrier. *Neuroscience* 56:529.

Ren, Y., Silverstein, R.L., Allen, J., Savill, J., 1995, CD36 gene transfer confers capacity for phagocytosis of cells undergoing apoptosis. *J. Exp. Med.* 181:1857.

Romani, N., Bhardwaj, N., Pope, M., 1997, Dendritic cells, in: *Weir's Handbook of Experimental Immunology*, 5th ed., vol IV, L.A. Herzenberg, D.M. Weir, L.A. Herzenberg and C. Blackwell, eds., Blackwell Science Inc, Cambridge, MA.

Suzuki, H., Kurihara, Y., Takeya, M., Kamada, N., Katoaka, M., Jishage, K., Ueda, O., Sakaguchi, H., Higashi, T., Suzuki, T., Takashima, Y., Kawabe, Y., Cynshi, O., Wada, Y., Honda, M., Kurihara, H., Aburatani, H., Doi, T., Matsumoto, A., Azuma, S., Noda, T., Toyoda, Y., Itakura, H., Ysazaki, Y., Horiuchi, S., Takahashi, K., Kar Kruijt, J., van Berkel, T., Steinbrecher, Urs P., Ishibashi, S., Maeda, N., Gordon, S., and Kodama, T., 1997, Resistance to atherosclerosis and susceptibility to infection in scavenger receptor knockout mice. *Nature* 386:292.

Van Rooijen, N., 1997, Selective depletion of macrophages by liposome-encapsulated drugs, in: *Weir's Handbook of Experimental Immunology*, 5th ed., vol IV, L.A. Herzenberg, D.M. Weir, L.A. Herzenberg and C. Blackwell, eds., Blackwell Science Inc, Cambridge, MA.

Walker, W.S., 1997, Mouse macrophage cell lines, in: *Weir's Handbook of Experimental Immunology,* 5th ed., vol IV, L.A. Herzenberg, D.M. Weir, L.A. Herzenberg and C. Blackwell, eds., Blackwell Science Inc, Cambridge, MA.

Wright, S.D., and Detmers, P.A., 1988, Adhesion-promoting receptors in phagocytes. *J. Cell Science supp.* 9: 99.

Wu, L., Li, C-L., and Shortman, K., 1996, Thymic dendritic cell precursors: relationship to the T lymphocyte lineage and phenotype of the dendritic cell progeny. *J.Exp Med.* 184:903.

STEALTH™ THERAPEUTIC SYSTEMS: RATIONALE AND STRATEGIES

Daan J.A. Crommelin and Gert Storm

Utrecht Institute for Pharmaceutical Sciences
Utrecht University
P.O.Box 80.082, 3508 TB Utrecht
The Netherlands

In this chapter the reasons to develop 'stealth' therapeutic systems for intravenous administration are discussed and strategies to achieve this goal are outlined. In particular, attention is paid to alternatives for polyethylene glycol as 'stealth' inducing agent.

WHY 'STEALTH' TECHNOLOGIES? WHAT ARE THE OBJECTIVES?

The question that is to be addressed concerns the rationale for the development of 'stealth' technologies in the field of advanced drug delivery. In other words, what should 'stealth' technology actually try to achieve? There are three reasons.

Distribution of Carrier Systems to non-MPS (Mononuclear Phagocyte System) Sites

Carrier systems play a major role in the present attempts to achieve site specific delivery of therapeutic agents. Carriers modulate the pharmacokinetics of the drug by changing its elimination and distribution profile. The success of 'conventional' active targeting strategies (using a homing device) is limited by the rapid and extensive uptake of these carrier systems by cells of the MPS, in particular Kupffer cells and spleen macrophages. Reduced uptake by the MPS (both in terms of rate and extent) and longer circulation times are desirable to increase the chance of the drug loaded carrier system to actually reach the target site. This is not only true for active targeting strategies where a homing device is coupled to the surface of the carrier. As will be explained later, it also holds for those passive targeting strategies aiming for delivery to target sites located outside the MPS.

The System Can Act as a Drug-Reservoir Circulating in the Blood Compartment

This means that the carrier provides a long-circulating drug reservoir from which the drug can be slowly released over a prolonged period of time. Candidate drugs of particular interest

Table 1. Drug carrier systems: options

Macromolecular	Cellular	Particulate
antibodies	erythrocytes	liposomes
hormones	leukocytes	nanoparticles
lectins	hepatocytes	microspheres
DNA	fibroblasts	emulsions
carbohydrates		LDL

(adapted from Gregoriadis, 1978)

are peptides and proteins with short elimination half lives for which a prolonged presence at the target site is desired.

Improvement of Immunospecific Targeting Strategies

Immunospecific targeting strategies with particulate systems (e.g., those using antibodies as homing devices) are expected to be more successful when the circulation time in blood is prolonged without jeopardizing the targeting potential.

CARRIER SELECTION

Many carrier systems have been proposed over the years. Table 1 (adapted from Gregoriadis, 1978) lists only a few. The different carrier systems can be roughly divided into three categories: macromolecular, cellular and particulate systems. Even systems in the same category can strongly differ in their behavior *in vivo*. Disposition characteristics strongly depend on physicochemical properties such as size, charge, 'hydrophilicity', and the presence of ligands attached for homing purposes. When confronted with the selection process for the most suitable carrier system to solve a particular drug delivery problem, criteria such as disposition in the body and safety issues should be taken into account (discussed in more detail in the following sections). But, also pharmaceutical issues such as upscaling, carrier quality and supply, reproducibility of the production process and shelf-life should be considered in the decision-taking process.

ANATOMICAL, PHYSIOLOGICAL AND PATHOLOGICAL CONSTRAINTS AND OPPORTUNITIES

Physiological Conditions

The body is a highly compartmentalized system. The extent to which a carrier system is confronted with this compartmentalization phenomenon depends on its size, but also, for instance, its charge. Water can easily move from one compartment to another through an endothelial or epithelial lining. For antibodies, with a molecular weight of 150 kD, distribution over the different body compartments becomes already more difficult. For colloids in the 1 μm size range there is hardly any escape possible from the blood compartment upon intravenous injection under physiological conditions, except through phagocytosis by the macrophages in liver and spleen that are in direct contact with the blood circulation.

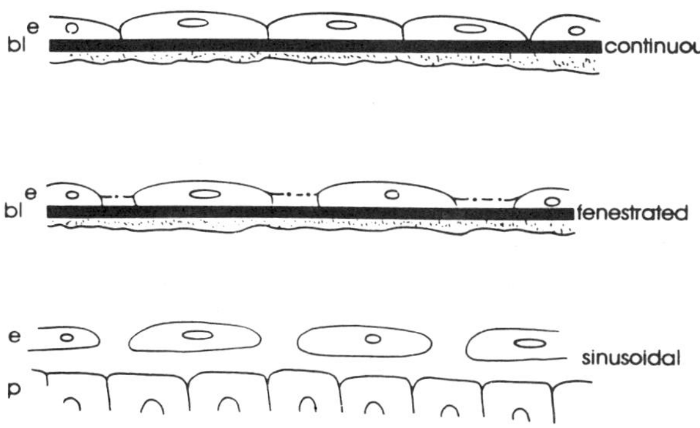

Figure 1. Schematic illustration of the structure of the wall of different classes of blood capillaries. A. Continuous capillary. The endothelium is continuous with tight junctions between adjacent endothelial cells. The subendothelial basement membrane is also continuous. B. Fenestrated capillary. The endothelium exhibits a series of fenestrae which are sealed by a membranous diaphragm. The subendothelial basement membrane is continuous. C. Discontinuous (sinusoidal) capillary. The overlying endothelium contains numerous gaps of varying size enabling materials in the circulation to gain access to the underlying parenchymal cells. The subendothelial basement is either absent (liver) or present as a fragmented interrupted structure (spleen, bone marrow). The fenestrae in the liver are about 0.1 - 0.2 μm in diameter; the pores between the endothelial cells and those in the basement membrane outside liver, spleen and bone marrow are much smaller. From Poste, 1985.

The body compartmentalization issue is illustrated in Figure 1 (taken from the work by Poste, 1985). It shows the difference in the endothelial lining of the blood vessels at different sites of the body under physiological conditions. The fenestrations in the continuous (endothelial cell layer plus basal membrane) endothelium have diameters up to about 20 nm in size. In sinusoid endothelium, such as that found in the liver and the spleen, where the basal membrane is interrupted, the fenestrae between endothelial cells are larger and reach dimensions in the 100 - 200 nm size range. This structural organization suggests that colloidal systems with particle sizes over 20 nm cannot leave the vascular bed readily in non-sinusoidal tissue. Thus, their therapeutic target should be circulating in the blood, or be present at the surface of the endothelial lining. Alternatively, macrophages in contact with blood can be the therapeutic target.

Pathological Conditions

The situation described in the previous section can change under pathological conditions. Maeda and collaborators (reviewed by Seymour, 1992) noticed that under pathological conditions the permeability of the vasculature in the diseased areas is often considerably enhanced. This effect was coined 'enhanced permeability and retention effect'. It is encountered in many tumors and inflammation sites. Under pathological conditions the vessel wall can have fenestrae as large as 150 nm.

Therefore, 'passive' targeting to these pathological sites, provided that the particle size of the carrier is between about 50 and 150 nm, may be a realistic option. Jain and collaborators studied the permeability of the vasculature in tumors in more detail and indeed found areas with a high permeability compared to vasculature in normal tissue (Jain, 1996). However, in these

Table 2. How to introduce 'stealth' effects into carrier systems

- Attach PEG (poly(ethylene glycol))
- Attach surface groups other than PEG; e.g., sialic acids GM1 polyglycerols
- Manipulate charge and/or molecular mass
- Change 'hydrophilicity'/'glycosylation' pattern

animal tumor models regions with poor penetration characteristics were also found. Thus, in principle, particulate systems with diameters below 150 nm can pass through the vasculature in areas with enhanced permeability under pathological conditions. But, the carrier systems should have enough time to escape from the circulation at the diseased site, i.e. they should be able to avoid the MPS long enough. 'Stealth' technology can provide the required long circulation times.

HOW CAN STABLE, LONG CIRCULATING 'STEALTH' SYSTEMS BE OBTAINED?

Table 2 lists several strategies to reduce the affinity of carrier systems for the MPS.

By far the most frequently used strategy is poly(ethylene glycol) (PEG) attachment to the carrier; this can be done by covalently attaching the PEG to the carrier. Alternatively, PEG can be attached non-covalently (through physical interactions) to the surface of colloidal particle systems such as polymeric nanoparticles by using surface active agents from the poloxamer or poloxamine series. Storm et al. (1995) provide a review on the publications that deal with different forms of pegylation of nanoparticles. There are two important issues that are usually not addressed in the literature and that are of utmost importance to obtain reproducible results: (a) there are several chemical binding strategies to attach PEG covalently to carrier systems. An overview of different options to couple PEG to phosphatidylethanolamine is given by Woodle and Lasic (1992). As the spacer group may play a role in the performance of the carrier system, the synthesis procedure should be carefully selected and described in detail in documents containing standard operating procedures; (b) polymers are heterogeneous macromolecules, for instance, in terms of their molecular weight. The characteristics and quality of polymers such as PEG and poloxamers, poloxamines and tweens have to be carefully checked and specifications should be met.

SELECTION OF NON-LIPOSOMAL SYSTEMS WITH 'STEALTH' CHARACTERISTICS

Polymeric Micelles

Polymeric micelles are made up of molecular units consisting of a hydrophilic (PEG) and a lipophilic section (e.g., polyaspartic acid). A cytostatic drug (e.g., doxorubicin) has been attached covalently to the lipophilic part. The schematic structure is given in Figure 2 (Okano et al., 1994). With these doxorubicin loaded micelles with diameters in the 50 - 100 nm range promising therapeutic results were obtained in mouse tumor model systems. An example of the superior therapeutic effect of the polymeric doxorubicin compared to free doxorubicin is presented in Figure 3 taken from Okano et al., (1994).

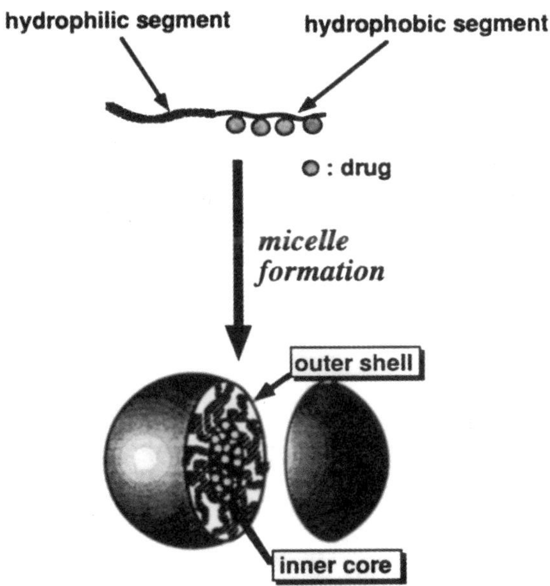

Figure 2. Concept of micelle-forming polymeric drug. The hydrophilic part can be PEG, the lipophilic segment, can be polyaspartic acid and the drug, the cytostatic doxorubicin.

Polymeric nanoparticles

Polymer-based particles in the nanometer range can be prepared with several different types of polymers (e.g., cyanomethacrylates, polylactic acid, polyglycolic acid). In the past these particles were converted into 'stealth' carriers by coating their surfaces with PEG containing copolymers of the poloxamer or poloxamine class (Storm et al., 1995). Recently, reports appeared where PEG was covalently bound to the nanoparticle surface. Veillard and co-workers (Bazile et al., 1995) showed that the presence of this covalently bound surface coating (PEG) dramatically changed the pharmacokinetic profile of the polylactic acid nanospheres in rats. The plasma half life increased from a few minutes to a number of hours.

SELECTION OF LIPOSOMAL SYSTEMS WITH 'STEALTH' COATINGS OTHER THAN PEG

At the present, PEG is the material of choice to prolong the circulation time of liposomes. A full issue of the J. Liposome Research (Huang, 1992) was devoted to long circulating liposomes and by far the most frequently used 'stealth' coating is PEG. GM1 or phosphatidylinositol in the bilayer are used to a much lesser extent. But, the question may be raised whether there are alternative coating materials in the pipeline with 'stealth' effects superior to PEG. Some publications have appeared where alternative coating strategies have been described and a selection of those is dealt with in this book. Sialic acids have been proposed (Schauer, 1985; Yamanauchi et al., 1995) and are indeed providing substantial prolongation of liposome half life compared to control liposomes containing GM1. Other examples are the polyoxazoline-DSPE (distearoylphosphatidylethanolamine) derivatives as developed by Zalipsky et al. (1996). Torchilin and co-workers (1995) described a number of different polymer-based 'stealth' approaches as well. And, finally, the idea of using phosphatidylpolyglycerols should be mentioned (Maruyama et al., 1995).

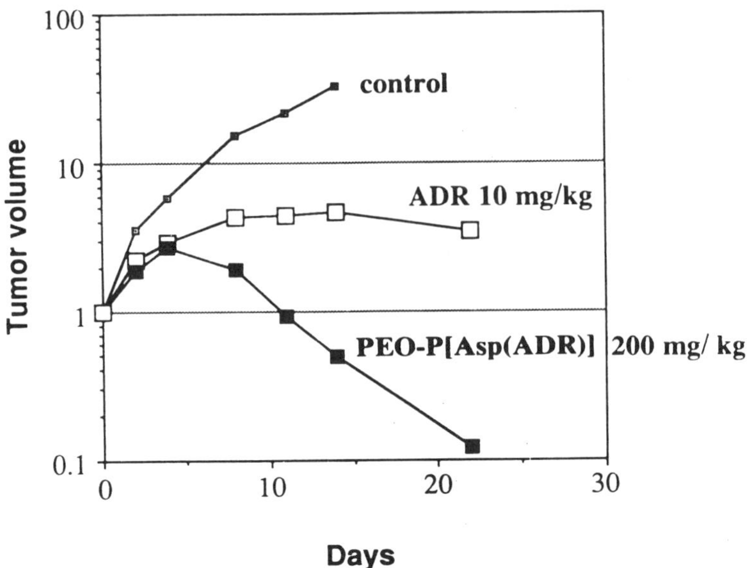

Figure 3. *In vivo* antitumor activity against murine C26. ADR: doxorubicin, PEO-P[Asp(ADR)]: polyethylene glycol - polyaspartic acid - doxorubicin conjugate. Taken from Okano et al., 1994; for details see Yokoyama et al., 1991.

In spite of all efforts made so far to develop coating materials that are superior to PEG as MPS-avoiding agents, the leading position of PEG as 'stealth' coating still does not seem to be seriously challenged.

DIRECT CHEMICAL CHANGES TO THE DRUG MOLECULE TO INCREASE CIRCULATION HALF LIFE

Apart from manipulating the disposition of colloidal particles by hydrophilic coatings, there is a completely different approach that has been successfully utilized in the past. Drug molecules can be structurally modified to prolong their circulation half life. A classical approach is the attachment of PEG to protein drugs such as asparaginase (MacEwen et al., 1987). An interesting example is the combination of PEG and interleukin-2 (IL-2) to form PEG-IL-2. The circulation half life of the pegylated protein is significantly prolonged compared to the native product; but, the question is whether that prolongation is indeed an advantage if one looks at the complicated PK/PD relationship for this drug (Braeckman, 1997). Recently, Gregoriadis and co-workers published studies on the sialylation of proteins to increase circulation half life and the therapeutic index of these protein drugs (Fernandes and Gregoriadis, 1996 and 1997)

Plasma clearance of protein drugs can also be reduced by removing clearance inducing amino acids by site directed mutagenesis, or by deleting sugar binding sites from the molecule (e.g., as done with tissue plasminogen activator) (cf. Braeckman, 1997).

Hashida and co-workers studied the effect of relatively small changes in the molecular structure of macromolecules on their renal and liver clearance in mice. The pronounced effects of changing molecular weight, charge and the attachment of galactose on the liver and renal clearances of a number of pharmaceutical proteins were recently reviewed and the reader is referred to this literature site for detailed information (Takakura and Hashida, 1996).

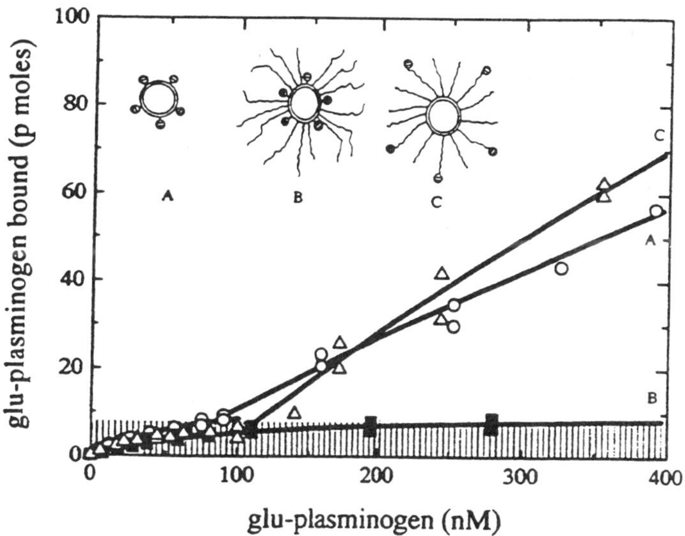

Figure 4a. Binding of liposome-associated glu-plasminogen to fibrin *in vitro*. (○) PEG-free proteoliposomes (type A); (■) PEG-coated proteoliposomes (type B); (△) proteo-PEG-coated liposomes (type C)(for further details see Blume et al., 1993).

ACTIVE TARGETING

For active targeting strategies with carriers with a homing device attached to their surface, long circulation half lives and MPS avoidance capability are highly preferred characteristics to maximize accumulation at the target site. Blume et al. (1993) and Torchilin et al. (1992) investigated whether long circulating liposomes with a PEG-coat could still be targeted to specific sites by using a homing device. Blume et al. (1993) designed three types of liposomes

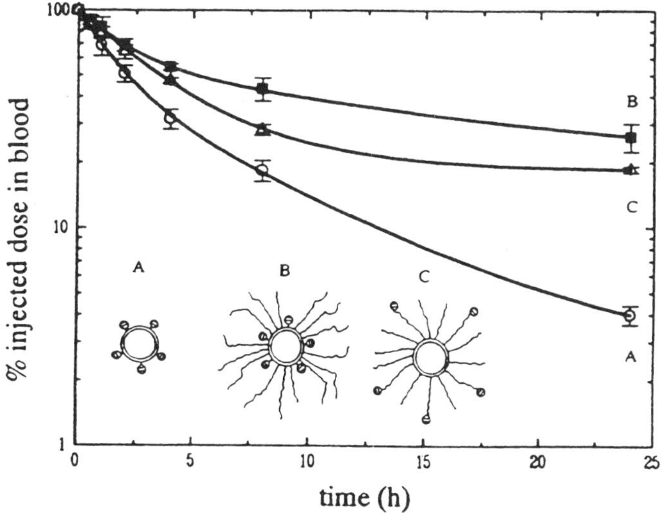

Figure 4b. Kinetics of elimination of glu-plasminogen proteoliposomes from the murine bloodstream after an i.v. injection of 2.5 μmoles of phospholipid. (○) PEG-free proteoliposomes (type A); (■) PEG-coated proteoliposomes (type B); (△) proteo-PEG-coated liposomes (type C)(for further details see Blume et al., 1993).

using plasminogen (plg) as a homing device. Plasminogen has an affinity for fibrin and the aim was to deliver these liposomes loaded with a thrombolytic specifically to fibrin clots in the blood circulation. The three liposome configurations are schematically shown in Figure 4. Liposomes were prepared with plasminogen covalently coupled to the liposome surface and no PEG present (see Figure 4, liposome A), liposomes with plasminogen covalently coupled directly to the liposome surface and PEG as well (Figure 4, liposome B), and liposomes with PEG coupled to the surface and plasminogen coupled to the end of the PEG chains sticking out in the aqueous medium (Figure 4, liposome C).

In Figure 4a the *in vitro* binding of the different plg-liposomes to fibrin is presented. It is clear that plg bound directly to the surface of long circulating liposomes with a PEG coat is no longer able to bind to fibrin. Figure 4b shows that when plasminogen is coupled to the PEG terminal ends, target binding ability is preserved while maintaining the long circulation characteristics in mice. These findings and observations with several other, different, experimental *in vivo* model systems (Torchilin et al., 1992; Maruyama et al., 1995; Emanuel et al., 1996a,b) lead to the conclusion that it is indeed possible to combine long circulation characteristics of pegylated liposomes with specific targeting strategies.

CONCLUSIONS

Some (proto)types of 'Stealth' liposome systems have reached the clinical stage of development and even entered the market place (Woodle and Storm, 1997). To provide the required hydrophilic surface characteristics for long blood circulation times, PEG (molecular weight range 2000 - 5000) either covalently bound to, or adsorbed onto the surface of nanoparticles has the most attractive features. But, at the present, alternative coatings such as sialic acids, poly(vinyl pyrrolidone) and polyglycerols are under investigation.

In conclusion, the rationale for the development of 'stealth' technologies is based on the following three concepts:

(i) **Passive targeting and MPS avoidance**: Passive targeting to non-blood pool and non-MPS target sites is possible, if long circulation times can be established and the carrier size does not exceed 100 - 150 nm. These are interesting options for targeted delivery of drugs or diagnostic agents (cf. Storm et al., 1997) to tumor and inflammation sites.

(ii) **Circulating microreservoir**: Liposomes have been used to provide a long circulating microreservoir for peptide delivery. Some sustained release effects have been found after intravenous administration of vasopressin encapsulated in long circulating liposomes (Woodle et al., 1992). Alternatively, one can refrain from using carrier systems and change the molecular structure of the active agent. Such a modification may reduce clearance rates while preserving pharmacological activity, if the position on the molecule responsible for fast clearance is identified and is known not to be involved in the desired activities of the molecule. It is always important to have insight into the PK/PD correlation of the drug in question, as this sheds light on the real need to prolong the blood circulation half-life of the drug (Braeckman, 1997).

(iii) **Active targeting**: Studies performed so far indicate that it is indeed possible to prolong the circulation time of targeted liposomes while preserving the homing capability of the homing device. This finding is of immense importance as it demonstrates that it is possible to combine two important features of the colloidal carrier system: long circulation times and escape possibilities from the blood circulation can go together with active targeting capabilities.

REFERENCES

Bazile, D., Prud'homme, C., Bassoullet, M-T., Marlard, M., Spenlehauer, G., and Veillard, M., 1995, Stealth Me.PEG-PLA nanoparticles avoid uptake by the mononuclear phagocyte system, *J.Pharm.Sci.*

84:493.

Blume, G., Cevc, G., Crommelin, D.J.A., Bakker-Woudenberg, I.A.J.M., Kluft, C., and Storm, G., 1993, Specific targeting with poly(ethylene glycol)-modified liposomes: coupling of homing devices to the ends of the polymeric chains combines effective target binding with long circulation times, *Biochim. Biophys. Acta.* 1149:180.

Braeckman, R., 1997, Pharmacokinetics and pharmacodynamics of peptide and protein drugs, in: *Pharmaceutical Biotechnology*, D.J.A. Crommelin and R. Sindelar, eds., Harwood Science Publ., Amsterdam.

Emanuel, E., Kedar, E., Bolotin, E.M., Smorodinsky, N.I., and Barenholz, Y., 1996a, Preparation and characterization of doxorubicin-loaded sterically stabilized immuno-liposomes, *Pharm. Research,* 13: 352.

Emanuel, E., Kedar, E., Bolotin, E.M., Smorodinsky, N.I., and Barenholz, Y., 1996b, Targeted delivery of doxorubicin via sterically stabilized liposomes: pharmacokinetics and biodistribution in tumor-bearing mice, *Pharm.Research,* 13:861.

Fernandes, A.I., and Gregoriadis, G., 1996, Synthesis, characterization and properties of sialylated catalase, *Biochim.Biophys. Acta,* 1293:90..

Fernandes, A.I., and Gregoriadis, G., 1997, Polysialylated asparaginase: preparation, activity and pharmacokinetics, *Biochim.Biophys. Acta,* 1341:26.

Gregoriadis, G., 1978, Liposomes in therapeutic and preventive medicine: the development of the drug-carrier concept. In: *Liposomes and their uses in Biology and Medicine*, D. Papahadjopoulos, ed., New York Academy of Sciences, New York

Huang, L., ed., 1992, Forum: Covalently attached polymers and glycans to alter the biodistributiom of liposomes, *J. Liposome Research,* 2:289.

Jain, R.K., 1996, Whitaker Lecture: delivery of molecules, particles and cells to solid tumors, *Ann Biomed. Eng.* 24:457.

MacEwen, E.G., Rosenthal, R., Matus, R., Viau, A.T., and Abuchowski, A., 1987, A preliminary study on the evaluation of asparaginase - polyethyleneglycol conjugate against canine malignant lymphoma. *Cancer Res.* 59:2011.

Maruyama, K., Takizawa, T., Yuda, T., Kennel, S.J., Huang, L., and Iwatsuru, M, 1995, Targetability of novel immunoliposomes modified with amphipatic poly(ethylene glycol)s conjugated at their distal terminals to monoclonal antibodies. *Biochim.Biophys. Acta* 1234:74.

Okano, T., Yui, N., Yokoyama, M., and Yoshida, R., 1994, Advances in Polymeric Systems for Drug Delivery, Gordon and Breach Science Publ., Switzerland.

Poste, G., 1985, Drug targeting in cancer therapy, in: *Receptor-Mediated Targeting of Drugs*, G. Gregoriadis, G. Poste, J. Senior, and A. Trouet, NATO ASI Series, Series A: Life Sciences, Plenum Press, New York

Schauer, R., 1985, Sialic acid and their role as biological masks, *TIBS* September 1985:357.

Seymour, L.W., 1992, Passive tumour-targeting of macromolecules and drug conjugates, *Crit.Rev.Ther. Drug Carrier Systems,* 9:135.

Storm, G., Belliot, S.H., Daemen, T., and Lasic, D.D., 1995, Surface modification of nanoparticles to oppose uptake by the mononuclear phagocyte system, *Adv. Drug Delivery Rev.* 17:31.

Storm, G., Bakker-Woudenberg, I.A.M.J., Schiffelers, R., Oyen, W.J.G., Crommelin, D.J.A., Corstens, F.H.M., and Boerman, O.C., Diagnostic and therapeutic targeting of infectious and inflammatory diseases using sterically stabilized liposomes, this book.

Takakura, Y., and Hashida, M., 1996, Macromolecular carrier systems for targeted drug delivery: pharmacokinetic considerations on biodistribution. *Pharm.Res.* 13:820.

Torchilin, V.P., Klibanov, A.L., Huang, L., O'Donnell, S., Nossif, N.D., and Khaw, B.A., 1992, Targeted accumulation of polyethylene glycol-coated immunoliposomes in infarcted rabbit myocardium, *FASEB J.*, 6:2716.

Torchilin, V.P., and Trubetskoy, V.S., 1995, Which polymers can make nanoparticulate drug carriers long-circulating?, *Adv. Drug Delivery Reviews*, 16:141.

Woodle, C., and Lasic, D.D., 1992, Sterically stabilized liposomes, *Biochim.Biophys. Acta* 1113:171.

Woodle, M.C., Storm, G., Newman, M.S., Jekot, J., Collins, L.R., Martin, F.J., and Szoka, F.C., 1992, Prolonged systemic delivery of peptide drugs by long circulating liposomes: illustration with vasopressin in BRATTLE BORO rat. *Pharm.Res.*, 9:260.

Woodle, M., and Storm, G., eds., 1997, *Long Circulating Liposomes. Old Drugs New Therapeutics*, Springer Verlag, NY.

Yamanauchi, H., Yano, T., Kato, T., Tanaka, I., Nakabayashi, S., Higashi, K., Miyoshi, S., and Yamada, H., 1995, Effects of sialic acid derivative on long circulation time and tumor concentration of liposomes. *J. Controlled Rel.* 113:141.

Yokoyama, M., Okano, T., Sakurai, Y., Ekimoto, H., Shibazaki, C., and Kataoka, 1991, Toxicity and antitumor activity against solid tumors of micelle-forming polymeric anticancer drug and its extremely long circulation in blood, *Cancer Res.* 51:3229.
Zalipsky, S., Hansen, C.B., Oaks, J.M., and Allen, T.A., 1996, Evaluation of blood clearance rates and biodistribution of poly(2-oxazoline)-grafted liposomes. *J.Pharm.Sci.* 85:133.

LONG CIRCULATING LIPOSOMES: EVOLUTION OF THE CONCEPT

Gregory Gregoriadis

Centre for Drug Delivery Research, The School of Pharmacy, University of London, 29-39 Brunswick Square, London, England

INTRODUCTION

The use of liposomes as a drug-carrier, proposed and tested in the early 1970s (Gregoriadis et al., 1971; Gregoriadis and Ryman, 1972a), has undergone a multitude of stages that have recently culminated in the licencing and marketing of several injectable pharmaceutical products (Gregoriadis, 1995). Comments on the impact of liposomes on drug delivery and targeting has been made elsewhere (Gregoriadis, 1998) . Here, I discuss some of the key developments which contributed to the understanding of vesicle fate in vivo and, eventually, control. The latter was achieved by tailoring the structural characteristics of liposomes in ways that ensure both quantitative retention of entrapped solutes during exposure of the carrier to blood en route to the target, and vesicle clearance rates from the circulation that are conducive to optimal pharmacokinetics.

Early work with solute-containing multilamellar vesicles (MLV) (Gregoriadis and Ryman, 1972a,b; Gregoriadis, 1973) revealed that vesicle rate of clearance from the blood of intravenously injected rats was rapid, dose-dependent and biphasic. Subsequently, it was observed (Gregoriadis and Neerunjum, 1974) that neutral MLV exhibited a slower rate of clearance than negatively charged MLV. The effect of neutral charge on vesicle clearance was also found (Juliano and Stamp, 1975) to apply to small unilamellar vesicles (SUV) which, however, exhibited a longer residence times than MLV. Another early observation (Gregoriadis, 1973) was that small, water-soluble drugs (eg. penicilin G) leaked considerably from circulating MLV. In view of what we know today, this is hardly surprising as MLV used in that study were cholesterol-poor. Perhaps more importantly, liposomes with materials that were still entrapped were shown to end up in the fixed macrophages of the reticuloendothelial system (RES), mainly in the liver and spleen, through the lysosomotropic pathway (Gregoriadis and Ryman, 1972b; Gregoriadis and Buckland, 1973; Black and Gregoriadis, 1974; Segal et al., 1974). Within lysosomes, vesicles were seen (Segal et al., 1974) to lose their organized structure (presumably through the action of phospholipases) and release their contents. Depending on molecular size and ability to withstand the hostile milieu of the organelles, drugs could then act either locally (Gregoriadis and Buckland, 1973) or, after diffusion through the lysosomal membrane, in other cell compartments (Black and Gregoriadis, 1976). Such knowledge of liposomal fate and behaviour enabled us (Gregoriadis, 1976a,b) to test, propose or anticipate a number of applications, including the treatment of certain inherited metabolic disorders, intracellular infections, cancer, gene therapy, immunopotentiation and vaccine delivery.

Another development in the 1970s was that liposomes coated with cell-specific ligands (eg. tumour cell-specific antibodies and the hepatocyte-specific asialoglycoproteins) could be targeted *in vivo* and internalized by cells expressing the appropriate antigens or receptors (Gregoriadis and Neerunjum, 1975; Gregoriadis et al., 1977) thus opening novel avenues of liposome research and potential uses. However, as initially predicted (Gregoriadis, 1974), success in this area, would not only require quantitative retention of entrapped drugs by vesicles but also sufficiently long time for circulating (targeted) vesicles to encounter and associate with, the target. Since the behaviour of vesicles *in vivo* emanates from their structural features and the environment in which the vesicles exist and interact with, knowledge of the influence of the biological milieu on both drug leakage and vesicle clearance rates appeared essential.

SOLUTE RETENTION BY LIPOSOMES

The first indication (Krupp et al., 1976; Scherpof et al., 1978) as to why liposomes might become unstable in the presence of blood thus allowing entrapped solutes to leak out, came from the observation that plasma high density lipoproteins (HDL) remove phospholipid molecules from the vesicle bilayer. It appeared reasonable then to assume that prevention of HDL action by, for example, rendering bilayers packed would improve vesicle stability. An earlier finding (Papahadjopoulos et al., 1973) that the presence of cholesterol in the bilayers reduces protein-induced vesicle permeability in buffer, was encouraging in this respect. Work to test this assumption were therefore carried out, first with MLVs (Gregoriadis and Davis, 1978) and later on with SUVs (Kirby et al., 1980a,b; Kirby and Gregoriadis, 1980; Tumer et al., 1983). With both types of vesicles, excess cholesterol (eg. a phospholipid to cholesterol molar ratio of 1:1) rendered bilayers more stable, with very little of the entrapped water-soluble marker or phospholipid being lost in the presence of blood *in vitro* or *in vivo* (blood) after intravenous, intraperitoneal or subcutaneous injection. Such a role for HDL in the destabilization of liposomes was further substantiated by (a) the demonstration (Senior et al., 1983) that cholesterol-free SUV exposed to blood plasma from lipoprotein deficient mice or a patient with congenital lecithin-cholesterol acyltransferase deficiency (characterized by low HDL levels), were more stable than when incubated with plasma from normal mice or humans; (b) the identification (Senior et al., 1983) of HDL as the only lipoprotein, among a number of lipoprotein species added to lipoprotein-deficient plasma, that was capable of restoring the vesicle-destabilizing activity of plasma.

Having established the role of excess cholesterol in promoting liposomal stability, it was of interest to see whether this could also be achieved with bilayers made rigid by substituting (egg) phosphatidylcholine (PC) with high-melting phospholipids. As anticipated (Kirby et al., 1980b), SUV or MLV made of hydrogenated lecithin, distearoyl phosphatidylcholine (DSPC) or sphingomyelin (SM) were less permeable to entrapped solutes in the presence of blood plasma, with bilayer stability becoming even greater when excess cholesterol was also present (Tumer et al., 1983; Senior et al., 1983; Gregoriadis and Senior, 1980; Senior and Gregoriadis, 1982; Senior et al., 1985; Wolff and Gregoriadis, 1984). For instance, liposomes composed of DSPC and equimolar cholesterol were able (Senior and Gregoriadis, 1982) to completely retain their solute (carboxyfluorescein; CF content) even after 48 hours of exposure to plasma at 37°C.

CLEARANCE OF LIPOSOMES FROM THE BLOOD CIRCULATION

Work on liposomal stability in the presence of blood plasma also showed that solute retention by cholesterol-rich SUV, although always pronounced during relatively short (eg. up to 2 h) periods of exposure to plasma, it was reduced significantly with some preparations (eg. SUV made of PC) on incubation for longer periods, whereas with others (eg. SM SUV) remained unchanged. It turned out (Gregoriadis and Senior, 1980) that this correlated with the residence time in the blood circulation of the corresponding vesicles. Thus, the greater the vesicle stability *in vitro* was the longer the vesicle half-life (Senior and Gregoriadis, 1982). To mention one example, SUV made of DSCP and equimolar cholesterol and shown to remain fully stable (in terms of CF release) on incubation with plasma for up to at least 48h, exhibited a half-life of 20h in the circulation of mice (Senior and Gregoriadis, 1982). A similarly long half life was also shown for cholesterol-rich sphingomyelin SUV (Gregoriadis and Senior, 1980), a finding observed independently by another group (Hwang et al., 1980). However, the relationship between vesicle stability and clearance observed for neutral SUV did not apply to negatively charged SUV or, more importantly, to larger neutral vesicles. In the case of large liposomes, clearance rates increased progressively with increasing size, even though bilayer stability remained high for prolonged periods in the presence of plasma (Senior et al., 1985). It thus appears that vesicle size and surface charge both override the state of bilayer stability in determining vesicle clearance from the circulation.

A tentative hypothesis (Gregoriadis, 1989) put forward, attempted to unify the role of liposomal phospholipids and cholesterol in controlling bilayer stability *in vitro* and half-lives of neutral vesicles *in vivo*. It was proposed that, for liposomes that survive complete disintegration following HDL attack, HDL-induced bilayer destabilization facilitates the adsorption onto or insertion into the bilayer of plasma proteins (eg. opsonins) thought to be responsible (Absolom, 1986) for the presentation of liposomes and other particles in a form recognizable by macrophages. The hypothesis predicts that the lesser the destabilization of the bilayer the fewer or smaller the "gaps" on it (and therefore the lesser the leakage of entrapped solutes) and, thus, the smaller the amount of opsonins capable of associating with the vesicles. Vesicles with decreasing amounts of opsonin on their surface would be then expected to exhibit a proportionally reduced uptake by the RES. Since 1976 a number of workers have shown that plasma components such as α_2-macroglobulin (Black and Gregoriadis, 1976), blood coagulation factors (Bonte and Juliano, 1986), components of the complement cascade (Loughrey et al., 1990), C reactive protein (Richards et al., 1979), and fibronectins (Rossi and Wallace, 1983) can bind to liposomes depending on vesicle lipid composition and surface properties. It has been recently claimed (Moghimi and Patel, 1993) that liposomes interact (or fail to interact) with tissue-specific proteins that are capable of facilitating (opsonins) or inhibiting (dysopsonins) uptake of vesicles with different lipid compositions by the macrophages of the liver, spleen or bone marrow. The ways by which the bilayer surface in its varying states of fluidity or destabilization encourages or discourages interaction with such proteins are unknown. However, the failure of a liver-specific opsonin to enhance uptake of liposomes rich in cholesterol or composed of high melting phospholipids by Kupffer cells, is compatible with the sequence of events in the proposed hypothesis .

LONG-CIRCULATING LIPOSOMES: FURTHER IMPROVEMENTS

Tailoring liposomes in ways that contribute to delaying their interception by the RES was early recognized (Gregoriadis, 1974, 1976a,b) as an essential component of any strategy for vesicle targeting to alternative sites. Such liposomes would be expected to circulate longer and interact effectively with a variety of cell targets intravascularly and, when possible, extravascularly (eg to treat tumors). To that end, the ability of long-lived cholesterol-rich SUV made of high melting phospholipids to deliver significant portions of the injected drug dose to solid tumours (Presant et al., 1988) or to cure solid-tumour bearing mice (Large and Gregoriadis, 1983) seemed promising. Indeed, the simplicity of the approach, coupled with novel technologies which achieve high (entrapped) drug to lipid mass ratios in small vesicles, has led to especially designed SUV for uses where a minimum vesicle size is a prerequisite for success, for instance the treatment of certain tumours (eg. Daunoxome™). Nonetheless, the challenge of long-lived larger vesicles, known to accommodate realistic quantities of drugs or molecules such as proteins (eg. haemoglobin and cytokines) and enzymes, still remains as strong as ever.

As manipulation of the lipid composition of large liposomes (over 100 nm in diameter) does not improve vesicle half-life in the circulation significantly (Senior et al., 1985), we (Senior et al., 1991) and others (see present book) have independently attempted to mimick Nature. Certain pathogenic bacteria with a highly hydrophilic coat, for instance, are capable of evading the immune system in animals, and the high concentration of sugars on the surface of circulating cells (eg. erythrocytes) is likely to contribute to their life span. It was thought conceivable that liposomes with a sufficiently hydrophilic surface would curtail opsonin adsorption and, thus, their interception by the RES. In this respect, the molecule of choice that eventually proved successful was polyethyleneglycol (PEG) and its various derivatives. As the present book attests, interest in this area is mushrooming with numerous applications in various stages of progress and a pegylated liposome-based formulation (Doxil™) already licenced. There is however room for improvement in terms, for instance, of further augmenting vesicle half-lives, applying the concept to even larger vesicles, or perhaps replacing PEG with a more natural, biodegradable molecule. PEG is non-biodegradable and the effect of its long term intracellular accumulation in the tissues (where the pegylated carrier, be it a liposome, a nanoparticle or a therapeutic protein, ends up), especially on chronic treatment, may be detrimental. There is evidence (Gregoriadis et al., 1993; Fernandes and Gregoriadis, 1996, 1997) to suggest that polysialic acids (easily harvested from the cultures of certain non-pathogenic bacteria) may be an attrative alternative.

Acknowledgements

I thank Mrs Concha Perring for excellent secretarial assistance.

REFERENCES

Absolom, D., 1986, Opsonins and dysopsonins - an overview, *Methods Enzymol.* 839:1

Black, C.D.V., and Gregoriadis, G., 1974, Intracellular fate and effect of liposome-entrapped actinomycin D injected into rats, *Biochem.Soc.Trans*, 2;869.

Black, C.D.V., and Gregoriadis, G., 1976, Interaction of liposomes with blood plasma proteins, *Biochem.Soc. Trans.* 4:253.

Bonte, F., and Juliano, R.L., 1986, Interaction of liposomes with serum proteins, *Chem.Phys.Lipids*, 40:359.

Fernandes, A., and Gregoriadis, G., 1996, Synthesis, characterization and properties of sialylated catalase, *Biophys. Biochim.Acta*, 1293:92.

Fernandes, A., and Gregoriadis, G., 1997, Polysialylated asparaginase: preparation, activity and pharmacokinetics, *Biophys.Biochim.Acta*, 1341:26.

Gregoriadis, G., 1973, Drug entrapment in liposomes, *FEBS Lett.* 36:292.

Gregoriadis, G., 1974, Structural requirements for the specific uptake of marcomolecules and liposomes by target tissues, in: *Enzyme Therapy in Lysosomal Storage Disease*, J.M. Tager, G.J.M. Hooghwinkel and W.Th. Daems, eds., North-Holland Publising Co., Amsterdam.

Gregoriadis, G., 1976a, The carrier potential of liposomes in biology and medicine, *New Engl.J.Med.* (Medical Progress article) 295:704.

Gregoriadis, G., 1976b, The carrier potential of liposomes in biology and medicine, *New Engl.J.Med.* 295:765

Gregoriadis, G., 1995, Engineering liposomes for drug delivery: progress and problems, *Trends in*

Biotechnology, 13:527.
Gregoriadis, G., 1998, Liposome research in drug delivery and targeting: Thoughts of an early participant, in: *Medical Applications of Liposomes*, D.D. Lasic and D. Papahadjopoulos, eds., Elsevier Science, BV Amsterdam.
Gregoriadis, G., and Ryman, B.E., 1972a, Fate of protein-containing liposomes injected into rats. An approach to the treatment of storage diseases, *Eur.J.Biochem.,* 24:485.
Gregoriadis, G., and B.E. Ryman, B.E., 1972b, Lysosomal localization of enzyme-containing liposomes injected into rats, *Biochem.J.* 128:142.
Gregoriadis, G., and Buckland, R.A., 1973, Enzyme-containing liposomes alleviate a model for storage disease, *Nature* (London) 244:170.
Gregoriadis, G., and Neerunjun, D.E., 1974, Control of the rate of hepatic uptake and catabolism of liposome-entrapped proteins injected into rats. Possible therapeutic applications, *Eur.J.Biochem.* 47:179.
Gregoriadis, G., and Neerunjun, E.D., 1975, Homing of liposomes to target cells *Biochem.Biophys.Res.Comm.* 65:537.
Gregoriadis, G., and Davis, C., 1979, Stability of liposomes *in vivo* and *in vitro* is promoted by their cholesterol content and the presence of blood cells, *Biochem. Biophys.Res.Comm.* 89:1287.
Gregoriadis, G., and Senior, J., 1980, The phospholipid component of small unilamellar liposomes controls the rate of clearance of entrapped solutes from the circulation, *FEBS Lett.* 119:43.
Gregoriadis, G., Leathwood, P.D., and Ryman, B.E., 1971, Enzyme entrapment in liposomes, *FEBS Lett.* 14:95.
Gregoriadis, G., Neerunjun, D.E., and Hunt, R., 1977, Fate of a liposome-associated agent injected into normal and tumour-bearing rodents. Attempts to improve localization in tumour tissues, *Life Sciences,* 21:357.
Gregoriadis, G., McCormack, B., Wang, Z., and Lifely, R., 1993, Polysialic acids: Potential in drug delivery, *FEBS Lett.*, 315:271.
Hwang, K.J., Luke, K.F.S., and Baumier, P.L., 1980, Hepatic uptake and degradation of unilamellar sphingomyelin/cholesterol liposomes: A kinetic study, *Proc.Natl.Acad.Sci.USA*, 77:4030.
Juliano, R.L., and Stamp, D., 1975, Effects of particle size and charge on the clearance of liposomes and liposome-encapsulated drugs, *Biochem.Biophys.Res.Commun.* 63:651.
Kirby, C., and Gregoriadis , G., 1980a, The effect of the cholesterol content of small unilamellar liposomes on the fate of their lipid components *in vivo, Life Sciences* 27:2223.
Kirby, C., Clarke, J., and Gregoriadis, G., 1980a, Effect of the cholesterol content of small unilamellar liposomes on their stability *in vivo* and *in vitro*, *Biochem.J.* 186:591.
Kirby, C., Clarke, J., and Gregoriadis, G., 1980b, Cholesterol content of small unilamellar liposomes controls phospholipid loss to high density lipoproteins in the presence of serum, *FEBS Lett.* 111:324.
Krupp, I., Chobanian, A.V., and Brecher, J.P., 1976, The *in vivo* transformation of phospholipid vesicles to a particle resembling HDL in the rat, *Biochem.Biophys.Res.Commun*, 72:1251.
Large, P., and Gregoriadis. G., 1983, Phospholipid composition of small unilamellar liposomes containing melphalan influences drug action in mice bearing PC6 tumours, *Biochem.Pharmacol* 32:1315.
Loughrey, H.C., Bally, M.B., Reinish, L.W., and Cullis, P.R., 1990, The binding of phosphatidylglycerol liposomes to rat platelets is mediated by complement, *Thromb.Haemost.* 64:172.
Moghimi, S.M., and Patel, H.M., 1993, Tecniques to study the opsonic effect of serum on uptake of liposomes by phagocytic cells from various organs of the RES, in: *Liposome Technology*, vol.3, G. Gregoriadis, ed., CRC Press, Boca Raton.
Papahadjopoulos, D., Cowden, M., and Kimelberg, H., 1973, Role of cholesterol in membranes. Effects of phospholipid-protein interactions, membrane permeability and enzyme activity, *Biophys.Biochim.Acta*, 310:8.
Presant, C.A., Proffitt, R.T., Turner, A.F., Williams, L.E., Winsor, D.W., Werner, J.L., Kennedy, P., Wiseman, C., Gala, K., McKenna, R.S., et al., Successful imaging of human cancer with indium^{111}-labelled phospholipid vesicles, *Cancer*, 62:905.
Richards, R.L., Gewurz, H., Siegel, J., and Alving, C.R., Interaction of C-reactive protein and complement with liposomes, *J.Immunol.*, 112:1185.
Rossi, J.D., and Wallace, B.A., 1983, Binding of fibronectin to phospholipid vesicles, *J.Biol.Chem.* 258:3327.
Segal, A.W., Wills, E.J., Richmond, J.E., Slavin, G., Black, C.D.V., and Gregoriadis, G., 1974, Morphological observations on the cellular and subcellular destination of intravenously administered liposomes, *Brit.J.Exp.Pathol.* 55:320.
Scherphof, G., Roerdink, G., Waite, M., and Parks, J., 1978, Disintegration of phosphatidylcholine liposomes inplasma as a result of interaction with high-density lipoproteins, *Biophys.Biochim.Acta*, 542:296.
Senior, J., and Gregoriadis, G., 1982, Is half-life of circulating small unilamellar liposomes determined by changes in their permeability?, *FEBS Lett* 145:109.
Senior, J., Gregoriadis, G., and Mitropoulos, K., 1983, Stability and clearance of small unilamellar liposomes: Studies with normal and lipoprotein-deficient mice, *Biochem.Biophys.Acta* 760:111.

Senior, J., Crawley, J.C.W., and Gregoriadis, G., 1985, Tissue distribution of liposomes exhibiting long half-lives in the circulation after intravenous injection, *Biochim.Biophys.Acta* 839:1.

Senior, J.H., Delgado, C., Fisher, D., Tilcock, C., and Gregoriadis, G., 1991, Influence of surface hydrophilicity of liposomes on their interaction with plasma proteins and clearance from the circulation: Studies with polyethylene glycol-coated vesicles, *Biochim.Biophys.Acta,* 1062:77.

Tumer, A., Kirby, C., Senior, J., and Gregoriadis, G., 1983, Fate of cholesterol-rich unilamellar liposomes containing ^{111}Inlabelled bleomycin after subcutaneous injection into rats, *Biochim.Biophys.Acta* 760:119.

Wolff, B., and Gregoriadis, G., 1984, The use of monoclonal anti-Thy$_1$ IgG$_1$ for the targeting of liposomes to AKR-A cells *in vitro* and *in vivo*, *Biochim.Biophys.Acta* 802:259.

STERICALLY STABILIZED IMMUNOLIPOSOMES: FORMULATIONS FOR DELIVERY OF DRUGS AND GENES TO TUMOR CELLS *IN VIVO.*

J.W. Park*, D. Kirpotin+^, K. Hong+^, W. Zheng+^ Y. Shao+^
O. Meyer+^, C.C. Benz* and D. Papahadjopoulos+^

Department of Cellular and Molecular Pharmacology[+] and
Division of Hematology/Oncology, Cancer Research Inst*, UCSF;
Liposome Research Laboratory, CPMCRI^, San Francisco, CA, USA

INTRODUCTION

Immunoliposomes represent a promising strategy to achieve targeted drug delivery for the treatment of cancers overexpressing specific surface receptors. Advances in immunoliposome design have been facilitated by independent progress in the areas of antibody-based therapeutics and liposomes, which can now be utilized for tumor-targeted drug delivery Park et al., 1997a).

A logical focus for the development of targeted cancer therapies is the HER2 receptor. The HER2 (C-ErbB-2, *neu*) protooncogene and its encoded p185^{HER2} (ErbB-2) receptor tyrosine kinase play an important role in the pathogenesis of breast and other cancers (for review, see Hynes and Stern, 1994). HER2 is overexpressed stably in a significant proportion of cancers, but in normal tissue it is expressed at much lower levels, if at all. As a cell surface receptor it is readily accessible to antibody-based therapeutics, and monoclonal antibodies directed against HER2 can inhibit tumor growth by affecting HER2 signal transduction. For example, muMAb4D5 and its humanized derivative, rhuMAbHER2, inhibit the growth of HER2-overexpressing breast cancer cells (Lewis et al., 1993) and enhance the efficacy of certain chemotherapy drugs (Shepard et al., 1991). In phase II clinical trials, treatment with rhuMAbHER2 alone (Baselga et al., 1996), or in combination with cisplatin chemotherapy (Pegram et al., 1996) was associated with encouraging antitumor activity against advanced breast cancer. rhuMAbHER2 is currently being evaluated in phase III clinical trials for the treatment of advanced breast cancer in conjunction with first-line chemotherapy.

Important new developments in liposomal drug delivery include "sterically stabilized" or "stealth" liposomes, which are more resistant to reticuloendothelial system (RES) clearance than so-called "conventional" liposomes (Papahadjopoulos et al., 1991). These developments

Targeting of Drugs 6: Strategies for Stealth Therapeutic Systems
Edited by Gregoriadis and McCormack, Plenum Press, New York, 1998

produced substantially prolonged drug circulation with enhanced tumor accumulation (Papahadjopoulos et al., 1991; Huang et al., 1992a; Lasic and Martin, 1995). The coupling of anti-HER2 antibody fragments to sterically stabilized liposomes (SIL) represents a more recent design modification to further increase the therapeutic index of an encapsulated drug via selective delivery to HER2-overexpressing cancer cells (Park et al., 1995). The receptor-mediated internalization property of this targeted delivery vehicle also affords an opportunity to achieve tumor-specific uptake of novel therapeutics such as genes and oligonucleotides.

FORMULATION OF ANTI-HER2 IMMUNOLIPOSOMES

Immunoliposomes (SIL) targeted to the HER2 receptor, incorporate multiple design elements to optimize intracellular delivery of encapsulated agent to tumor cells. These include: (a) rhuMAbHER2 Fab' for reduced clearance and immunogenicity, and to provide internalization and antiproliferative activities; (b) PEG-Fab' linkage to facilitate immunoliposome binding and internalization; (c) sterically stabilized immunoliposomes for prolongation of circulation and selective tumor extravasation; and (d) encapsulated therapeutic agents (e.g.doxorubicin, nucleic acids) for enhanced therapeutic index via targeted intracellular delivery.

These designs were tested by constructing multiple versions of anti-HER2 immunoliposomes containing covalently linked rhuMAbHER2 Fab' (Genentech, Inc.) and poly(ethylene glycol) (PEG)-phosphatidylethanolamine (PE) in varying proportions (0-12 mol % of total phospholipid). SIL were prepared by conjugation of "conventional" (phosphatidylcholine + cholesterol; PC/Chol) or "sterically stabilized" (PC/Chol/PEG-PE) small unilamellar liposomes with rhuMAbHER2-Fab'. Initially, Fab' was conjugated to maleimido-phosphatidylethanolamine (M-PE), resulting in Fab' directly linked to the liposome surface, and hence in parallel with PEG. Recently, we have also prepared SIL with Fab' conjugated to maleimide-terminated PEG-PE (MMC-PEG-DSPE or MP-PEG-DSPE) (Kirpotin et al, 1997; O'Connell et al., 1993), resulting in Fab' linked to the distal end of PEG chains. Both procedures were highly efficient, typically yielding 50-100 Fab' fragments per liposome particle.

INTERACTION WITH CANCER CELLS *IN VITRO*

Studies of SIL binding, internalization and intracellular drug delivery were performed quantitatively with SIL containing a pH-sensitive fluorescent probe (1-hydroxypyrene-3,6,8-trisulfonic acid (HPTS/pyranine). These studies demonstrated rapid uptake of SIL by SK-BR-3 cells into a neutral environment, with subsequent accumulation in an acidic intracellular compartment, consistent with surface binding followed by receptor-mediated endocytosis (Kirpotin et al., 1997a). Total uptake of SIL in SK-BR-3 cells reached 23,000 SIL/cell; while total uptake of non-targeted control liposomes was essentially undetectable. In addition, total uptake of SIL in non-HER2-overexpressing MCF-7 cells was 700-fold lower than in SK-BR-3 cells. With liposomes Fab'-linked SIL, but not with PEG-Fab'-linked SIL, high PEG content was associated with reduced binding affinity and endocytosis. Binding and endocytosis also depended on the quantity of conjugated Fab', reaching a plateau at ~40 Fab'/immunoliposome for binding, and ~10 Fab'/immunoliposome for internalization.

The tumor cell uptake of SIL containing colloidal gold particles was studied by electron microscopy (Park et al., 1995). SK-BR-3 cells treated with anti-HER2 SIL (0 mol% PEG) showed gold-loaded SIL at the cell surface and intracellularly in coated pits, coated vesicles, endosomes, multivesicular bodies, and lysosomes, consistent with internalization occurring via the coated pit pathway. In addition, gold particles were noted free within the cytoplasm and not associated with membrane-bound organelles, indicating that delivery outside the endolysosomal pathway had also occurred.

PHARMACOKINETICS FOLLOWING I.V. INJECTION

Pharmacokinetic studies of doxorubicin (dox)-loaded anti-HER2 immunoliposomes following single i.v. injection were performed in normal adult rats. Plasma levels of dox following i.v. injection with either SIL or control (no Fab') sterically stabilized liposomes (SL) were similar, and indicated biexponential rate of clearance with terminal plasma half-lives of greater than 10 h and mean residence times of 16-24 h (Park et al., 1996). By comparison, levels of dox following injection of free dox at the same dose (800 mg) were undetectable at 5 min. The integrity of dox-loaded anti-HER2 SIL *in vivo* was assayed by two-component pharmocokinetic studies, in which plasma pharmacokinetics of dox and of rhuMAbHER2-Fab' were independently determined on identical plasma samples following single i.v. injection. The terminal plasma half-lives of rhuMAbHER2-Fab'(as measured by ELISA) and that of dox (as measured fluorimetrically) were both approximately 10 h, indicating negligible dissociation or drug leakage.

LOCALIZATION IN HUMAN TUMOR XENOGRAFTS IN MICE

Tumor localization of dox-loaded anti-HER2 SIL were evaluated in HER2-overexpressing tumor xenograft models, in which nude mice carrying established subcutaneous (s.c.) BT-474 tumor xenografts received a single i.v. injection of SIL (Park et al., 1996). The concentration of dox in tumor tissue still exceeded 1%/g of tissue 67 h after injection, which was significantly higher than dox levels in all other tissues except liver, the major site of liposome clearance. Tumor/blood and tumor/muscle ratios were both greater than 22-fold.

SIL localization in tumors was studied histologically in two tumor xenograft models, BT-474 and MCF7/HER2 (MCF7 cells stably transfected with HER2), following administration of SIL containing colloidal gold (Park et al., 1997). Gold-loaded SIL were administered i.v. every 48 h for 3 doses and subsequently visualized by light microscopy within tumor tissue using a silver enhancement technique (Huang et al., 1992). Silver grains indicating the presence of gold particles were observed throughout tumor tissue, both in perivascular areas and within cellular regions of the tumor. Importantly, silver grains were frequently observed within the cytoplasm of individual tumor cells, indicating intracellular delivery of gold particles. In contrast, treatment with control (no Fab') sterically stabilized liposomes (SL) showed silver grains accumulating in predominantly extracellular and perivascular spaces, consistent with previous reports of the tumor interstitial localization of sterically stabilized liposomes (Huang et al., 1992b). These control liposomes were not observed within individual tumor cells.

Intracellular localization of anti-HER2 SIL *in vivo* was confirmed using encapsulated ^{67}Ga-DTPA chelate administered by single i.v. injection in the MCF7/HER2 tumor xenograft model (Park et al., 1996). After 48 h, excised tumors were analyzed by electron microscopic autoradiography. Autoradiographic grains signifying ^{67}Ga emission resulting from immunoliposome delivery were frequently detected within tumor cells, at or near the cell surface, within the cytoplasm, and within the nucleus. In contrast, tumors from mice treated with ^{67}Ga-loaded control liposomes showed no intracellular localization of ^{67}Ga.

ANTI-TUMOR EFFICACY

Delivery of dox by anti-HER2 SIL may represent a particularly advantageous strategy for the treatment of HER2-overexpressing breast cancers, since these cancers appear to possess an especially steep dose-response relationship to dox-based therapy (Muss et al., 1994). In addition, the significant clinical problem of dox toxicity to myocardium and hematopoietic cells may be largely mitigated by anti-HER2 immunoliposome delivery, as HER2 expression is negligible in these cell types (Press et al., 1990). Finally, the antiproliferative effect of the anti-HER2 immunoliposome vehicle (i.e., the rhuMAbHER2 effect) may work synergistically with dox, since it is has been shown that rhuMAbHER2 augments the efficacy of dox in animal models (Baselga et al., 1994).

Studies with dox-loaded anti-HER2 SIL *in vitro* showed efficient and specific cytotoxicity against HER2-overexpressing breast cancer cells (Park et al., 1995). Treatment of SK-BR-3 cells for 1 h with dox-loaded anti-HER2 SIL yielded dose-dependent cytotoxicity (IC50 = 0.3 mg/ml) equivalent to that of free dox, indicating that SIL delivered dox as efficiently as the rapid diffusion of free dox into cells *in vitro*, and were up to 30-fold more cytotoxic than dox-loaded SIL bearing irrelevant Fab'. The specificity of targeting was further confirmed by treatment of WI-38 cells, a nonmalignant lung fibroblast cell line expressing minimal levels of HER2, wherein dox-loaded anti-HER2 SIL demonstrated cytotoxicity that was 20-fold less than free dox and equivalent to SIL with irrelevant Fab'.

Anti-tumor efficacy studies using dox-loaded anti-HER2 SIL in multiple HER2-overexpressing tumor xenograft-nude mice models showed a significant increase in the therapeutic index of dox due to targeted delivery (Park et al., 1996). In these studies, experimental treatment was initiated 1-2 wks after tumor implantation, at which time tumors were 200-1000 mm^3 in volume. SIL were administered at a total dox dose of 15 mg/kg, divided over three weekly i.v. injections. Therapeutic effects were compared to those of dox-loaded control (no Fab') SL at the same dose and schedule. Additional control arms include treatment with free dox, which was given at its maximum tolerated dose (MTD) in these animals (7.5 mg/kg), free rhuMAbHER2, and saline. In each of the four models (two independent sublines of BT-474, MCF7/HER2, and MDA-MB-453), treatment with SIL resulted in marked tumor growth inhibition and/or regression (including some cured animals) and was significantly superior to all other treatment conditions. Furthermore, SIL did not produce any apparent acute toxicities or significant weight loss in the nude mice. The MTD determined for SIL represents a 2.5-fold increase over that of free dox.

DELIVERY OF OLIGONUCLEOTIDES AND PLASMIDS

The results described above suggest that anti-HER2 SIL may be particularly advantageous for tumor-specific treatment with agents that require intracellular delivery. For example, anti-HER2 SIL could represent a gene therapy vector system that provides targeted delivery of therapeutic genes or oligonucleotides to tumor cells. To overcome existing obstacles for successful implementation of this strategy, SIL must be reconstructed to have the following properties: efficient packaging of DNA, favorable retention of drug delivery properties (such as stability, long circulation, minimal non-specific reactivity, and reduced immunogenicity), selective and efficient delivery of nucleic acids to tumor cells, and finally intracellular delivery that results in adequate gene expression.

A modification of anti-HER2 SIL for targeted gene therapy has been to include cationic lipids for efficient and high capacity packaging of DNA. Cationic liposomes, which readily form complexes with DNA molecules via electrostatic interactions and can mediate gene transfer to a variety of cell types, have previously attracted much attention as a gene therapy vector system. However, cationic liposomes developed to date have been limited by poor stability, high non-specific reactivity, and lack of targeting. We recently reported cationic liposome-plasmid DNA complexes that are significantly more stable by manipulation of their lipid composition, inclusion of PEG-PE, and condensation of plasmid DNA with polyamines (Hong et al., 1997). These stable constructs show high gene transfer efficiency *in vitro* and in multiple tissue sites *in vivo* following i.v. administration. For targeting, rhuMAbHER2-Fab' fragments have recently been conjugated covalently to PEG-containing cationic liposomes, thus generating anti-HER2 cationic immunoliposome-DNA complexes. In these cationic SIL, PEG-PE provides the attachment site for Fab', and when present at appropriate concentration, minimizes non-specific reactivity. By this design, anti-HER2 cationic SIL mediate efficient transfection of SK-BR-3 cells *in vitro*: the addition of Fab' is associated with a 20-fold increase in reporter (firefly luciferase) gene expression as compared to cationic liposomes lacking Fab', with minimal transfection observed in non-HER2-overexpressing MCF7 cells (W. Zheng and K. Hong, unpublished results).

Anti-HER2 cationic SIL are also being developed for targeted delivery of oligonucleotides. In recent studies, we have been able to show that a new formulation of these SIL were found to be internalized in SK-BR-3 cells, resulting in rhodamine-labeled liposome accumulation within the cytosol, while the released FITC-labeled oligonucleotides accumulated within the nucleus (Meyer et al., 1998).

ENHANCED RATES OF RELEASE AND EXTRAVASATION

In an attempt to stimulate the extravasation and release of liposome material after localization in tumors, we have developed several different approaches, which are described below. Although these formulations are not targeted to tumor cells specifically, their application to this field could be substantial and could certainly include targeting.

Thermosensitive liposomes have been formulated to respond to an increase in temperature from 37 to 42 degrees C, by an increase in the rate of release of encapsulated drugs in the presence of plasma or serum (Gaber et al., 1995). *In vivo* studies with such liposomes have shown that externally applied local hyperthermia can not only increase the extravasation of intact lipoosomes into the tumor mass, but also increase drastically the rate of release of the encapsulated drug, doxorubicin (Gaber et al., 1996).

Increased extravasation of liposomes into specific tissues has also been observed after systemic administration of vasoactive agents such as substance P (Rosenecker et al., 1996). In such a system, there is a very large increase in the accumulation of liposomes along with their contents into the bronchial airways, the oesophagus and the urinary bladder (Rosenecker et al., 1996). This indicates the potential of increasing the permeability of specific tissues to liposomes by selecting specific ligands that have effects on the endothelial cells of such tissues. Recently we have been able to show that when such liposomes encapsulate β-adrenergic agonists, they can inhibit the effect of subsequent injections of substance P, indicating an anti-inflammatory effect (Zhang et al., 1998).

Finally, we have reported recently the synthesis of new conjugates of PEG-PE where PEG can be detached from liposomes under the influence of mild reducing agents (Kirpotin et al., 1997c). Such detachment is followed by aggregation and fusion, which may be useful for applications requiring intracellular delivery (Park et al., 1997b; Kirpotin et al., 1997a).

DISCUSSION

Anti-HER2 SIL represent a potentially powerful strategy for the treatment of HER2-

overexpressing cancers because of their ability to provide tumor-targeted intracellular drug delivery. Similarly to non-targeted sterically stabilized liposomes, SIL are stable and long circulating *in vivo*. Unlike liposomes, SIL bind to and internalize in target cells but are not reactive with normal cells. In HER2-overexpressing tumor xenograft-nude mouse models, anti-HER2 SIL localize selectively in tumor tissue, where they deliver encapsulated agents intracellularly. In each animal model tested, dox-loaded anti-HER2 SIL greatly extend the therapeutic index of dox, both by increasing its antitumor efficacy and by reducing the systemic host toxicity; SIL exhibited significantly superior efficacy as compared with either dox-loaded sterically stabilized liposomes or free dox.

The anti-HER2 immunoliposome strategy may also enable new therapeutic applications, such as gene therapy. Compared with cationic liposomes, anti-HER2 cationic SIL provided significantly higher levels of reporter gene expression in target cells, but not in non-target cells. Anti-HER2 cationic SIL have also been developed to deliver oligonucleotides to target cells, resulting in the internalization of the complex and nuclear accumulation of oligonucleotides.

The properties of anti-HER2 SIL described here make them a promising and novel strategy for tumor-targeted therapy. In particular, the development of immunoliposome constructs which are stable, long circulating, and able to deliver encapsulated agents intracellularly in tumor cells appears to be important for the success of this strategy. These properties may confer sufficient pharmacokinetic and pharmacodynamic advantages to overcome the previous limitations of monoclonal antibody therapy and liposomal drug delivery (Park et al, 1997b; Kirpotin et al, 1997b).

Acknowledgments

We sincerely thank Paul Carter, Ph.D., Gilbert-Andre Keller, Ph.D., and William I Wood, Ph.D. of Genentech, Inc.(So.S.F., CA). We also thank Drs. Gail T. Colbern, D.V.M. of the Geraldine Brush Cancer Research Instititute/California Pacific Medical Center (S.F., CA), and Jose Baselga, M.D. and John Mendelsohn, M.D. of Memorial Sloan-Kettering Cancer Center (N.Y., N.Y.) for their contributions relating to the tumor xenograft-nude mouse studies. This work was partially supported by grants from the Breast Cancer S.P.O.R.E. Program of the National Cancer Institute and National Institutes of Health (P50-CA 58207-01); the U.S. Army Medical Research and Materiel Command (DAMD17-94-J-4195); the California Breast Cancer Research Program (2CB-0250); and the American Society of Clinical Oncology Young Investigator Award (J.W.P.) sponsored by the Don Shula Foundation, and a grant from Sequus Pharmaceuticals, Menlo Park, CA.

REFERENCES

Baselga, J., Tripathy, D., Mendelsohn, J., Baughman, S., Benz, C.C., Dantis, L., Sklarin, N.T., Seidman, A.D., Hudis, C.A., Moore, J., Rosen, P.P., Twaddell, T., Henderson, I.C. and Norton, L., 1996, Phase II study of weekly intravenous recombinant humanized anti-p185^{HER2} monoclonal antibody in patients with HER/*neu*-overexpressing metastatic breast cancer, *J. Clin. Onc.*. 14:737.

Baselga, J., Norton, L., Shalaby, R. and Mendelsohn, J., 1994, Anti-HER2 humanized monoclonal antibody (Mab) alone and in combination with chemotherapy against breast carcinoma xenografts, *Proc. Amer. Soc. Clin. Oncol.* 13:63.

Gaber, M.H.., Hong, K., Huang, S-K. and Papahadjopoulos, D., 1995, Thermosensitive sterically stabilized liposomes: Formulation and *in vitro* studies on mechanism of doxorubicin release, *Pharmaceut Res.* 2:1407.

Gaber, M.H., Wu, N., Hong, K., Huang, S., Dewhirst, M. and Papahadjopoulos, D., 1996, Thermosensitive liposomes: extravasation and release of contents in tumor microvascular networks, *Intl J Rad Oncol Biol Phys* .36:1177.

Hong, K., Zheng, W., Baker, A. and Papahadjopoulos, D., 1997, Stabilization of cationic liposome-plasmid DNA complexes by polyamines and poly(ethylene glycol)-phospholipid conjugates for efficient *in vivo*

gene delivery, *FEBS Lett.* 400:233.
Huang, S.K., Mayhew, E.M., Gilani, S., Lasic, D.D., Martin, F.J. and Papahadjopoulos, D., 1992a, Pharmacokinetics and therapeutics of sterically stabilized liposomes in mice bearing C-26 colon carcinoma, *Cancer Res.* 52:6774.
Huang, S.K., Lee, K.D., Hong, K., Friend, D.S. and Papahadjopoulos, D., 1992b, Microscopic localization of sterically stabilized liposomes in colon carcinoma-bearing mice, *Cancer Res.* 52:5135.
Hynes, N.E., and Stern, D.F., 1994, The biology of *erb*B-2/*neu*/HER-2 and its role in cancer, *Biochim. Biophys. Acta* . 1198:165.
Kirpotin, D., Park, J.W., Hong, K., Zalipsky, S., LI, W.L., Carter, P., Benz, C.C. and Papahadjopoulos, D., 1997a, Sterically stabilized anti-HER2 immunoliposomes: design and targeting to human breast cancer cell *in vitro*, *Biochemistry.* 36:66.
Kirpotin, D., Park, J.W., Hong, K., Shao, Y., Shalaby, R., Cobern, G., Benz, C. and Papahadjopoulos, D., 1997b, Targeting of liposomes to solid tumors: the case of sterically stabized anti-HER2 immunoliposomes, *Forum on "Liposomes Targeting In Animal Models", J Liposome Res.* 7(4):391.
Kirpotin, D., Hong, K., Mullah, N., Papahadjopoulos, D., and Zalipsky, S., 1997c, Liposomes with detachable polymer coating: destabilization and fusion of dioleylphosphatidylethanolamine vesicles trigered by cleavage of surface grafted PEG, *FEBS Lett.* 388:115.
Lasic, D.D. and Martin, F.J., eds, 1995, Chapters 1, 19, 11, 12, 14, 21, 22, in: *Stealth Liposomes*. CRC, Boca Raton.
Lewis, G.D., Figari, I., Fendly, B., Wong, W.L., Carter, P., Gorman, C., and Shepard, H.M., 1993, Differential responses of human tumor cell lines to anti-p185HER2 monoclonal antibodies, *Cancer Immunol. Immunother.* 37:255.
Meyer, O., Kirpotin, D., Hong, K., Sternberg, B., Park, J.W., Woodle, M.C., and Papahadjopoulos, D., 1998, Cationic liposomes coated with poly(ethylene glycol) as carriers for oligonucleotides, *J Biol Chem.* submitted.
Muss, H.B., Thor, A.D., Berry, D.A., Kute, T., Liu, E.T., Koerner, F., Cirrincione, C.T., Budman, D.R., Wood, W.C., Barcos, M. and Henderson, I.C., 1994, c-*erb*B-2 expression and response to adjuvant chemotherapy in women with node-positive early breast cancer, *N. Engl. J. Med.* 330:1260.
O'Connell, R.P., Carter, P., Presta, L., Eigenbrot, C., Covarrubias, M., Snedecor, B., Speckart, R., Blank, G., Vetterlein, D. and Kotts, C., 1993, Characterization of humanized anti-p185HER2 antibody Fab fragments produced in *E.coli.* in: *Protein Folding In Vivo and In Vitro*, J. L. Cleland, ed., American Chemical Society, Washington, D.C.
Papahadjopoulos, D., Allen, T.M., Gabizon, A., Mayhew, E., Matthay, K., Huang, S.K., Lee, K.D., Woodle, M.C., Lasic, D.D., Redemann, C. and Martin, F.J., 1991, Sterically stabilized liposomes: improvements in pharmacokinetics and antitumor therapeutic efficacy, *Proc. Natl. Acad. Sci. USA.* 88:11460.
Park, J.W., Hong, K., Kirpotin, D.B., Papahadjopoulos, D., and Benz, C.C., 1997a, Immunoliposomes for cancer treatment, in: Gene Therapy, J. T. August, M. W. Anders, F. Murad, and J. T. Coyle, eds, Academic Press, San Diego
Park, J.W., Hong, K., Carter, P., Asgari, H., Guo, L.Y., Keller, G.A., Wirth, C., Shalaby, R., Kotts, C., Wood, W.I., Papahadjopoulos, D. and Benz, C.C., 1995, Development of anti-p185HER2 immunoliposomes for cancer therapy, *Proc. Natl. Acad. Sci. USA* . 92:1327.
Park, J.W., Colbern, G., Baselga, J., Hong, K., Shao, Y., Kirpotin, D., Nuijens, A., Wood, W., Papahadjopoulos, D. and Benz, C., 1996, Antitumor efficacy of anti-p185HER2 immunoliposomes: enhanced therapeutic index due to targeted delivery, *Proc. Amer. Soc. Clin. Oncol.*. 15:501.
Park, J., Hong, K., Kirpotin, D., Meyer, O., Papahadjopoulos, D. and Benz, C., 1997b, Anti-HER2 immunoliposomes for targeted therapy of human tumors, *Cancer Lett.* 188:153.
Pegram, M., Lipton, A., Pietras, R., Hayes, D., Weber, B., Baselga, J., Tripathy, D., Twaddell, T., Glaspy, J. and Slamon, D., 1996, Phase II study of intravenous recombinant humanized anti-p185 HER-2 monoclonal antibody (rhuMAb HER-2) plus cisplatin in patients with HER-2/*neu* overexpressing metastatic breast cancer, *Proc. ASCO.* 14:106.
Press, M.F., Cordon-Cardo, C. and Slamon, D.J., 1990, Expression of the HER-2/*neu* proto-oncogene in normal human adult and fetal tissues, *Oncogene.* 5:953.
Rosenecker, J., Zhang, W., Hong, K., Lausier, J., Geppetti, P., Yoshihara, S., Papahadjopoulos, D. and Nadel, J., 1996, Increased liposome extravasation in selected tissues: effect of substance, P. *Proc Natl Acad Sci USA.* 93:7236.
Shepard, H.M., Lewis, G.D., Sarup, J.C., Fendly, B.M., Maneval, D., Mordenti, J., Figari, I., Kotts, C.E., Palladino, M.A., Ullrich, A. and Slamon, D., 1991, Monoclonal antibody therapy of human cancer: taking the HER2 protooncogene to the clinic, *J. Clin. Immunol.* 11:117.
Zhang, W., Guo, L., Nadel, J. and Papahadjopoulos D., 1998, Inhibition of tracheal veascular extravasation by liposome-encapsulated albuterol in rat, *Pharm. Res.* in press.

PEGYLATION OF LIPOSOMES IN CELL-SPECIFIC TARGETING:

IT DOES NOT ALWAYS MAKE SENSE

Gerrit L. Scherphof, Gerben A. Koning,
Jan A.A.M. Kamps, Toos Daemen

Department of Physiological Chemistry, Groningen Institute
for Drug Studies,University of Groningen, Medical Sciences,
Antonius Deusinglaan 1, 9713 AV Groningen, The Netherlands

INTRODUCTION

Four studies on the use of PEG-ylated liposomes in delivery to liver-associated cells are presented in this paper: (1) hepatocyte-directed delivery; (2) alleviation of Kupffer cell toxicity caused by liposomal doxorubicin; (3) immunospecific delivery to intrahepatic colon tumor cells; (4) scavenger receptor-mediated delivery to hepatic endothelial cells.

Study 1 demonstrates that PEGylation results not only in reduced uptake by (hepatic) macrophages but at least as much in reduced uptake by hepatocytes. Because it is generally accepted that the effect of PEG on the circulation time of liposomes is caused by reduced binding of opsonizing proteins, this result led us to postulate the existence of "hepatocyte-specific opsonins". We have now obtained convincing evidence that apolipoprotein E can serve as such.

In study 2 we showed that incorporation of PEG-PE in the liposomal bilayer significantly alleviates the toxicity of doxorubicin-loaded liposomes towards Kupffer cells. Study 3 describes attempts to achieve increased delivery of tumor-cell specific immunoliposomes to metastatically growing colon tumor cells in the rat liver. Although high liposome concentration in metastatic nodules was only observed occasionally with immunoliposomes carrying PEG, we found that this accumulation is mostly due to macrophage uptake rather than tumor cell uptake. Also, coupling the antibody to the distal end of the PEG chain does not significantly alter the affinity of the immunoliposomes towards macrophages. In contrast to the preceding three studies, in study 4 we demonstrate that rapid and efficient targeting of surface-modified liposomes can be achieved to a cell population other than macrophages without the need to equip the liposomes with a "long-circulation" device such as PEG. In this case, surface-grafted PEG is even counter-productive. It is concluded that, although prolonged circulation times may be beneficial for improving targeting of liposomes to cells not readily accessible from the circulation or to

Targeting of Drugs 6: Strategies for Stealth Therapeutic Systems
Edited by Gregoriadis and McCormack, Plenum Press, New York, 1998

decrease macrophage toxicity, measures to achieve this may be detrimental to the targeting of liposomes to readily accessible cells possessing a high affinity towards the carrier.

RESULTS AND DISCUSSION

PEG and hepatocytic delivery

We have elaborately shown in the past that systemically administered liposomes of sufficiently small size can interact with and be processed by hepatocytes (Roerdink et al., 1984; Spanjer et al., 1986; Scherphof and Kamps, 1998). Fig. 1 shows that there is a clear-cut relationship between liposome size and the extent to which they are taken up by this hepatic cell population: the smaller the liposomes, in this case consisting of egg PC and cholesterol (1:1 molar ratio), the larger the contribution of the hepatocytes in total hepatic

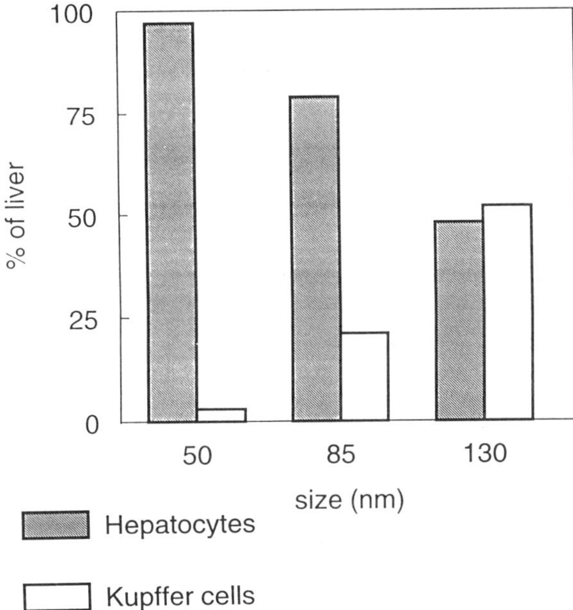

Figure 1. Effect of liposome size on intrahepatic distribution of eggPC/Chol liposomes. Radioactively labeled liposomes were prepared from PC/Chol (1:1) either by sonication (50 nm) or by extrusion (85 and 125 nm). They were injected i.v. into rats (5 μmol lipid per animal) and after 20 h hepatocytes and Kupffer cells were isolated and their radioactivity contents determined.

uptake. This is likely to be a reflection of the size of the fenestrations in the hepatic sinusoidal endothelium, which are approximately between 100 and 150 nm in diameter. Although one would thus expect that particles with a diameter larger than this would be denied access to the hepatocytes, we recently reported on observations indicating that phosphatidylserine-containing liposomes of up to 400 nm in diameter can be taken up in vivo by hepatocytes to a considerable extent (Daemen et al. 1997b).

Other factors, in addition to the size of the fenestrations, apparently determine the extent of participation of the hepatocyte population in liposome clearance from the blood. Since it is widely accepted that the effect of PEG on liposome circulation time has to be ascribed to the interference of the polymer chains with the adsorption of proteins on the liposomal surface, which in turn decreases its affinity towards macrophages, we set out to determine whether this polymer also has an effect on the hepatocyte-mediated elimination of liposomes. Fig. 2 demonstrates that this is the case. The left panel represents the amounts of liposomal lipid recovered in blood, spleen and liver as well as the intrahepatic distribution between hepatocytes and Kupffer cells, 20 h after injection of PC/chol (1:1) liposomes (5 μmol of liposomal lipid). The liver accumulates almost 2 μmol of this dose of which 75% is in the hepatocytes. The presence of PEG-DSPE in the eggPC/chol liposomes (right panel) caused an almost 90% reduction in the amount of liposomes taken up by the hepatocytes, even more than the reduction in Kupffer cell uptake. This observation led us to postulate the existence of hepatocyte-specific opsonins (Scherphof et al. 1994).

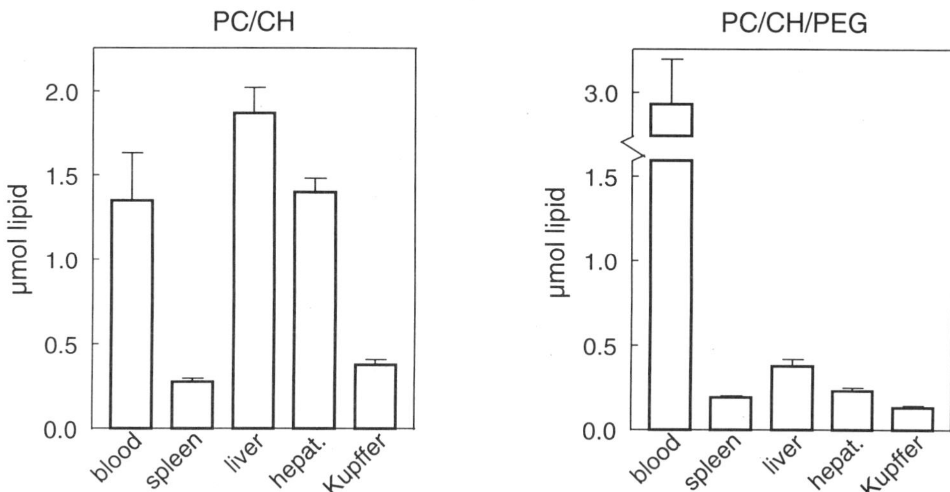

Figure 2. Effect of PEG on uptake of eggPC/chol liposomes by Kupffer cells and hepatocytes following intravenous administration.
Radioactively labeled liposomes with or without 5 mol% PEG-DSPE were injected i.v. into rats; 20 h after injection radioactivity in isolated hepatocytes and Kupffer cells was determined.

Recently we have obtained evidence that apolipoprotein E (apoE) may serve as a hepatocyte-specific opsonin. By immunoblotting we showed that during the incubation of liposomes with rat or human plasma this protein is one of the proteins that becomes adsorbed to the liposomal surface. In addition, we found that hepatic uptake of eggPC/chol liposomes in apoE-deficient mice is reduced by about 75% as compared to control mice, confirming the role of apoE in the clearance of these liposomes. Remarkably, hepatic uptake of 10% PS-containing liposomes was not reduced at all in the apoE-deficient mice, clearly indicating that neutral and negatively charged liposomes are cleared by different mechanisms by the hepatocytes, one apoE-mediated and the other not. Based on experiments in which we up-regulated the hepatic LDL receptor density by estradiol treatment we conclude that the apoE-mediated mechanism does not involve the LDL receptor. It is therefore likely to be the remnant receptor that is to be held responsible for the elimination of neutral liposomes (Ishibashi et al., 1994). From earlier experiments we know that PS-liposomes on the one hand are processed within the hepatocytes by an endo-lysosomal route (Roerdink et al., 1984) and on the other hand may only transiently interact with these cells allowing selective uptake of certain liposomal constituents such as the fluorescent marker rhodamine-PE (Verkade et al., 1992). The putative HDL receptor, a scavenger B-I type receptor which is known to be down-regulated by estradiol treatment (Landschulz et al., 1996), is a potential candidate for this interaction mechanism since we found that estradiol treatment substantially reduces uptake of liposomal cholesterol-oleyl ether by hepatocytes. Table 1 summarizes our current ideas about the possible interaction mechanisms of liposomes with hepatocytes: neutral PC/chol vesicles are (predominantly) taken up by an apoE-mediated mechanism, most likely a remnant receptor; negatively charged PS-containing liposomes interact by at least two mechanisms with the hepatocyte, one leading to complete endocytic internalization and processing of liposomes in the lysosomal compartment and the other involving a transient surface interaction resulting only in transfer of selective bilayer constituents; this might be an HDL receptor, recently identified as a scavenger receptor of the B-I type.

In conclusion, the use of surface-exposed PEG in these experiments allowed us to gain further insight into the way in which liposomes, once gaining access to the hepatocytes, may interact with these cells. The different interaction mechanisms might be exploited to deliver various substances to different intracellular compartments in this cell type, as we showed for example for (liposomal) cholesterol entering different pools of bile acids depending on the carrier used (Kuipers et al., 1986).

Effect of PEG on Kupffer cell toxicity of liposomal doxorubicin

During a study on combination therapy of hepatic metastases with chemo- and immunotherapy we observed that liver macrophages are extremely sensitive to doxorubicin, particularly when this is encapsulated in liposomes (Daemen et al., 1995). Treatment of rats with liposomal doxorubicin at a dose of 2 x 1 mg/kg with a 3-day interval caused a severe reduction in the Kupffer cell population of these animals. Not only was the phagocytic capacity of the liver almost completely lost, but also the cells themselves disappeared from the liver. After dosing with 2x 5 mg/kg both phagocytic capacity and cells were virtually completely lost. The effect was shown to be reversible: within two weeks phagocytic potential was restored to pre-treatment levels while the Kupffer cell population was even

Table 1. Putative interaction mechanisms of liposomes with hepatocytes

	ApoE-mediated		Not apoE-mediated	
Liposomes	LDL receptor	remnant receptor	HDL receptor (not endocytosed)	unknown (endocytosed)
neutral	no	yes	no	no
negative	no	no	yes	yes

Table 2. Kupffer cell toxicity of doxorubicin in PEGylated liposomes

Treatment[a])	Phagocytic capacity [b]) (nmol lipid per 10^6 cells)	number of Kupffer cells isolated per liver x 10^{-6}
none	13.5 ± 1.5	18.5 ± 1.5
1x 5 mg dox/kg in PEG-lip	11.4 ± 2.2	14.9 ± 2.0
2x 5 mg dox/kg in PEG-lip	2.5 ± 0.5	9.5 ± 2.1
3x 5 mg/kg in PEG-lip	2.0 ± 1.2	3.2 ± 1.5

a) Rats received 1, 2 or 3 times 5 mg/kg doxorubicin i.v. in 25 μmol of liposomal lipid per kg (PC/Chol/PEG-DSPE 55:45:5 molar ratio; liposome diameter 125 nm) with a 3-day interval.
b) 24 h after the last lip-dox injection rats received ^{3}H-cholesterololeylether-labeled test liposomes (PC/Chol/PS 4:5:1 molar ratio; 25 μmol lipid per kg). Two hours later Kupffer cells were isolated and the amount of radioactivity per 10^6 cells was determined.

somewhat larger by that time. Possible harmful effects of this treatment include the persistence of bacteriemia during the period of Kupffer cell deficiency (Daemen et al., 1995) and the excessive growth of hepatic tumors under those conditions (Heuff et al., 1993).

In a follow-up study (Daemen et al., 1997a) we showed that PEG incorporation in the liposomes significantly attenuated the toxic effects, compatible with the notion that PEG slows down the uptake of liposomes by macrophages (and hepatocytes! See above). Nonetheless, with PEG liposomes also the reduction in phagocytic activity and number of Kupffer cells can still be quite substantial. The use of small liposomes in combination with conservative dosage regimens, as are regularly applied in current-day patient treatment protocols with PEGylated liposomal doxorubicin (Uziely et al., 1995; Goebel et al., 1996), will possibly keep Kupffer cell toxicity within acceptable limits.

Immunospecific delivery of liposomes to intrahepatic tumor cells

The effect of PEGylation on intratumoral localization of immunoliposomes in a hepatic colon-derived metastasis model in the rat was studied by us. Liposomes carrying on their surface a covalently linked monoclonal antibody against CC531 colon carcinoma cells and containing the fluorescent dye DiI were shown to be able to concentrate to an appreciable extent in tumor nodules in the liver when PEG-DSPE was present in the bilayers (Scherphof et al., 1997), while this was not observed for PEG liposomes without antibody. However, accumulation of fluorescence in excess of that in the surrounding liver tissue was not observed in all tumor nodules: some displayed little or even no fluorescence. Average accumulation of liposomes in tumor areas as compared to that in the surrounding tissue is shown in Fig. 3 which summarizes data obtained with a radioactive liposome label, ^{3}H-cholesteryloleylether. Although there is a tendency for immunoliposomes to accumulate to a somewhat higher extent in the tumor nodules than in liver (filled bars in Fig. 3), this difference is not significant. Also the difference between immunoliposomes and control liposomes is insignificant when measured with this assay. By coupling the antibody via its peroxidized Fc fragment to the distal end of the PEG chains (Zalipsky et al., 1995), in close collaboration with Dr. Samuel Zalipsky, SEQUUS Pharmaceuticals, accumulation in the tumor was even somewhat lower than in the liver tissue, although, again, this was not significant (hatched bars).

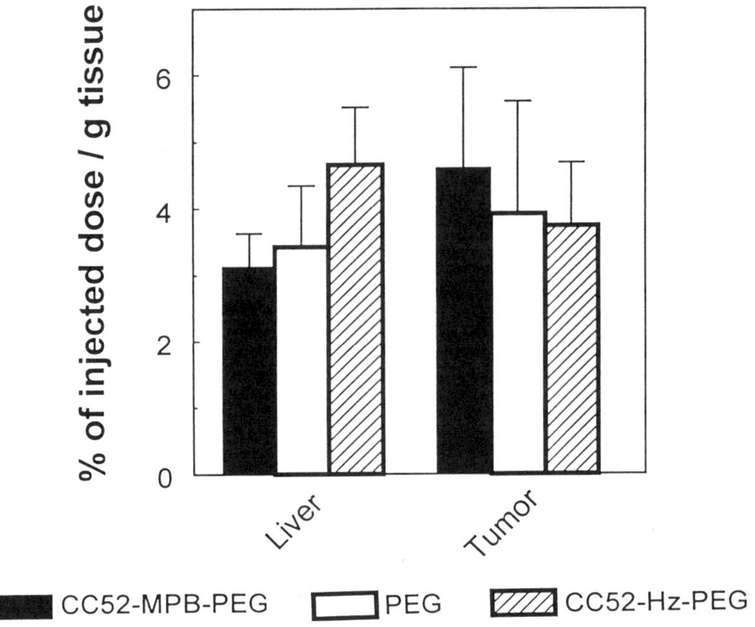

Figure 3. Specific uptake of PEG-DSPE immunoliposomes by metastatic tumor nodules in the liver as compared to liver tissue.
The monoclonal antibody CC52, specific for the CC531 colon adenocarcinoma cells, was covalently coupled (Derksen and Scherphof, 1986) to ^{3}H- cholesteryloleylether-labeled PEG-DSPE containing liposomes. These immunoliposomes were injected i.v. (2 μmol liposomal lipid per 100 g) into male Wag/Rij rats carrying hepatic metastases of the syngeneic CC531 colon tumor cells. After 24 h the liver and individual tumor nodules were dissected and weighed pieces of liver and tumor tissue were assayed for radioactivity. Open bars: PEG-DSPE/MPB-PE liposomes without antibody; filled bars: PEG-DSPE/MPB-PE immunoliposomes with antibody coupled via MPB-PE; hatched bars: PEG-DSPE immunoliposomes with antibody coupled at distal end of the PEG chains

Experiments with isolated Kupffer cells and the distally coupled immunoliposomes confirmed the slightly increased affinity of these liposomes for the hepatic macrophages. Upon electron microscopic examination of metastatic nodules, applying encapsulated colloidal gold as a liposomal marker, we observed that the gold particles in the tumor areas were concentrated, disappointingly, mostly in macrophage-like cells rather than in tumor cells. It is not unlikely that the preferential uptake of our immunoliposomes by tumor-associated macrophages rather than by tumor cells is a consequence of the antibody coupling procedure we used. This procedure, which is based on a random thiolation of the antibody molecule with succinimidyl-S-acetylthioacetate followed by coupling to MPB-PE in the liposomes (Derksen and Scherphof, 1985), results in random coupling of the antibody to the liposomal surface and thus in exposure of the Fc portion of at least a fraction of the antibody molecules. Conceivably this will, in turn, lead to interaction with Fc receptors on the macrophages. Possibly, the tumor macrophages display even a higher affinity towards the Fc portion of this antibody than the competing Kupffer cells in the surrounding liver tissue. The conspicuous heterogeneity in liposome distribution between different tumor nodules may be caused by histological differences (Dingemans et al., 1994). We also observed a substantial variation in the histological appearance of tumor nodules within one liver: the nodules may contain different numbers of macrophages and tumor cells within the nodules appear both as easily accessible single cells and as clusters of tightly packed cells.

In conclusion, although we confirm observations made by others that PEGylation tends to enhance tumor localization of immunoliposomes, we want to emphasize that this does not necessarily mean that such liposomes do actually interact with the tumor cells. Inaccessibility of the tumor cells combined with severe competition by tumor-associated macrophages, in addition to that by Kupffer cells, cause the immunoliposomes to concentrate at best to the same extent in the tumor as in the surrounding liver tissue but, even so, only marginally in the tumor cells.

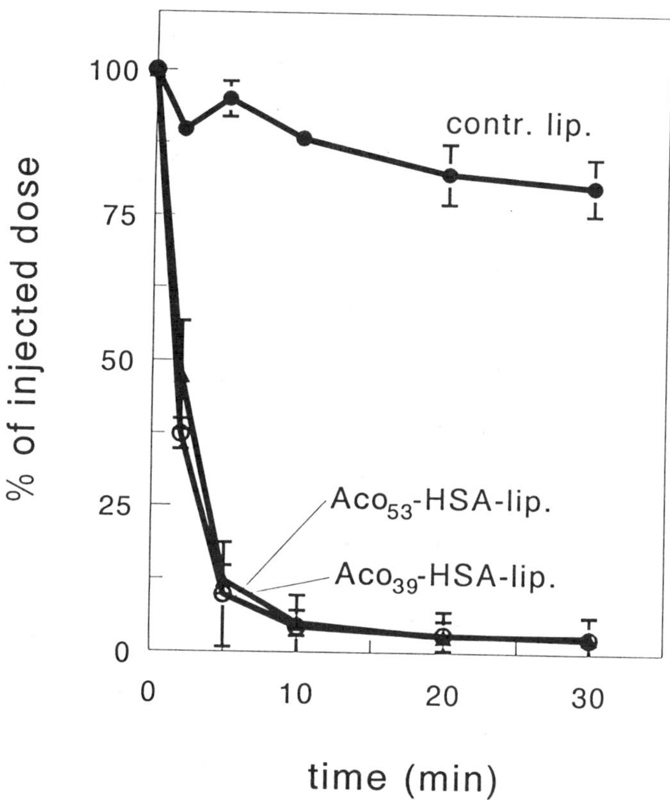

Figure 4. Plasma elimination of polyanionic albumin (Aco-HSA) liposomes. Radiolabeled Aco-HSA liposomes (see Table 3 for details), 90 nm in diameter, were injected i.v. and plasma was sampled for radioactivity after 30 min. Control liposomes were of the same composition but without Aco-HSA modification. Aco39 and Aco53 refer to the number of NH_2 groups substituted by aconitic acid.

Scavenger receptor-mediated targeting to hepatic endothelial cells

We wil conclude with an example of very successful active targeting of liposomes without the use of PEG or of any other means of slowing down liposome elimination; in this case the presence of PEG was shown to be even inhibitory. The target cell in these experiments was the hepatic sinusoidal endothelial cell. This cell type is known or suspected to be involved in a variety of important hepatic functions and pathological conditions; it is also the first site of interaction of tumor cells invading the liver. This cell shares an obvious advantage as a target for site-specific drug delivery with all vascular endothelial cells: it is immediately and unrestrictedly accessible from the blood stream. However, in contrast to the other major cell types in the liver, the Kupffer cells and the hepatocytes, the liver endothelial cells, in our hands, have consistenly been virtually inert to interaction with liposomes. Recently, D.K.F. Meijer and co-workers showed that poly-anionized albumins display a high affinity *in vivo* to liver endothelial cells, presumably involving scavenger receptors (Jansen et al., 1993).

In collaboration with D.K.F. Meijer we coupled such albumins, *i.e.* poly-aconitylated human serum albumin (Aco-HSA), to liposomes and observed that these liposomes are cleared extremely rapidly from the circulation as compared to control liposomes without the Aco-HSA (Kamps et al., 1996) (Fig. 4). Inclusion of PEG-DSPE in the bilayers of these liposomes led to a substantial decrease in the rate of elimination and thus an enhanced chance for interaction with cells other than the target cell. As shown in Fig. 5, the liver accounts for nearly all liposomes eliminated from the blood and within the liver it is predominantly the endothelial cell population that is responsible for this elimination (about 70%), Kupffer cells and hepatocytes accounting for about 20% and 10%, respectively (Kamps et al., 1997). The liposomes are readily internalized by a scavenger receptor-mediated process, as indicated by the strong inhibitory effect of polyinosinic acid, and degraded in the endo-lysosomal compartment. However, when using pH-sensitive liposomes containing cholesterol-hemisuccinate (Chu and Szoka, 1994), we found a model drug, the strongly pH-sensitive fluorophore hydroxypyridinetrisulfonate (HPTS) {Daleke et al., 1990), to be released to an appreciable extent in a non-acidic compartment of the cell, presumably the cytosol. This indicates that at least a significant fraction of the liposomes entering the endothelial cells in this way may escape intralysosomal degradation and be able to deliver their contents to the cytoplasmic compartment, a prerequisite for meaningful intracellular delivery of lysosome-sensitive drugs such as, for example, anti-sense oligonucleotides.

An important aspect of our observations was also that the great majority of the endothelial cell population participates in the elimination process, as demonstrated by fluorescence microscopy and flow cytometry using DiI-labeled liposomes (Kamps et al., 1997). Thus, with this approach we are able to deliver a liposomal drug to virtually all of the endothelial cells in the liver.

Aco-HSA-liposome uptake by the hepatic endothelial cells is strongly dependent on liposome size. Gradually increasing the liposomal diameter caused a shift in ratio of endothelial cell total uptake to Kupffer cell total uptake from 2.6 for 92-nm liposomes to 0.2 for 590-nm liposomes in favor of the Kupffer cells (calculated from Table 3, taking into consideration that endothelial cells in the liver outnumber the Kupffer cells by a factor of 3).

GENERAL CONCLUSIONS

In our laboratory, surface modification of liposomes with PEG has led us to propose the existence of hepatocyte-specific factors in plasma ("opsonins") and subsequently to

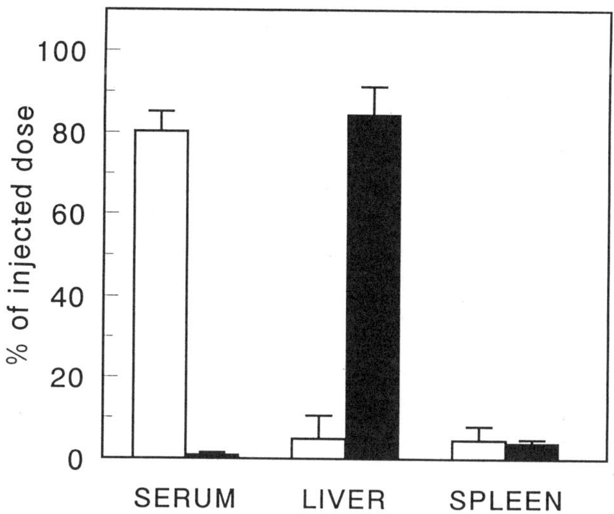

Figure 5. Preferential uptake of polyanionic liposomes by the liver.
Data present the amount of control (open bars) and Aco-HSA (filled bars) liposomes remaining in circulation 30 min after i.v. injection. (For details see Table 3)

Table 3. Effect of size of Aco-HSA liposomes on endothelial cell uptake

Liposome size (nm)	Liposome uptake (nmol lipid per 10^6 cells)	
	Endothelial cells	Kupffer cells
92	2.8 ± 0.5	3.3 ± 0.5
153	1.9 ± 0.1	7.4 ± 0.0
263	1.0 ± 0.1	15.7 ± 1.9
590	1.0 ± 0.1	17.6 ± 2.2

^{3}H-Cholesteryloleylether-labeled liposomes, consisting of eggPC, cholesterol and MPB-PE and surface-modified with poly-aconitylated human serum albumin (Aco-HSA), were obtained by extrusion through filters with pore sizes of 50, 100, 200 or 400 nm and injected i.v. into rats at a dose of 1 µmol of liposomal lipid per 100 g body weight. After 2h liver endothelial cells and Kupffer cells were isolated and liposome uptake was calculated from their radioactivity content.

demonstrate that apolipoprotein E may serve as one such hepatocyte-directed opsonin for neutral liposomes. Furthermore we obtained evidence in support of the view that liposomes of different composition may interact with the hepatocyte by means of different factors adsorbed at their surface and thus by different mechanisms. Our results even lend support to the notion that one type of liposome may interact by more than one mechanism with the hepatocyte, leading liposomal constituents into different intracellular routes and compartments.

We have also found PEG to be advantageous in attenuating the toxic effect of liposomal doxorubicin towards (hepatic) macrophages. Currently available commercial formulations of liposomal doxorubicin combine both small size and the presence of PEG in order to achieve long circulation times and thus better and prolonged exposure of tumor cells to the drug. Fortunately, obviously and even unavoidably, this simultaneously serves to diminish the accumulation of the drug in the macrophages and thus to keep the toxic effect within acceptable limits. We also observed, like several others, that PEGylation of (immuno)liposomes helps to achieve liposome accumulation in tumor tissue, in our case specifically in intrahepatic micrometastatic lesions originating from colon adenocarcinoma cells. However, we found, first of all, that only part of the tumor nodules accumulated such liposomes in excess of the surrounding liver tissue and that, secondly, most of the liposomes thus accumulating in the metastases were in macrophages rather than tumor cells. Apparently, although the PEG chains successfully provided conditions favorable for accumulation in the tumor lesions, macrophages in the tumor area were still able to compete effectively with the tumor cells, many of which appeared microscopically to be virtually inaccessible because of tight mutual interactions.

Finally, we emphasize that the use of PEG should not be considered a *conditio sine qua non* for successful targeting purposes. It should be appreciated that under conditions where the target cells are readily accessible and the affinity of the carrier for the target is sufficiently high, there is no need to prolong carrier circulation time. Such measures may even be counterproductive in achieving fast and efficient interaction with the target cells. These considerations are derived from our observations on the rapid and very efficient targeting of liposomes surface-modified with poly-anionic albumin to the scavenger receptor of hepatic endothelial cells.

Acknowledgements

The autors gratefully acknowledge stimulating and fruitful collaborations with Dr. Folkert Kuipers, Dr. Dirk Meijer and Dr. Piet Swart, all in Groningen, Dr. Irma Bakker, Rotterdam, Dr. Louis Havekes, Leiden, Dr. Samuel Zalipsky, Menlo Park, Dr Maria Velinova in Sofia and Dr. Terry Allen, Edmonton and the excellent technical assistance of Joke Regts, Henriëtte Morselt and Bert Dontje in the various studies summarized in this paper.

The work described received support from the Dutch Cancer Society (grant G-UKC 94-767) and the Netherlands Organization for Scientific Research, NWO (grant 902-23-187).

REFERENCES

Chu, C.-J. and Szoka, F.C., 1994, pH-Sensitive liposomes, *J. Liposome Res.* 4:361.

Daemen, T., Hofstede, G., Ten Kate, M.T., Bakker-Woudenberg, I.A.J.M. and Scherphof, G.L., 1995, Liposomal doxorubicin induced toxicity: impairment of phagocytic activity and depletion of liver macrophages, *Int. J.Cancer*, 61:716.

Daemen, T., Regts, J., Meesters, M., Ten Kate, M.T., Bakker-Woudenberg, I.A.J.M. and Scherphof, G.L., 1997a, Toxicity of doxorubicin entrapped within long-circulating liposomes, *J. Control. Release*, 44:1.

Daemen, T., Velinova, M., Regts, J., De Jager, M., Kalicharan, R., Donga, J., Van der Want, J.J.L. and Scherphof, G.L., 1997b, Different intrahepatic distribution of phosphatidylglycerol and phosphatidylserine liposomes in the rat, Hepatology, 26:416

Daleke, D.L., Hong, K. and Papahadjopoulos D., 1990, Endocytosis of liposomes by macrophages: binding, acidification and leakage of liposomes monitored by a new fluorescence assay, *Biochim. Biophys. Acta*, 1024:352

Derksen, J.T.P. and Scherphof G.L., 1985, An improved method for the covalent coupling of proteins to liposomes, *Biochim. Biophys. Acta*, 814:151

Dingemans, K.P., Van den Berg Weerman, M.A., Keep, R.F. and Das P.K., 1994, Developmental stages in experimental liver metastases: relation to invasiveness, *Int. J.Cancer*, 57:433

Goebel, F.-D., Goldstein, D., Goos, M., Stewart, J.S., 1996, Efficacy and safety of Stealth liposomal doxorubicin in AIDS-related Kaposi's sarcoma, *Brit. J. Cancer*, 73:989

Heuff, G., Oldenburg, H.S.A., Boutkan, H., Visser, J.J., Beelen, R.H.J., Van Rooijen, N., Dijkstra, C.D. and Meyer, S., 1993, Enhanced tumor growth in the rat liver after selective elimination of Kupffer cells, *Cancer Immunol. Immunother.*, 37:125

Ishibashi, S., Herz, J. Maeda, N., Goldstein, J.L. and Brown, M.S., 1994, The two-receptor model of lipoprotein clearance: tests of the hypothesis in "knock-out" mice lacking the low density lipoprotein receptor, apolipoprotein E or both proteins, *Proc. Natl. Acad. Sci. USA,* 89:4431

Jansen, R.W., Olinga, P., Harms, G. and Meijer, D.K.F., 1993, Pharmacokinetic analysis and cellular distribution of the anti-HIV compound succinylated human serum albumin *in vivo* and in the isolated perfused rat liver, *Pharmaceut. Res.* 10:1611

Kamps, J.A.A.M., Morselt, H.W.M., Swart, P.J. Meijer, D.K.F. and Scherphof G.L., 1997, Massive targeting of liposomes, surface-modified with anionized albumins, to hepatic endothelial cells, *Proc. Natl. Acad. Sci. USA,* 94:11681

Kamps, J.A.A.M., Swart, P.J., Morselt, H.W.M., Pauwels, R., De Béthune, M.P., De Clercq, E., Meijer, D.K.F. and Scherphof, G.L., 1996, Preparation and characterization of conjugates of (modified) human serum albumin and liposomes: drug carriers with an intrinsic anti-HIV activity. *Biochim. Biophys. Acta*, 1278:183

Kuipers, F., Spanjer, H.H., Havinga, R., Scherphof, G.L. and Vonk, R.J., 1986, Lipoproteins and liposomes as *in vivo* cholesterol vehicles in the rat: preferential use of cholesterol carried by small unilamellar liposomes for the formation of muricholic acids. *Biochim. Biophys. Acta,* 876:559

Landschulz, K.T., Pathak, R.K., Rigotti, A., Krieger, M. and Hobbs H.H., 1996, Regulation of a scavenger receptor class B, type I, a high density lipoprotein receptor in liver and steroidogenic tissues of the rat, *J. Clin. Invest.*, 98:984

Roerdink, F.H., Regts, J., Van Leeuwen, B. and Scherphof G.L., 1984, Intrahepatic uptake and processing of small unilamellar vesicles in rats, *Biochim. Biophys. Acta,* 770:195

Scherphof, G.L. and Kamps, J.A.A.M.,1998, Receptor vs. non-receptor mediated clearance of liposomes, *Adv. Drug Deliv. Rev.*, in press

Scherphof, G.L., Morselt, H. and Allen, T.M., 1994, Intrahepatic distribution of long-circulating liposomes containing poly(ethyleneglycol) distearoylphosphatidylethanolamine. *J. Liposome Res.*, 4:213

Scherphof, G.L. and Crommelin D.J.A., 1996, Cells involved in removing liposomes from the blood circulation: why are they so special?, *J. Liposome Res.*, 6:19

Scherphof, G.L., Kamps, J.A.A.M. and Koning, G.A., 1997, *In vivo* targeting of surface-modified liposomes to metastatically growing colon carcinoma cells and sinusoidal endothelial cells in the rat liver, *J. Liposome Res.*, 7:419

Spanjer, H.H., Van Galen, M., Roerdink, F.H., Regts, J . and Scherphof, G.L., 1986, Intrahepatic distribution of small unilamellar liposomes as a function of liposomal lipid composition, *Biochim. Biophys. Acta*, 863:224

Steinberg, D., 1996, A docking receptor for HDL cholesterol ester, *Science*, 271:460

Uziely, B., Jeffers, S., Isacson, R., Kutsch, K., Wei-Tsao, D., Yehoshua, Z., Libson, E., Muggia, F.M. and Gabizon, A., 1995, Liposomal doxorubicin: antitumor activity and unique toxicities during two complementary phase I studies, *J. Clin. Oncol.*, 13:1775

Verkade, H.J., Zaal, K.J., Derksen, J.T.P., Vonk, R.J., Hoekstra, D., Kuipers, F. and Scherphof G.L., 1992, Processing of the phospholipid analog Rhodamine-PE by rat hepatocytes *in vitro* and *in vivo*, *Biochem. J.*, 284:259

Zalipsky, S., Puntambekar, B., Boulikas, P., Engbers, C.M. and Woodle, M.C., 1995, Peptide attachment to extremities of liposomal surface-grafted PEG chains: preparation of the long-circulating form of laminin pentapeptide YIGSR, *Bioconjug. Chem.*, 6:705

STEALTH™ LIPOSOMES FOR THE TARGETING OF DRUGS IN CANCER THERAPY

Theresa M. Allen, Daniel Lopes de Menezes,
Christian B. Hansen, Elaine H. Moase

Department of Pharmacology, University of Alberta,
Edmonton, AB, Canada T6G 2H7

INTRODUCTION

Anticancer drugs have a variety of dose-limiting side effects as a consequence of their distribution to normal, sensitive tissues as well as to cancerous tissues. The goal for the targeting of liposomal drugs in cancer therapy is to increase the localization of the anticancer drugs to the diseased tissue, while at the same time decreasing their localization to normal tissue, i.e. to increase the selective toxicity of the drugs. Put another way, by the entrapment of anticancer drugs in liposomes we are aiming to increase the efficacy of the drugs, while reducing their side effects (toxicity), resulting in an increase in the therapeutic indices for the drugs.

Two types of targeting have been achieved for liposomal anticancer drugs: passive targeting and ligand-mediated ('active') targeting. Several recent reviews are available (Allen et al. 1995; Allen and Moase 1996; Allen et al. 1997; Maruyama et al. 1997). A number of basic developments in the liposome field were necessary before either of these types of targeting could be realized, as outlined in Table 1. Developments 1, 3 and 4 have largely been achieved (Olson et al. 1979; Mayer et al. 1986; Mayer et al. 1986; Klibanov et al. 1990; Papahadjopoulos et al. 1991; Haran et al. 1993; Allen 1994), while there is room for considerable research and improvement to achieve developments 5 and 6. Development 2 has been achieved for some weak base or weak acid anticancer drugs such as doxorubicin, daunorubicin and vincristine (Mayer et al. 1986; Haran et al. 1993; Boman et al. 1994), but for many other anticancer drugs efficient loading remains to be achieved. Development 7, which has been achieved, applies only to ligand-mediated targeting, and is not relevant to passive targeting.

A discussion of each type of targeting, of coupling methods, as well as a discussion of liposome pharmacokinetics as it relates to anticancer drug, occurs below.

Targeting of Drugs 6: Strategies for Stealth Therapeutic Systems
Edited by Gregoriadis and McCormack, Plenum Press, New York, 1998

Table 1. Some basic developments necessary to the achievement of liposome targeting.

1. Small populations of liposomes of homogeneous size distribution, preferably around 50-100 nm in diameter.
2. Efficient loading of the anticancer drugs into liposomes.
3. Circulation times sufficiently long to enable the liposomal drug to localize to the diseased cells, preferably at least several hours.
4. Retention of the majority of anticancer drug within the liposomes until the liposome localizes to its target population of cells.
5. Ability of the liposome to release its associated drug once the target is reached.
6. Delivery of the liposomal drug to the cell in a usable form, or to the appropriate cell compartment once the target is reached.
7. Efficient techniques for coupling ligands, including antibodies, to the liposome surface which result in long circulation times, and good target recognition.

PHARMACOKINETICS OF LIPOSOMAL ANTICANCER DRUGS

Association of drugs, including anticancer drugs, with liposomes can result in substantial alterations in the pharmacokinetics (PK) and biodistribution of the drugs. An understanding of these alterations will help to understand how and why liposomes can be useful for the targeting of anticancer drugs.

Entrapping anticancer drugs in liposomes

The ease of association of anticancer drugs with liposomes depends on the solubility of the drug(Defrise-Quertain et al. 1984). Anticancer drugs of high hydrophilicity, e.g. cytosine arabinoside, are readily encapsulated in, and retained in, the aqueous internal compartment of liposomes, albeit with low trapping efficiency, and are slowly released from the liposomes over several hours to several days (Defrise-Quertain et al. 1984; Allen et al. 1992). Drugs of high hydrophobicity, basically non-existent among approved clinically anticancer drugs, are easily and efficiently inserted into the fatty acyl chain region of the lipid bilayer, and are not readily released from the liposome. Many anticancer drugs, e.g. doxorubicin, are of intermediate solubility and readily partition between the liposome bilayer and the exterior or interior aqueous phase, resulting in their rapid release from liposomes. This would prevent these drugs from being targeted, as they would leave the liposomes before the liposomes reached their intended target. However, manipulation of the interior pH of the liposomes, or formation of molecular complexes of these partially soluble drugs within the liposomes can result in excellent retention of weak bases such as doxorubicin or daunorubicin in liposomes (Mayer et al. 1986; Haran et al. 1993; Forssen 1997) which allows the entrapped drugs to reach their *in vivo* targeting.

Pharmacokinetics of the liposome carrier

The degree to which the PK of the drug-liposome package is different from that of the free drug depends on the liposome size, composition, surface charge, administered dose, route of administration, the release rate of the drug from the liposmes and the PK properties of the free drug (Abra and Hunt 1981; Hwang 1987; Allen et al. 1995). A basic assumption is that high retention of drugs in liposomes is essential if the drugs are to be carried to their intented target cancer cells by the liposomes. This being the case, then an analysis of the PK of liposome anticancer drugs becomes equivalent to an analysis of the PK of the liposome carrier. Obviously, if the drug is rapidly released from the liposomes, the PK approaches that of the free drug. As a general rule, unlike free (non-entrapped) drugs, for liposome-associated drugs the biological activity of the drug will be delayed until the drug is released from the liposome. The liposome-associated drug generally has a lower rate of clearance that the free drug and a biodistribution restricted primarily to the vasculature and those few tissues like the mononuclear phagocyte system (MPS) which accumulate liposomes. The metabolism and elimination of the liposome-associated drug will also be altered compared to the free drug (Hwang 1987; Allen and Hansen 1991; Allen and Stuart 1998).

Pharmacokinetics of Stealth versus classical liposomes

As was mentioned in point 3 of Table 1, for liposomes to achieve targeting to cancerous tissues, long circulation times are necessary. Sterically stabilized (Stealth, SL) liposomes have two important PK advantages over classical liposomes (Allen and Hansen 1991; Papahadjopoulos et al. 1991; Allen et al. 1995). They have decreased rates of clearance, and dose-independent PK. SL are formed by grafting hydrophilic polymers such as polyethylene glycol, by means of a lipid anchor, to the liposome surface (Blume and Cevc 1990; Klibanov et al. 1990; Allen et al. 1991; Senior et al. 1991; Woodle et al. 1991; Torchilin and Trubetskoy 1995; Zalipsky et al. 1996). This results in an inhibition of the binding of plasma protein opsonins to the liposome surface and, as a consequence, significantly decreases the uptake of the SL by cells of the MPS (Senior et al. 1991; Chonn and Cullis 1992). For SL in the size range of $\leq$100 nm in diameter, circulation half-lives of 16-24 hours have been described in rodents, and as high as 45 hours in humans (Allen and Hansen 1991; Woodle et al. 1991; Northfelt et al. 1993; Gabizon et al. 1994). These half-lives are long enough to cause increased localization of liposomes to solid tumours (Gabizon and Papahadjopoulos 1988; Northfelt et al. 1993; Zu et al. 1993; Gabizon et al. 1994).

The dose-independent PK of SL also has consequences for their ability to be targeted to tumours. Unlike SL, there is a dose-dependent saturation of the uptake of classical liposome by the MPS (Abra and Hunt 1981; Hwang 1987; Allen and Hansen 1991). The consequence of this is that, at low concentrations, they experience rapid clearance into the MPS and do not circulate long enough to localize in solid tumours. At much higher concentrations, saturation of the MPS uptake mechanism leads to significant increases in circulation half-lives, and therefore increased localization to solid tumours (Abra and Hunt 1981; Hwang 1987; Allen and Hansen 1991). Thus, the ability of classical liposomes to localize to solid tumours would be expected to be disproportiate to dose. SL have long circulation times, independent of concentration, and this would be expected to lead to tumour uptake which is proportionate to dose.

PASSIVE TARGETING

Passive targeting refers to the opportunistic tendency of liposomes, if they circulate long enough, to localize in tissues with increased vascular permeability, which includes many solid tumours and regions of infection (Jain 1987; Bakker-Woudenberg et al. 1992; Zu et al. 1993; Forssen et al. 1996; Unezaki et al. 1996). As solid tumours grow, they require additional blood vessels to supply oxygen and nutrients. As new capillaries sprout into the growing tumour, in a process called angiogenesis, they are leaky relative to normal vasculature.(Jain 1987). Long-

Table 2. Examples of liposomal anticancer drugs approved for clinical use, or in clinical trials.

Drug	Indications	Status	Commerical Name & Company
Doxorubicin	First line & refractory Kaposi's sarcoma	approved	Doxil® (US), Caelyx (Eur) SEQUUS Pharmaceuticals, Inc.
Daunorubicin	First line & refractory Kaposi's sarcoma	approved	DaunoXome® (US & Eur) NeXstar Inc.
Doxorubicin	Advanced breast, ovarian, lung cancers	Phase II	Doxil® SEQUUS Pharmaceuticals, Inc.
Doxorubicin	Advanced breast cancer	Phase II	TLC-D99® The Liposome Company
Vincristine	Pancreatic and colon cancer	Phase II	Onco TCS® Inex Pharmaceuticals

circulating liposomes extravasate through these leaky capillaries into the tumour interstitium in greater quantities than in regions with normal vasculature, which is essentially impermeable to liposomes (Jain 1987; Bakker-Woudenberg et al. 1992; Zu et al. 1993; Forssen et al. 1996; Unezaki et al. 1996). The result is that concentrations of liposomal anticancer drugs can be increased in solid tumours several-fold compared to those observed with the free drug.

Several liposomal anticancer drugs, relying on passive targeting mechanisms, have either been approved for clinical use, or are in clinical trials (Table 2). The two approved liposomal anticancer drug formulations for passive targeting are both for the treatment of Kaposi's sarcoma (Gill et al. 1995; Harrison et al. 1995; Goebel et al. 1996; Northfelt et al. 1996; Northfelt et al. 1997). Kaposi's sarcoma lesions are characterised by greatly increased vascular permeability within the tumour. Doxil®, a long-circulating PEG-containing liposomal formulation of doxorubicin, was able to deliver between 5 and 11 times more doxorubicin to Kaposi's sarcoma lesions than to normal skin (Northfelt et al. 1996), leading to an overall response rate of as high as 80% (Goebel et al. 1996; Northfelt et al. 1997). DaunoXome® is a formulation of daunorubicin into small, solid liposomes with no PEG, and is also effective in the treatment of Kaposi's sarcoma (Gill et al. 1995). Both formulations resulted in decreased uptake of the anticancer drug by the heart, a site of major dose-limiting toxicity of the free drug (Gill et al. 1995; Uziely et al. 1995). The toxicity profiles of the two preparations differ somewhat (Gill et al. 1995; Uziely et al. 1995).

The passive localisation of long-circulating liposomes to sites with leaky vasculature indicates that their use might be expanded to include other solid tumours. Ongoing clinical trials are examining the efficacy of Doxil® in advanced breast and ovarian cancers (Ranson et al. 1996; Muggia et al. 1997), and spontaneously arising malignant tumours in dogs (Vail et al. 1997). DaunoXome® is being tested for efficacy in the treatment of adenocarcinoma of the colon (Eckardt et al. 1994).

LIGAND-MEDIATED TARGETING

Ligand-mediated, or 'active', targeting is being investigated as a means of increasing the site-specific delivery of liposomes to cancer cells. Unlike passively targeted liposomes, in this case, ligands which specifically binds to surface epitopes on the target cancer cells, are coupled to the surface of the liposomes (see below). It is proposed that this will lead to the accumulation of the liposomal drug at/in the target tissue, further increasing the concentration of drug to the intented target relative to free drug or passively targeted liposomal drug. This topic has been recently reviewed (Allen et al. 1995; Allen and Moase 1996; Allen et al. 1997; Maruyama et al. 1997).

Coupling ligands to Stealth liposomes

Prior to the development of SL, the greatest limitation to targeted liposomal drug delivery was the rapid clearance of ligand (antibody)-bearing liposomes into the MPS (Aragnol and Leserman 1986; Debs et al. 1987). The discovery that PEG inhibited the uptake of ligand-liposomes by the MPS has revived interest in targeted drug delivery (Blume et al. 1993; Hansen et al. 1995; Maruyama et al. 1995; Torchilin et al. 1996).

Several different molecules have been attached to liposomes for the purpose of targeting to specific cells. Monoclonal antibodies (mAb) and antibody fragments have been used extensively *in vitro* as well as *in vivo* (Maruyama et al. 1990; Ahmad and Allen 1992; Torchilin et al. 1992; Ahmad et al. 1993; Maruyama et al. 1995; Goren et al. 1996; Vingerhoeds et al. 1996; Kirpotin et al. 1997). In addition, various endogenously occuring ligand have been attached to liposomes for the purpose of targeting specific cells. These include peptides (DeFrees et al. 1996; Zalipsky et al. 1997), proteins and lipoproteins (Brown and Silvius 1990; Blume et al. 1993; Lundberg et al. 1993), growth factors and vitamins (Rosenberg et al. 1987; Ishii et al. 1989; Lee and Low 1994; Lee and Low 1995) and carbohydrates (Zalipsky et al. 1997). The main advantages and disadvantages of using antibodies versus ligands is summarized in Table 3.

A variety of different coupling chemistries have been developed for the attachment of mAb and ligands to PEG-containing liposomes. Two main strategies have been used: those in which the ligand is attached at the liposome surface (Fig. 1A), and those in which the ligand is attached to the PEG terminus (Fig. 1B) (Loughrey et al. 1990; Maruyama et al. 1990; Loughrey et al. 1993; Maruyama et al. 1995; Zalipsky et al. 1995; Kirpotin et al. 1996; Kirpotin et al. 1996; Klibanov et al. 1996; Maruyama et al. 1997).

Table 3. Some advantages and disadvantages of antibodies versus ligands.

Targeting moiety	Advantages	Disadvantages
Whole antibodies	High degree of specificity possible	Immune reactions may occur. Expensive & time-consuming to produce. Target eptiope may be down-regulated or shed.
Antibody fragments	High degree of specificity possible. Orientation of binding region preserved. Immune reactions may be reduced.	Binding avidity may be lost. Expensive and time-consuming to produce. Target epitope may be down-regulated of shed.
Ligands (e.g. to folate, transferrin, LDL receptors, gowth factor receptors etc.)	Decreased chance of immune reactions. Widely available, and often inexpensive.	Target cell specificity may be reduced relative to mAb as normal cells often express the same receptors.

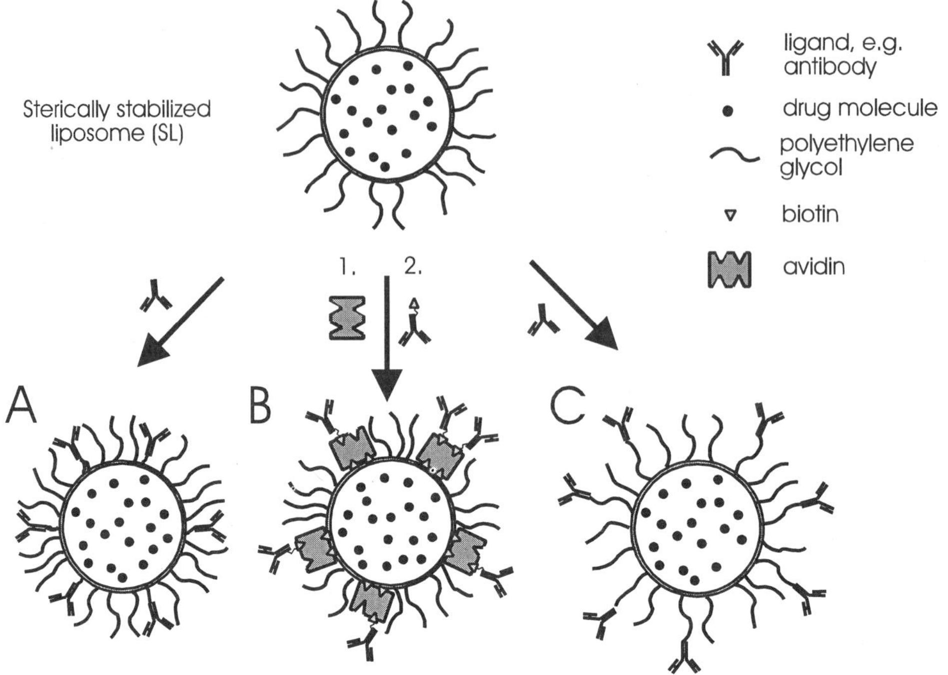

Figure 1. Strageties for the formation of targeted sterically stabilized liposomes. A. Covalent attachment of mAb or ligand to the surface of SL. B. Non-covalent attachment of biotinylated mAb to the surface of SL using an avidin linker. C. Covalent attachment of mAb or ligand via end-functionalized PEG.

For attaching ligands at the surface of SL (Fig. 1A,B), much of the coupling chemistry has been adapted from procedures for conjugating mAb or Ab fragments to the liposomes surface (Heath and Martin 1986; Weiner 1990; Torchilin and Klibanov 1993). Coupling lipids such as N-pyridyldithiopropionyl-phosphatidylethanolamine (PDP-PE) or maleimidophenylbutyrate-PE) (MPB-PE), respectively, are incorporated into the liposomes during formation. A thiolether bond is then formed with maleimide-derivitized or thio-derivatized mAb, respectively (Heath and Martin 1986; Allen and Moase 1996). In another method, antibodies have been coupled to a lipid anchor in detergent solution and incorporated into liposomes by dialysis (Holmberg et al. 1989; Klibanov and Huang 1992). Another variation involved coupling mAb to non-PEG liposomes and importing the PEG into the liposomes after conjugation from PEG micelles (Suzuki et al. 1995; Uster et al. 1996). The non-covalent, but high affinity interaction of avidin or streptavidin with biotin has also been adapted for coupling ligands to the liposome surface (Loughrey et al. 1987; Loughrey et al. 1990; Corley and Loughrey 1994; Longman et al. 1995).

A problem which has arisen when attaching the ligands at the liposome surface, is the interference of the hydrophilic PEG chains with the approach of the ligand to it epitope, with a consequent reduction (or elimination) of binding (Mori et al. 1991; Allen et al. 1994). A solution to this problem has been the development of coupling chemistries to attach the ligand to the functionalised terminal ends of the PEG (Fig. 1C). The ligand-PEG-lipid conjugate may be constructed either before (Blume et al. 1993; Allen et al. 1995; Hansen et al. 1995;

Maruyama et al. 1995; Zalipsky et al. 1997) or after (Suzuki et al. 1995; Ansell et al. 1996; DeFrees et al. 1996) liposome formation, and proceeds unencumbered by the steric barrier of the polymer. The various coupling chemistries involve thiol-reactive (Martin and Papahadjopoulos 1982; Hansen et al. 1995; Suzuki et al. 1995), carboxyl(Blume et al. 1993; Maruyama et al. 1995; Takizawa et al. 1996), hydrazide (Zalipsky 1993; Zalipsky et al. 1993; Hansen et al. 1995; Zalipsky et al. 1995; Ansell et al. 1996) and amino-succinimidyl (Zalipsky et al. 1994; Lee and Low 1995; Klibanov et al. 1996) functional groups. These coupling methods have also been reviewed recently (Allen et al. 1995; Allen and Moase 1996; Allen et al. 1997).

Pharmacokinetics of ligand-liposomes

The clearance rate of mAb-liposomes formed from SL appears to be dependent on mAb density at the liposome surface (Maruyama et al. 1990; Allen et al. 1995). At lower densities (below approximately 50 μg mAb/μmol phospholipid), the mAb-liposomes are cleared at rates only slightly more rapid than those seen for mAb-free SL (Allen et al. 1995). At densities above approximately 100 μg mAb/μmol phospholipid, clearance rates become rapid, with half-lives of only a few minutes (Allen et al. 1995). The clearance is likely mediated by the Fc region of the whole antibody (Aragnol and Leserman 1986).

Non-antibody ligands, when coupled to SL, result in only slight reductions in circulation times relative to ligand-free SL, even at high densities (Blume et al. 1993; Zalipsky et al. 1995; DeFrees et al. 1996; Zalipsky et al. 1997). The use of mAb fragments, like the Fab fragment, which lacks the Fc region of the mAb, decreases the clearance of mAb-liposomes relative to whole mAb (Shahinian and Silvius 1995; Maruyama et al. 1996; Harding et al. 1997; Maruyama et al. 1997).

Applications of targeted Stealth liposomes

Several models have been developed for examining the potential for targeted liposomes. In some of these models, only *in vitro* targeting has been examined. In other models, *in vivo* targeting has been measured but no therapeutic endpoint has been examined, and in a few models therapeutic endpoints have been measured. Various types of targeted liposomes and two different strategies have been used in these experiments. These are presented in a cartoon in Fig. 2.

Many cancer cells overexpress ligand receptors which can be exploited in targeting applications. The folate receptor is overexpressed on cancers of epithelial origin. Folate-liposomes, with entrapped doxorubicin, were shown to be more toxic to target cells than free drug (Lee and Low 1995; Lee and Huang 1996). Selective cytotoxicity could be demonstrated in co-cultures of the target, HeLa cells with human lung fibroblasts not expressing the folate receptor (Lee and Low 1995). Another receptor which can be overexpressed by cancer cells is the low density lipoprotein (LDL) receptor. This receptor can be targeted by attaching apolipoprotein-B (Apo-B) at the liposome surface. Apo-B-liposomes loaded with hydromycin B resulted in selective cytotoxicity to leukemic L2C lymphocytes *in vitro* (Vidal et al. 1985). Likewise, the transferrin receptor is expressed at high densities on some malignant cells. Transferrin-liposomes loaded with cytosine arabinoside were shown to result in enhanced toxicity against CV-1 cells compared to free drugs (Brown and Silvius 1990). Binding of folate, apo-B or transferrin to their respective receptors results in internalization of the liposome and its contents into the target cells, which may result in increased drug delivery to the cells (Vidal et al. 1985; Brown and Silvius 1990; Lee and Low 1995; Lee and Huang 1996). Other targeting molecules which could be exploited in the treatment of cancer include nerve growth factor and epidermal growth factor (Rosenberg et al. 1987; Ishii et al. 1989).

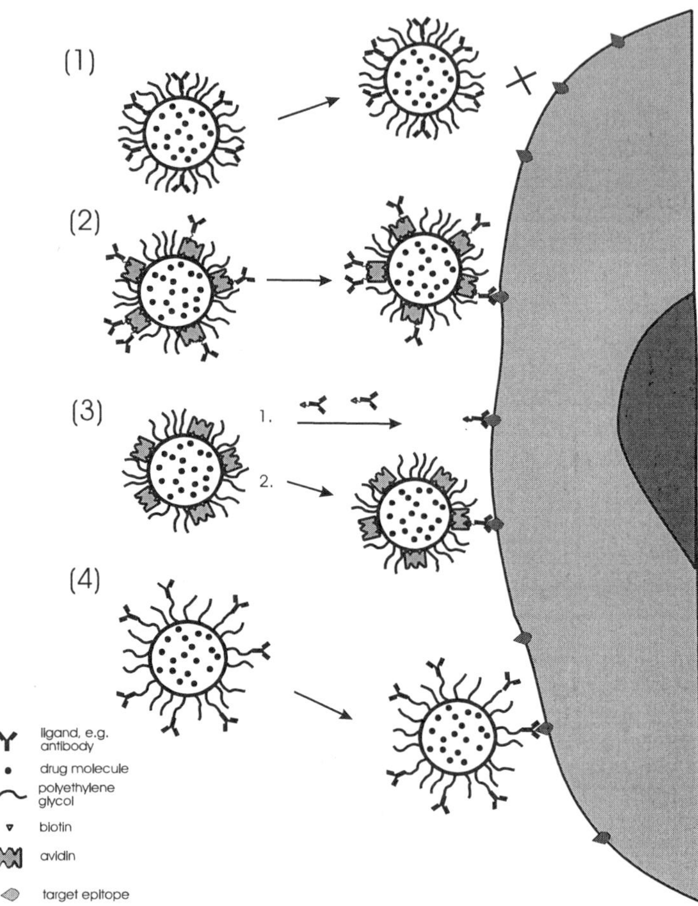

Figure 2. Types of liposomes and targeting strategies used in applications of targeted SL. 1) Antibody or ligand attached at the liposome surface. PEG can interfere with the binding of the Ab-liposomes with their target epitope. 2) Use of spacers, in this case avidin, to extend the ligand or antibody above the surface of the PEG hydrophilic barrier can increase target binding. 3) Two-step strategy which involves binding of a biotinylated ligand to the epitope, followed by binding of avidin- or streptavidin-liposomes. 4) Liposomes with ligand of antibody attached at the PEG terminus have efficient target binding. (copied with permission)

There have been a few demonstrations of *in vivo* targeting of long-circulating liposomes. Liposomes coupled to a mAb against myosin could be targeted to experimentally infarcted rabbit myocardium (Torchilin et al. 1992). Also, liposomes coupled to a mAb against the lung endothelial anticoagulant protein thrombomodulin resulted in significant increases in the binding and retention of these liposomes in lung of mouse (Maruyama et al. 1990; Maruyama et al. 1995). These experiments were extended to examine a therapeutic endpoint by delivery of the lipophilic prodrug dipalmitoyl-fluoro-deoxyuridine (dpFUdR) to EMT-6 mouse mammary tumours seeded into the lung of mice. A significant increase in therapeutic effect was observed when the mAb-liposomes were injected into mice at days 1 and 3 after tumour inoculation (Mori et al. 1995).

Other experiments have examined therapeutic endpoints for targeted liposomes. SL coupled to a mAb recognizing a carbohydrate at the surface of murine lung squamous carcinoma cells were loaded with doxorubicin and used to treat a murine lung cancer model. Treatment of the mice with the mAb-liposomes at three days post-injection of tumour resulted in significant decreases in the number and size of tumours and significant increases in survival times, relative to free drug or mAb-free liposomes. Some long-term survivors were observed (Ahmad et al. 1993). Treatment of more advanced tumours was unsuccessful using this approach, possibly as a result of receptor down-regulation or lack of tumour penetration of the mAb-liposomes (Allen et al. 1995).

Liposomes conjugated to a Fab' fragment against the OA3 antigen, and loaded with doxorubicin, were targeted against an ascitic ovarian cancer. No therapeutic benefit resulted, possibly due to lack of internalization of the drug-liposome package, or to rapid release rates of the drug from the liposomes (Vingerhoeds et al. 1996). In other experiments, using liposomal doxorubicin to treat subcutaneous Caov.3 human ovarian cancer xenografts in mice, we observed that mAb-free SL were more effective than mAb-liposomes, which in turn were more effective than free drug in reducing the rate of tumour growth (Allen et al. 1995). These tumours more advanced, being larger than 1mm in diameter, and differences in the rate of penetration of the targeted liposomes versus the non-targeted liposomes into the tumour might account for these observations (Allen et al. 1995).

Cells within the vasculature are postulated to be an easily accessible target for ligand-liposomes injected by the intravenous route. Haematological malignacies such as multiple myeloma, B and T cell lymphomas, etc. are logical models to examine this hypothesis. In a model of haematological maliganacy, anti-CD19 mAb coupled to doxorubicin-containing SL were targeted to a human B lymphoma xenograft in mice. A significant increase in life span was observed for the mAb-liposomes compared to Ab-free liposomes or free drug (Allen et al. 1995). The success of this model is attributed in part to the readily accessible target cells and in part to the use of a mAb against an internalizing epitope on the cell surface which increases the efficiency of drug delivery to the cell (Allen et al. 1995; Lopes de Menezes et al. 1995).

In another targeting model, anti-HER2/*neu* Fab' fragments coupled to doxorubicin-containing SL were targeted to breast tumour xenografts overexpressing the $p185^{HER2}$ receptor. Cell binding, internalization and antiproliferative activity of the mAb-liposomes were readily demonstrated when the fragments were coupled at the PEG terminus (Kirpotin et al. 1996; Kirpotin et al. 1997). Increased antitumour activity of targeted versus non-targeted doxorubicin-containing liposomes has been reported in a preliminary study (Kirpotin et al. 1996). However, in a similar model, targeted doxorubicin-containing liposomes were equal in efficacy to non-targeted liposomes, although both had superior therapeutic efficacy to free drug (Goren et al. 1996).

BARRIERS TO THE TARGETING APPROACH FOR THE DELIVERY OF ANTICANCER DRUGS

Several problems can be identified for the ligand-mediated targeting approach for the site-specific delivery of anticancer drugs to cancer cells *in vivo*. One of the primary problems is the occurrence of heterogeneity. Cancer cells, with a few exceptions, are notorious for their heterogeneity. Some cells in the targeted population, or at an extreme, most or all of the cells in the cancer, will not express the epitope being targeted. Furthermore, cancer cells will down-regulate or shed epitopes from the cell surface. These characteristics of cancer will make it very difficult to deliver drug to all of the target cells. This problem may be partially overcome by the 'bystander effect' where liposomes localizing in the interstitial fluid of solid tumours can release drug which will diffuse to cells not containing the target epitope.

Another problem which has been identified is the 'binding site barrier' (Osdol et al. 1991). This hypothesis states that ligands will bind to the first target they encounter, which will tend to be cells at the periphery of solid tumours. This will retard or prevent the penetration of ligand-liposomes into the tumour interior. Indeed, non-targeted liposomes may have greater penetrability than targeted liposomes. The binding site barrier is not expected to be a problem for single cell haematological malignancies, for metastatic cells migrating in blood of lymph, or for micrometastases a few cells in size.

Antibodies or antibody fragments have advantages for targeting applications because of their excellent selectivity, but they bring with them the problem of immune reactions against the foreign protein. The production of human anti-mouse antibodies (HAMA) has been shown to occur when mouse mAbs are administered to humans patients (Schroff et al. 1985; Courtenay-Luck et al. 1986; Phillips and Dahman 1995). The use of humanized or chimerized antibodies or antibody fragments may reduce this immunogenicity (Harding et al. 1997). Alternatively, liposomes can be targeted by means of ligand against receptors which are either overexpressed or uniquely expressed on the target cells, but because ligands tend to be less selective than mAbs for the target, this may lead to an increase in non-selective toxicity.

CONCLUSIONS

The development of both sterically stabilized (Stealth) liposomes and new coupling chemistries for attaching antibodies and ligands to these liposomes has led to a revival of interest in *in vivo* targeting applications. Many recent experiments have help us to appreciate both the promise and the challenges inherent in this approach to targeting of drugs in cancer therapy.

REFERENCES

Abra, R.M., and C.A., Hunt, 1981, Liposome disposition in vivo. III. Dose and vesicle-size effects. *Biochim. Biophys. Acta,* 666:493.

Ahmad, I., and Allen, T.M., 1992, Antibody-mediated specific binding and cytotoxicity of liposome-entrapped doxorubicin to lung cancer cells *in vitro*. *Cancer Res.* 52:4817.

Ahmad, I., Longenecker, M., Samuel, J., and Allen, T.M., 1993, Antibody-targeted delivery of doxorubicin entrapped in sterically stabilized liposomes can eradicate lung cancer in mice. *Cancer Res.* 53:1484.

Allen, T.M., 1994, The use of glycolipids and hydrophilic polymers in avoiding rapid uptake of liposomes by the mononuclear phagocyte system. *Adv. Drug Del. Rev.* 13:285.

Allen, T.M., Agrawal, A.K., Ahmad, I., Hansen C.B., and Zalipsky, S., 1994, Antibody-mediated targeting of long-circulating (Stealth®) liposomes. *J. Liposome Res.* 4:1.

Allen, T.M., Ahmad, I., Lopes de Menezes, D.I., and Moase, E.H., 1995, Immunoliposome-mediated targeting of anti-cancer drugs *in vivo*. *Biochem. Soc. Trans.* 23:1073.

Allen, T.M., Brandeis, E., Hansen, C.B., Kao. G.Y., and Zalipsky, S., 1995, A new strategy for attachment of antibodies to sterically stabilized liposomes resulting in efficient targeting to cancer cells. *Biochim.*

Biophys. Acta, 1237:99.
Allen, T.M. and Hansen, C.B., 1991, Pharmacokinetics of Stealth versus conventional liposomes: effect of dose. *Biochim. Biophys. Acta,* 1068:133.
Allen, T.M., Hansen, C.B., and Lopes de Menezes, D.E., 1995, Pharmacokinetics of long circulating liposomes. *Adv. Drug Del. Rev.* 16:267.
Allen, T.M., Hansen, C.B., Martin, F., Redemann, C., and Yau-Young, A., 1991, Liposomes containing synthetic lipid derivatives of poly(ethylene glycol) show prolonged circulation half-lives *in vivo. Biochim. Biophys. Acta,* 1066:29.
Allen, T.M., Hansen, C.B., and Stuart, D.D., 1998, Targeted sterically stabilized liposomal drug delivery. in: *Medical Applications of Liposomes.* D.D. Lasic and D. Papahadjopoulos, eds. Elsevier Science Publishers, Amsterdam, (in press).
Allen, T.M., Hansen, C.B., and Zalipsky, S., 1995, Antibody-targeted Stealth® Liposomes, in: *Stealth Lipsomes.* D.D. Lasic and F. Martin, eds., CRC Press, Inc., Boca Raton.
Allen, T.M., Mehra, T., Hansen, C.B., and Chin, Y.C., 1992, Stealth liposomes: an improved sustained release system for 1-β-D-arabinofuranosylcytosine. *Cancer Res.* 52:2431.
Allen, T.M. and Moase, E.H., 1996, Therapeutic opportunities for targeted liposomal drug delivery. *Adv. Drug Del. Rev.* 21:117.
Allen, T.M. and Stuart, D., Eds., 1998, Liposome Pharmacokinetics: Classical, Sterically Stabilized, Cationic Liposomes and Immunoliposomes, in: *Liposomes: Rational Design.* A.S. Janoff, ed. Marcel Dekker, Inc., New York, (in press)
Ansell, S.M., Tardi, P.G., and Buchkowsky, S.S., 1996, 3-(2-pyridyldithio)propionic acid hydrazide as a cross-linker in the formation of liposome-antibody conjugates. *Bioconjug. Chem.* 7:490.
Aragnol, D., and Leserman, L., 1986, Immune clearance of liposomes inhibited by an anti-Fc receptor antibody in vivo. *Proc. Natl. Acad. Sci. USA* 83:2699.
Bakker-Woudenberg, I.A.J.M., Lokerse, A.F., ten Kate, M.T., and Storm, G., 1992, Enhanced localization of liposomes with prolonged blood circulation time in infected lung tissue. *Biochim. Biophys. Acta,* 1138:318.
Blume, G., and Cevc, G., 1990, Liposomes for the sustained drug release in vivo. *Biochim. Biophys. Acta,* 1029:91.
Blume, G., Cevc, G., Crommelin, M.D., Bakker-Woudenberg, L.A., Kluft, C., and Storm, G., 1993, Specific targeting with poly(ethylene glycol)-modified liposomes: coupling of homing devices to the ends of the polymeric chains combines effective target binding with long circulation times. *Biochim. Biophys. Acta,* 1149:180.
Boman, N.L., Masin, D., Mayer, L.D., Cullis, P.R., and Bally, M.B.,1994, Liposomal vincristine which exhibits increased drug retention and increased circulation longevity cures mice bearing P388 tumors. *Cancer Res.* 54:2830.
Brown, P.M., and Silvius, J.R., 1990, Mechanisms of delivery of liposome-encapsulated cytosine arabinoside to CV-1 cells in vitro. Fluorescence-microscopic and cytotoxicity studies. Biochim. Biophys. Acta, 1023:341.
Chonn, A., and Cullis, P.R., 1992, Ganglioside GM_1 and hydrophilic polymers increase liposome circulation times by inhibiting the association of blood proteins. *J. Liposome Res.* 2:397.
Corley, P., and Loughrey, H.C., 1994, Bonding of biotinated-liposomes to streptavidin is influenced by liposome composition. *Biochim. Biophys. Acta,* 1195:149.
Courtenay-Luck, N.S., Epenetos, A.A., Moore, R., Larche, M., Pecatasides, D., and Ritter, M.A., 1986, Development of primary and secondary immune responses to mouse monoclonal antibodies used in the diagnosis and therapy of malignant neoplasms. *Cancer Res.* 46:6489.
Debs, R. J., Heath, T.D., and Papahadjopoulos, D., 1987, Targeting of anti-Thy 1.1 monoclonal antibody conjugated liposomes in Thy 1.1 mice after intravenous administration. *Biochim. Biophys. Acta,* 901:183.
DeFrees, S.A., Phillips, L., Guo, L., and Zalipsky, S., 1996, Sialyl Lewis X liposomes as a multivalent ligand and inhibitor of E-selectin mediated cellular adhesion. *J. Am. Chem. Soc.* 118:6101.
Defrise-Quertain, F., Chatelain, P., Delmelle, M., and Ruysschaert, J., 1984, Model Studies for Drug Entrapment and Liposome Stability. in: *Liposome Technology.* G. Gregoriadis, ed., CRC Press, Inc. 2:1-17.
Eckardt, J.R., Campbell, E., Burries, H.A., Weiss, G.R., Rodriguez, G.I., Fields, S.M., Thurman, A.M., Peacock, N.W., Cobb, P., Rothenberg, M.L., Ross, M.E., and Von Hoff, D.D., 1994, A Phase II trial of DaunoXome, liposome encapsulated daunorubicin, in patients with metastatic adenocarcinoma of the colon. *Am Journal Clin Oncol Cancer Clin Trials* 17:498.
Forssen, E.A., 1997, The design and development of DaunoXome^R for solid tumor targeting in vivo. *Adv. Drug Del. Rev.* 24:133.
Forssen, E.A., Male-Brune, R., Adler-Moore, J.P., Lee, M.J.A., Schmidt, P.G., Kraieva, T.B., Shimizu, S., and Tromberg, B.J., 1996, Fluorescence imaging studies for the disposition of daunorubicin liposomes

(DaunoXome) in tumor tissue. *Cancer Res.* 56:2066.
Gabizon, A., Catane, R., Uziely, B., Kaufman, B., Safra, T., Cohen, R., Martin, F., Huang, A., and Barenholz, Y., 1994, Prolonged circulation time and enhanced accumulation in malignant exudates of doxorubicin encapsulated in polyethylene-glycol coated liposomes. *Cancer Res.* 54:987.
Gabizon, A., and Papahadjopoulos, D., 1988, Liposome formulations with prolonged circulation time in blood and enhanced uptake by tumors. *Proc. Natl. Acad. Sci. USA,* 85:6949.
Gill, P.S., Espina, B.M., Muggia, F., Cabriales, S., Tulpule, A., Esplin, J.A., Liebman, H.A., Forssen, E., Ross, M.E., and Levine, A.M., 1995, Phase I/II clinical and pharmacokinetic evaluation of liposomal daunorubicin. *J. Clin. Oncol.* 13:996.
Goebel, F.-D., Goldstein, D., Goos, M., Jablonowski, H., and Stewart, J.S., 1996, Efficacy and safety of Stealth^R lipsomal doxorubicin in AIDS-related Kaposi's sarcoma. *Brit. J. Cancer,* 73:989.
Goren, D., Horowitz, A.T., Zalipsky, S., Woodle, M.C., Yarden, Y., and Gabizon, A., 1996, Targeting of stealth liposomes to erB-2 (Her/2) receptor: *in vitro* and *in vivo* studies. *Br. J. Cancer* 74:1749.
Hansen, C.B., Kao, G.Y., Moase, E.H., Zalipsky, S. and Allen, T.M., 1995, Attachment of antibodies to sterically stabilized liposomes: evaluation, comparison and optimization of coupling procedures. *Biochim. Biophys. Acta,* 1239:133.
Haran, G., Cohen, R., Bar and, L,K., and Barenholz, Y., 1993, Transmembrane ammonium sulfate gradients in liposomes produce efficient and stable entrapment of amphipathic weak bases. *Biochim. Biophys. Acta,* 1151:201.
Harding, J.A., Engbers, C.M., Newman, M.S., Goldstein, N.I., and Zalipsky, S., 1997, Immunogenicity and pharmacokinetic attributes of poly(ethylene glycol)-grafted immunoliposomes. *Biochim. Biophys. Acta* (in press).
Harrison, M., Tomlinson, D., and Stewart, S., 1995, Liposomal- entrapped doxorubicin: an active agent in AIDS-related Kaposi's sarcoma. *J. Clin. Oncol.* 13:914.
Heath, T.D., and Martin, F.J., 1986, The development and application of protein-liposome conjugation techniques. *Chem. Phys. Lipids,* 40:347.
Holmberg, E., Maruyama, K., Litzinger, D.C., Wright, S., Davis, M., Kabalka, G.W., Kennel, S.J., and Huang, L., 1989, Highly efficient immunoliposomes prepared with a method which is compatible with various lipid compositions. *Biochem. Biophys. Res. Commun.* 165:1272.
Hwang, K.J., 1987, Liposome Pharmacokinetics. in: *Liposomes: from Biophysics to Therapeutics.* M. J. Ostro, ed., Marcel Dekker, New York.
Ishii, Y., Aramaki, Y., Hara, T., Tsuchiya, S., and Fuwa, T., 1989, Preparation of EGF labeled liposomes and their uptake by hepatocytes. *Biochem. Biophys. Res. Commun.* 160:732.
Jain, R.K., 1987, Transport of molecules across tumor vasculature. *Cancer Metastasis Rev.* 559.
Kirpotin, D., Hong, K., Mullah, N., Papahadjopoulos, D., and Zalipsky, S., 1996, Liposomes with detachable polymer coating: destabilization and fusion of dioleoylphosphatidylethanolamine vesicles triggered by cleavage of surface-grafted poly(ethylene glycol). *FEBS Lett.* 388:115.
Kirpotin, D., Park, J.W., Hong, K., Keller, G., Benz, C., and Paphadjopoulos, D., 1996, Binding and endocytosis of sterically stabilized anti-HER2 immunoliposomes by human breast cancer cells. *Proc. Am. Assoc. Cancer Res.* 37:3186.
Kirpotin, D., Park, J.W., Hong, K., Zalipsky, S., Li, W.-L., Carter, P., Benz, C.C., and Papahadjopoulos, D., 1997, Sterically stabilized anti-HER2 immunoliposomes: design and targeting to human breast cancer cells *in vitro. Biochemistry* 36:66.
Klibanov, A.L. and Huang, L., 1992, Long-circulating liposomes: development and perspectives. *J. Liposome Res.* 2:321.
Klibanov, A.L., Maruyama, K., Torchilin, V.P., and Huang, L., 1990, Amphipathic polyethyleneglycols effectively prolong the circulation time of liposomes. *FEBS Lett.* 268:235.
Klibanov, A.L., Serbina, N., Torchilin, V.P., and Huang, L., 1996, Attachment of ligands to liposomes via PEG spacer for prolonged liposome circulation and targeting. *J. Liposome Res.* 6:195.
Lee, R.J., and Huang, L., 1996, Folate-targeted, anionic liposome-entrapped polylysine-condensed DNA for tumor cell-specific gene transfer. *J. Biol. Chem.* 271:8481.
Lee, R.J. and Low, P.S., 1994, Delivery of liposomes into cultured KB cells via folate receptor-mediated endocytosis. *J. Biol. Chem.* 269:3198.
Lee, R.J. and Low, P.S., 1995, Folate-mediated tumor cell targeting of liposome-entrapped doxorubicin *in vitro. Biochim. Biophys. Acta,* 1233:134.
Longman, S.A., Cullis, P.R., Choi, L., de Jong, G., and Bally, M.B., 1995, A two-step targeting approach for delivery of doxorubicin-loaded liposomes to tumour cells in vivo. *Cancer Chemother. Pharmacol.* 36:91.
Lopes de Menezes, D.E., Pilarski, L.M., and Allen, T.M., 1995, Selective cytoxicity of immunoliposomal to B lymphocytes. *Proc. Am. Assoc. Cancer Res.* 36:307.
Loughrey, H., Bally, M.B., and Cullis, P.R., 1987, A non-covalent method of attaching antibodies to liposomes. *Biochim. Biophys. Acta,* 901:157.

Loughrey, H.C., Choi, L.S., Cullis, P.R., and Bally, M.B., 1990, Optimized procedures for the coupling of proteins to liposomes. *J. Immunol. Methods,* 132:25.
Loughrey, H.C., Choi, L.S., Wong, K.F., Cullis, P.R., and Bally, M.B., 1993, Preparation of streptavidin-liposomes for use in ligand specific targeting applications. in: *Liposome Technology,* 2nd ed. G. Gregoriadis, ed. CRC Press. Boca Raton.
Lundberg, B., Hong, K., and Papahadjopoulos, D., 1993, Conjugation of apolipoprotein B with liposomes and targeting to cells in culture. *Biochim. Biophys. Acta,* 1149:305.
Martin, F.J., and Papahadjopoulos, D., 1982, Irreversible coupling of immunoglobulin fragments to preformed vesicles. An improved method for liposome targeting. *J. Biol. Chem.* 257:286.
Maruyama, K., Kennel, S.J., and Huang, L., 1990, Lipid composition is important for highly efficient target binding and retention of immunoliposomes. *Proc. Natl. Acad. Sci. USA*, 87:5744.
Maruyama, K., Kennel, S.J., and Huang, L., 1990, Lipid composition is important for highly efficient target binding and retention of immunoliposomes. *Proc. Natl. Acad. Sci. USA* 87:5744.
Maruyama, K., Takahashi, N., Tagawa, T., Nagaike, K., and Iwatsuru, M., 1997, Immunoliposomes bearing polyethyleneglycol-coupled Fab' fragment show prolonged circulation time and high extravasation into targeted solid tumors *in vivo. FEBS Lett.* 413:177.
Maruyama, K., Takizawa, T., Takahashi, N., Tagawa, T., Nagaike, K., and Iwatsuru, M., 1996, Factors influencing longevity and target binding of PEG-immunoliposomes conjugated antibodies at PEG's terminals. *J. Liposome Res.* 6:206.
Maruyama, K., Takizawa, T., Takahashi, N., Tagawa, T., Nagaike, K., and Iwatsuru, M., 1997, Targeting efficiency of PEG-immunoliposome-conjugated antibodies at PEG terminals. *Adv. Drug Del. Rev.* 24:235.
Maruyama, K., Takizawa, T., Yuda, T., Kennel, S.J., Huang, L., and Iwatsuru, M., 1995, Targetability of novel immunoliposomes modified with amphipathic poly(ethylene glycol)s conjugated at their distal terminals to monoclonal antibodies. *Biochim. Biophys. Acta,* 1234:74.
Mayer, L.D., Bally, M.B., and Cullis, P.R., 1986, Uptake of adriamycin into large lunilamellar vesicles in response to a pH gradient. *Biochim. Biophys. Acta,* 857:123.
Mayer, L.D., Hope, M.J., and Cullis, P.R., 1986, Vesicles of variable sizes produced by a rapid extrusion procedure. *Biochim. Biophys. Acta* 858:168.
Mori, A., Kennel, S.I., Waalkes, M.V.B., Scherphof, G.L., and Huang, L., 1995, Characterization of organspecific immunoliposomes for delivery of 3', 5'-O-dipalmitoyl-5-fluoro-2'-deoxyuridine in a mouse lung-metastasis model. *Cancer Chemother. Pharmacol.* 35:447.
Mori, A., Klibanov, A.L., Torchilin, V.P., and Huang, L., 1991, Influence of the steric barrier of amphipathic poly(ethyleneglycol) and ganglioside GM_1 on the circulation time of liposomes and on the target binding of immunoliposomes *in vivo.* FEBS Lett. 284:263.
Muggia, F., Hainsworth, J.D., Jeffers, S., Miller, P., Groshen, S., Tan, M., Roman, L., Uziely, B., Muderspach, L., Garcia, A., Burnett, A., Greco, F.A., Morrow, C.P., Paradiso, L.J., and Liang, L-J., 1997, Phase II study of liposomal doxorubicin in refractory ovarian cancer: antitumor activity and toxicity modification by liposomal encapsulation. *J. Clin. Oncol.* 15.
Northfelt, D.W., Dezube, B.J.,Thommes, J.A., Levine, R., Von Roenn, J.H., Dosik, G.M., Rios, A., Krown, S.E., DuMond, C., and Mamelok, R.D., 1997, Efficacy of pegylated-liposomal doxorubicin in the treatment of AIDS-related Kaposi's sarcoma after failure of standard chemotherapy. *J. Clin. Oncol.* 15:653.
Northfelt, D.W., Martin, F.J., Kaplan, L.D., Russell, J., Andersen, M., Lang, J., and Volberding, P.A., 1993, Pharmacokinetics, tumour localization and safety of Doxil (liposomal doxorubicin) in AIDS patients with Kaposi's sarcoma. *Proc. Am. Soc. Clin. Oncol.* 12:51.
Northfelt, D.W., Martin, F.J., Working, P.,Volberding, P.A., Russell, J., Newman, M., Amantea, M.A., and Kaplan., L.D., 1996, Doxorubicin encapsulated in liposomes containing surface-bound polyethylene glycol: pharmacokinetics, tumour localization, and safety in patients with AIDS-related Kaposi's sarcoma. *J. Clin. Pharmacol.* 36:55.
Olson, F., Hunt, C.A., Szoka, F.C., Vail, W.J., and P.D., 1979, Preparation of liposomes of defined size distribution by extrusion through polycarbonate membranes. *Biochim. Biophys. Acta,* 557:9.
Osdol, W.V., Fujimori, K., and Weinstein, J.N., 1991, An analysis of monocolonal antibody distribution in microscope tumour nodules: consequences of a "binding site barrier". *Cancer Res.* 51:4776.
Papahadjopoulos, D., Allen, T.M., Gabizon, A., Mayhew, E., Matthay, K., Huang, S.K., Lee, K.D., Woodle, M.C., Lasic, D.D., Redemann, C., and Martin, F.J., 1991, Sterically stabilized liposomes: improvements in pharmacokinetics and antitumor therapeutic efficacy. *Proc. Natl. Acad. Sci. USA* 88:11460.
Phillips, N.C. and Dahman, J., 1995, Immunogenicity of immunoliposomes: reactivity against species-specific IgG and liposomal phospholipids. *Immunol. Lett.* 45:149.
Ranson, M., O'Bryne, K., Carmichael, J., Smith, D., Stewart, S., and Howell, A., 1996, Phase II dose-finding trial of DOX-SL (Stealth® liposomal doxorubicin HCl) in the treatment of advanced breast cancer.

Proc. Am. Soc. Clin. Oncol. 15:124.
Rosenberg, M. B., Breakefield, X.O., and Hawrot, E., 1987, Targeting of liposomes to cells bearing nerve growth factor receptors mediated by biotinylated nerve growth factor. *J. Neurochem*. 48:865.
Schroff, R. W., Foon, K.A., Beatty, S.M., Oldham, R.K., and Morgan, A.C., 1985, Human anti-mouse immunoglobulin responses in patients receiving monoclonal antibody therapy." *Cancer Res.* 48:879.
Senior, J., Delgado, C., Fisher, D., Tilcock, C., and Gregoriadis, G., 1991, Influence of surface hydrophilicity of liposomes on their interaction with plasma protein and clearance from the circulation: studies with the poly(ethylene glycol)-coated vesicles. *Biochim. Biophys. Acta*, 1062:77.
Shahinian, S. and Silvius , J.R., 1995, A novel strategy affords high-yield coupling of antibody Fab' fragments to liposomes. *Biochim. Biophys. Acta,* 1239:157.
Suzuki, S., Watanabe, S., Masuko, T., and Hashimoto, Y., 1995, Preparation of long-circulating immuno-liposomes containing adriamycin by a novel method to coat immunoliposomes with poly(ethylene glycol). *Biochim. Biophys. Acta* 1245:9.
Suzuki, S., Watanabe, S., Masuko, T., and Hashimoto, Y., 1995, Preparation of long-circulating immuno-liposomes containing adriamycin by a novel method to coat immunoliposomes with poly(ethylene glycol). *Biochim. Biophys. Acta,* 124:9.
Takizawa, T., Maruyama, K., and Iwatsuru, M., 1996, Novel immunoliposomes modified with amphipathic polyethyleneglycols conjugated at their distal terminals to monoclonal antibodies. *J. Liposome Res.* 6:261.
Torchilin, V.P., and Klibanov, A.L., 1993, Coupling of ligands with liposome membranes. *Drug Target. Del.* 2:227.
Torchilin, V.P., Klibanov, A.L., Huang, L., O'Donnell, S., Nossiff, N.D., and Khaw, B.A., 1992, Targeted accumulation of polyethylene glycol-coated immunoliosomes in infarcted rabbit myocardium. *FASEB J.* 6:2716.
Torchilin, V.P., Narula, J., Halpern, E., and Khaw, B.A., 1996, Poly(ethylene glycol)-coated anti-cardiac myosin immunoliposomes: factors influencing targeted accumulation in the infarcted myocardium. *Biochim. Biophys. Acta,* 1279:75.
Torchilin, V.P., and Trubetskoy, V.S., 1995, New synthetic amphilphilic polymers for steric protection of liposomes in vivo. *J. Pharm. Sci.* 85:85.
Unezaki, S., Maruyama, K., Hosoda, J.-I., Nagae, I., Koyanayi, Y., Nakata, M., Ishida, O., Iwatsuru, M., and Tsuchiya, S., 1996, Direct measurement of the extravasation of polyethyleneglycol-coated liposomes into solid tumor tissue by in vivo fluorescence microscopy. *Int. J. Pharmaceut.* 144:11.
Uster, P.S., Allen, T.M., Daniel, B.E., Mendez, C.J., Newman, M.S., and Zhu, G.Z., 1996, Insertion of poly-(ethylene glycol) derivatized phospholipid into preformed liposomes results in prolonged in vivo circulation time, *FEBS Lett.* 386:243.
Uziely, B., Jeffers, S., Isacson, R., Kutsch, K., Wei-Tsao, D., Yehoshua, Z., Libson, E., Muggia, F.M., and Gabizon, A., 1995, Liposomal doxorubicin: antitumor activity and unique toxicities during two complementary phase I studies. *J. Clin. Oncol.* 13:1777.
Vail, D.M., Kravis, L.D., Cooley, A.J., Chun, R., and MacEwan, E.G., 1997, Preclinical trial of doxorubicin entrapped in sterically stabilized liposomes in dogs with spontaneously arising malignant tumors. *Cancer Chemother. and Pharmacol.* 39:410.
Vidal, M., Sainte-Marie, J., Philippot, J.R., and Bienvenue, A., 1985, LDL-mediated targeting of liposomes to leukemic lymphocytes in vitro. *EMBO J.* 4:2461.
Vingerhoeds, M.H., Steerenberg, P.A., Hendriks, J.J.G.W., Kekker, L.C., van Hoesel, Q.G.C.M., Crommelin, D.J.A., and Storm, S., 1996, Immunoliposome-mediated targeting of doxorubicin to human ovarian carcinoma *in vitro* and *in vivo*. *Br. J. Cancer* 74:1023.
Weiner, A.L., 1990, Chemistry and Biology of Immunotargeted Liposomes. Targeted Therapeutic Systems. P. Tyle and B. P. Ram, eds. Marcel Dekker, Inc., New York.
Woodle, M.C., Lasic, D.D., Redemann, C., Newman, M., Babbar, S., and Martin, F.J., 1991, In vivo studies of long circulating (Stealth) liposomes in rats. *Periodicum Biologorum.* 93:349.
Zalipsky, S., 1993, Synthesis of end-group functionalized polyethylene glycol-lipid conjugates for preparation of polymer-grafted liposomes. *Bioconj. Chem.* 4:296.
Zalipsky, S., Brandeis, E., Newman, M., and Woodle, M.C., 1994, Long circulating, cationic liposomes containing amino-PEG-phosphatidylethanolamine. *FEBS Lett.* 353:71.
Zalipsky, S., Hansen, C.B., Oaks, J.M., and Allen, T.M., 1996, Evaluation of blood clearance and biodistribution of poly(2-oxazoline)-grafted liposomes. *J. Pharm. Sci.* 85:133.
Zalipsky, S., Mullah, N., Harding, J.A., Gittelman, J., Guo, L., and DeFrees, S.A., 1997, Poly(ethylene glycol)-grafted liposomes with oligopeptide or oligosaccharide ligands appended to the termini of the polymer chains. *Bioconj. Chem.* 8:111.
Zalipsky, S., Newman, M., Punatambekar, B., and Woodle, M.C., 1993, Model ligands linked to polymer chains on liposomal surfaces: application of a new functionalized polyethylene glycol-lipid conjugate. *Polym. Materials: Sci. Eng.* 67:519.

Zalipsky, S., Puntambekar, B., Bolikas, P., Engbers, C.M., and Woodle, M.C., 1995, Peptide attachment to extremities of liposomal surface grafted PEG chains: Preparation of the long-circulating form of laminin pentapetide, YIGSR. *Bioconj. Chem.* 6:705.

Zu, N.Z., Da, D., Rudoll, T.L., Needham, D., Whorton, A.R., and Dewhirst, M.W., 1993, Increased microvascular permeability contributes to preferential accumulation of Stealth liposomes in tumor tissue. *Cancer Res.* 53: 3765.

STEALTH™ LIPOSOMES AS CARRIERS OF DOXORUBICIN

Dorit Goren[1], Samuel Zalipsky[2], Aviva T. Horowitz[1], and Alberto Gabizon[1]

[1]Department of Oncology, Hadassah Hebrew University Hospital, Jerusalem, Israel, and [2]Sequus Pharmaceuticals Inc., Menlo Park, California, USA

INTRODUCTION

Liposomes, as non-covalently bound carriers, biocompatible and biodegradable, have raised considerable interest as a drug delivery system in cancer chemotherapy (Gregoriadis, 1988). Most applications of liposomes in cancer chemotherapy are directed at altering tissue distribution and various pharmacokinetic parameters of the drug in question in such a way that toxicity can be reduced and/or efficacy increased (Mayhew and Papahadjopoulos, 1983). Reduced toxicity may be gained through site circumvention of drug sensitive tissues and by slow release of the cytotoxic agent from the carrier, avoiding peak plasma concentrations after bolus injection of free drug. Liposome- mediated decrease in toxicity could enable escalation of dose, which will result in increased tumor exposure to the drug.

One of the most encouraging areas in the liposome-anticancer drug field is the work with anthracyclines. Doxorubicin (DOX), a major anti-neoplastic anthracycline and one of the drugs most widely used in cancer chemotherapy, has a broad spectrum of anti-tumor activity against solid tumors and leukemias (Young et al., 1981). Anthracycline-induced cardiotoxicity (Minow et al., 1975), a severe cumulative chronic toxicity, may be attenuated by a carrier system that decreases drug uptake by the heart muscle without lessening the anti-tumor activity, thereby improving the therapeutic index. Liposomes can adequately fulfill this task, given their relative inability to cross continuous capillaries and the lack of cells of the reticulo-endothelial system (RES) in the myocardial tissue. In addition, the slow-release effect of liposomal delivery of anthracyclines may reduce the peak plasma concentration of free drug, a factor which is directly correlated with cardiotoxicity (Legha et al., 1982). Various liposome formulations have been tested as carriers of DOX in the early 80's (Gabizon et al., 1982; Forssen and Tokes, 1981; Forssen and Tokes, 1983; Olson et al., 1982; Van Hoesel et al., 1984; Rahman et al., 1985; Gabizon et al., 1986a). Examination of their pharmacologic properties, indicates that liposome entrapment of DOX decreases drug distribution to the heart, thus reducing the cardiotoxic effect. Furthermore, liposomal encapsulation of DOX enhances its therapeutic activity in a limited number of experimental tumor models (Gabizon et al., 1985; Gabizon et al., 1986b;

Targeting of Drugs 6: Strategies for Stealth Therapeutic Systems
Edited by Gregoriadis and McCormack, Plenum Press, New York, 1998

Gabizon et al., 1983; Rahman et al., 1986).

A remarkable progress in liposome engineering in recent years has resulted in the development of liposomal drug delivery systems with valuable pharmacologic properties: inhibition of the rapid clearance from circulation by the RES and reduction of the rate of drug leakage, leading to stable long-circulating liposome formulations. In particular, the coating of liposomes with polyethylene-glycol (PEG), a hydrophilic polymer, that generates a steric barrier preventing the hydrophobic interactions of plasma opsonins with the vesicle surface, has a major impact in conferring protection to the vesicles from RES-mediated clearance (Torchilin, 1995). Moreover, bilayer rigidification using high Tm phospholipids reduces the rate of leakage of liposome content (Gabizon et al., 1993a). The term "Stealth liposomes" has been coined to designate these long-circulating, PEG-coated liposomes (Lasic and Martin, 1995).

In the design of stable liposome formulations for drug delivery the process of drug encapsulation is an essential determinant of the carrier properties. An efficient remote loading technique of preformed liposomes with doxorubicin, compatible with Stealth formulations was developed (Haran et al., 1993). By means of transmembrane pH and ammonium sulfate gradients that drive cationic amphiphiles such as DOX, into the liposome water compartment (Haran et al., 1993; Mayer et al., 1990), extremely high drug load (150-200μg DOX/μmol phospholipid) is obtained. Once inside, the drug becomes ionized, thus preventing its escape through the lipid bilayer. This is followed by the formation of a gel-like precipitate, as the drug in liposomes reaches a concentration exceeding its aqueous solubility (Lasic et al., 1992). For intact gradient maintenance, high phase-transition temperature phospholipids in combination with cholesterol are essential components. This type of formulation can remain for long periods in circulation with minimal leakage of drug (Gabizon et al., 1993a, b). A Stealth liposome formulation of DOX, known as Doxil (a registered trademark of Sequus, Menlo Park, CA), has been approved for clinical use, underscoring the validity of this approach.

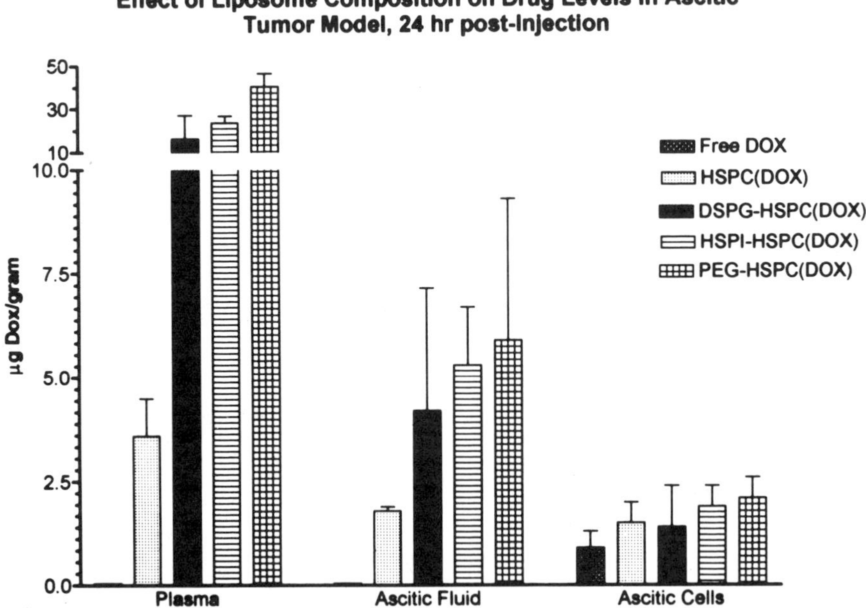

Figure 1. The effect of liposome composition on DOX levels in a murine ascitic tumor model (J-6456 lymphoma), 24h post injection of 10mg Doxil/kg

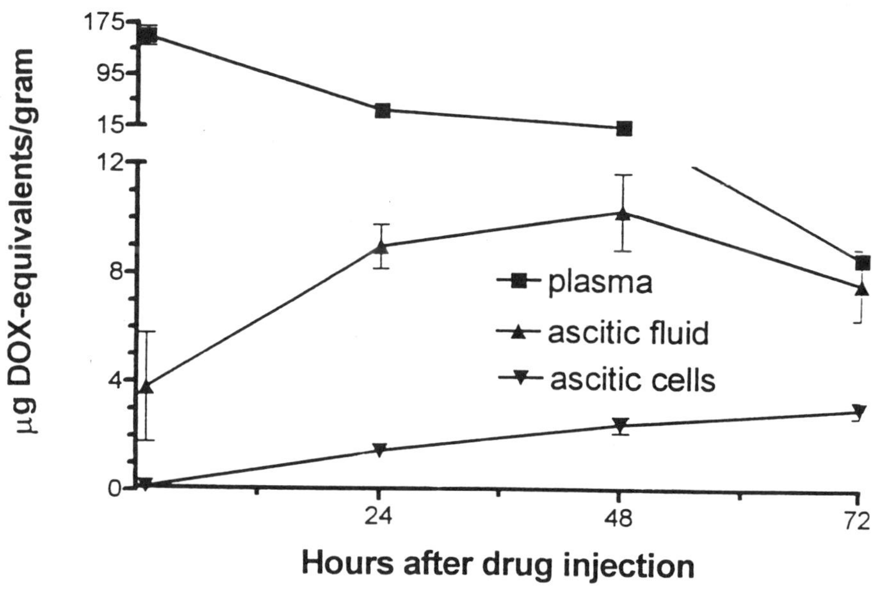

Figure 2. Kinetics of DOX delivery to the J6456 ascitic tumor. Mice previously inoculated i.p. with J6456 tumor cells received i.v. 10mg/kg of DOX encapsulated in PEG-HPC-Chol liposomes. Note that the drug peak in the ascitic fluid is obtained 48 h after drug injection. Drug penetration in the cellular compartment is slower and equilibration with the extracellular compartment is still not reached 72h after injection. Each time point is the mean of 3 to 4 mice. (Reproduced with permission from Gabizon et al., 1996)

LIPOSOME LONGEVITY IN CIRCULATION CORRELATION WITH TUMOR ACCUMULATION

A key issue in cancer drug delivery is enhancing selectively drug accumulation in tumors. The relevance of the Stealth liposome formulation to this favorable pharmacologic property was investigated in several murine and human tumor models (Gabizon et al., 1996). An example of an interesting tumor model examined is the J-6456 murine lymphoma inoculated intraperitoneally. A biodistribution study of a variety of liposomal-DOX formulations administered intravenously, in mice with an ascitic J-6456 lymphoma tumor, indicated that PEG-containing liposomes is probably the most advantageous formulation (Gabizon et al., 1997). As seen in Fig. 1, 24 h after injection, the highest plasma DOX levels as well as highest peak levels in the ascitic fluid are obtained with a PEG-hydrogenated soybean phosphatidylcholine (HSPC)-cholesterol (Chol) formulation, a result which is evidently due to their long circulation time (Gabizon et al., 1996; Gabizon et al., 1997). Interestingly, the drug levels in mice injected with free DOX are undetectable in plasma and ascitic fluid (Gabizon et al., 1996; Gabizon, 1992), but only slightly lower than for liposome-encapsulated DOX in the case of ascitic cells (Gabizon et al., 1996). Despite this small difference in cell-associated DOX, it should be stressed that the reservoir of drug in plasma is huge in the case of formulations like PEG-HSPC-Chol. Through a gradual process of extravasation of liposomal DOX from plasma into the ascitic fluid, followed by efflux of drug from liposomes, the ascitic cellular compartment is gradually fed, resulting ultimately in a greater cell exposure to drug. To illustrate this point, the kinetics of the process of accumulation in ascites using PEG-HPC-Chol(DOX) liposomes is shown in Fig. 2. We found that the DOX levels in the ascitic fluid gradually rise reaching a peak at 48h after injection, and equilibrating with plasma levels at 72h after injection. This is indicative of a slow equilibration of liposomal drug between two compartments (intravascular and extracellular), probably mediated by the extravasation of circulating liposomes. Another important observation in Fig. 2 is the late but steady increase in the levels of cell-associated drug, which have not yet reached their peak at 72h after injection. Clearly, the pattern of drug accumulation in cells results from the slow efflux of DOX from liposomes present in the surrounding ascitic fluid. Since, even at 72h after injection there is still a high concentration of liposomal drug in the ascitic fluid, it is likely that drug levels in cells will further increase. However, due to the advanced stage of the tumor and the deterioration of the animals, we did not examine the drug levels beyond 72h. Interestingly, no significant accumulation of liposomal DOX was observed in peritoneal washes from tumor-free mice (data not shown), indicating that extravasation depends on the increased microvascular permeability caused by the ascitic tumor (Gabizon, 1992).

Another relevant aspect to the pharmacology of PEG-HSPC-Chol liposomes is the route of injection. Allen et al. have reported a rapid entry of liposomes into the circulatory system after subcutaneous or intraperitoneal injection, followed by a plasma clearance curve similar to that observed after direct i.v. injection (Allen et al., 1993). To examine this point, we did a comparative study of the pharmacokinetics of PEG-HSPC-Chol(DOX) administered by the i.v. or i.p. route into J-6456 ascites bearing mice. As seen in Fig. 3, high plasma levels were rapidly reached after i.p. injection of liposomal DOX and the ensuing plasma clearance was similar to that of i.v.-injected liposomal DOX. However, both, peak plasma levels and AUC values (area

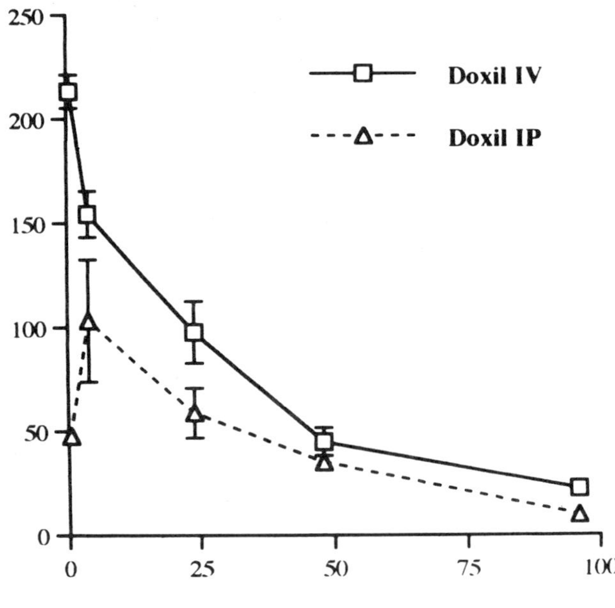

Figure 3. DOX levels in plasma after i.v. or i.p. injection of 10mg/kg Doxil dose, into BALB/c female mice bearing J-6456 lymphoma, inoculated 5 days previously.

under the concentration versus time curve) are significantly greater when the drug is given intravenously (Fig. 3), suggesting that a sizable fraction of the liposome dose is trapped in the peritoneal cavity or in draining lymph nodes. The route of injection is highly relevant to the therapeutic efficacy (Cabanes et al., 1998). In the ascitic J6456 lymphoma, the i.v. route is more effective than the i.p. route for PEG-liposomal DOX, while the opposite is true for free DOX and DOX encapsulated in egg-derived phosphatidylglycerol (EPG)-phosphatidylcholine(EPC)-Chol oligolamellar liposomes (Gabizon et al., 1988a).

LIPOSOME COMPOSITION /FORMULATION AND THERAPEUTIC EFFICACY

Evidently, the critical issue is to look into the relevance of differences in circulation longevity and drug delivery to tumors and their implications in anti-tumor activity. The effect of bilayer fluidity of PEG-coated liposomes on the therapeutic activity of encapsulated DOX (Fig. 4) was examined, using the ascitic J6456 tumor model. Clearly, PEG- HPC-Chol- (DOX) and PEG-dipalmitoyl-PC(DPPC)-Chol(DOX) were more effective than PEG-EPC-Chol(DOX) in prolonging the survival of tumor-bearing mice, an observation that underscores the relevance of a high T_m bilayer for an efficient drug delivery to the tumor. This result is similar to a previous observation in the same tumor model in which a hydrogenated phosphatidylinositol-HPC-Chol formulation of DOX was found to be therapeutically superior to a formulation of EPG-EPC-Chol liposomes (Gabizon, 1992). Despite the longer circulation time of PEG-HPC-Chol(DOX) over PEG-DPPC-Chol(DOX), there was no additional gain in

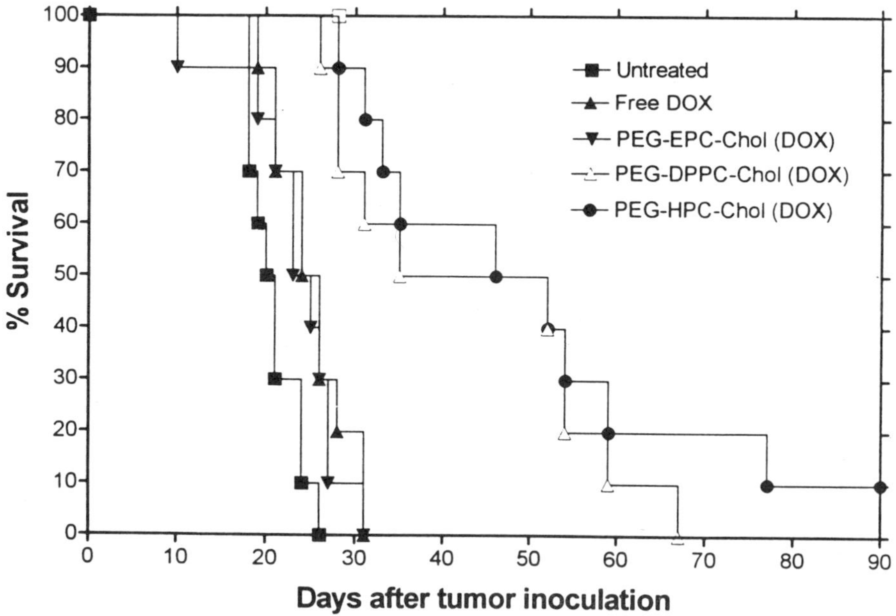

Figure 4: Therapeutic efficacy of liposome-encapsulated DOX against the ascitic J-6456 lymphoma. BALB/c mice were inoculated i.p. with 106 J-6456 cells and treated 5 days later i.v. with 10mg/kg of free or liposome-encapsulated DOX. There were 10 mice in each experimental group. The survival of PEG-HPC(DOX)-treated mice was significantly longer than that of free DOX, and PEG-EPC(DOX)-treated mice (Log-rank test, $p \leq 0.0006$). There was no significant difference between the PEG-HPC(DOX) and PEG-DPPC(DOX) groups. (Reproduced with permission from Gabizon et al., 1996)

anti-tumor activity. This suggests that there is a balance between stability in circulation and drug release rate in the tumor fluid that determines optimal antitumor activity. At any rate, the HPC- and DPPC-based formulations show an improvement in anti-tumor activity against ascites J6456 lymphoma as compared with free-DOX, as manifested by an increase in median survival and even a few cases of cures (Gabizon et al., 1993b; Gabizon et al., 1996). Liposomal DOX-increased anti-tumor activity is not related to a buffering of toxicity conferred to the drug through encapsulation, since it is obtained at sub-toxic dose levels, although the therapeutic effect is more pronounced at levels that are toxic for free DOX, i.e., 15mg/kg (Gabizon, 1992). A recently published study of Unezaki et al in which PEG-distearoyl-PC (DSPC)-Chol liposomes are compared directly with DSPC-Chol liposomes for delivery of DOX in the mouse C-26 tumor model, demonstrates an unequivocal advantage for the PEG formulation in terms of drug tumor localization and therapeutic efficacy (Unezaki, et al., 1995).

CLINICAL TRIALS WITH PEGYLATED LIPOSOMAL DOX (DOXIL)

Doxil was the first liposomal drug to be approved by the FDA for the treatment of AIDS-related Kaposi's Sarcoma (Harrison et al., 1995). In a pilot clinical study with Doxil in patients with solid tumors, the pharmacokinetic parameters were found to be remarkably different from those of free DOX and in agreement with expectations from preclinical findings in rodents and dogs (Gabizon et al., 1993a; Gabizon et al., 1994). A long circulation half-life ($t_{1/2}$ ~45h), slow clearance rate, small volume of distribution and reduced urinary excretion of DOX and metabolites are all characteristic of Doxil (Gabizon et al., 1994). This pharmacokinetic pattern may allow for enhanced tumor exposure to the drug. Subsequent phase I and phase II studies with Doxil have established that the toxicity profile is significantly different from that of free doxorubicin with a lower incidence of acute toxicities, myelosuppression, and alopecia (Uziely et al., 1995). There are also results suggesting that cardiac toxicity, a well-known cumulative damage of anthracyclines, is attenuated although longer follow-up is needed before drawing conclusions.

Analysis of anti-tumor responses to Doxil in the two institutions performing the phase I study indicated significant and durable responses with 33% (15/45) of evaluable patients showing partial response or improvement. Responses were noticed in patients with tumors considered as "DOX-sensitive" such as breast and ovarian cancer, as well as with tumors seldom responsive to DOX, such as head and neck, mesothelioma, prostate, and renal cell carcinoma (Uziely et al., 1995). More information on anti-tumor activity is available from phase II studies (Muggia et al., 1997; Ranson et al., 1997) which indicate that Doxil is definitely an active agent in advanced breast and ovarian cancer. A direct comparison with conventional chemotherapy in the frame of phase III studies is still needed before the role of Doxil as standard therapy can be established.

Based on these pharmacokinetic and pharmacodynamic observations, Doxil should be seen as a new tool in cancer chemotherapy, and not mistaken as one of the many newly developed anthracycline analogs.

ACTIVE TARGETING OF STEALTH LIPOSOMES FOR DRUG DELIVERY

The proven performance of PEGylated Stealth liposomes as long-circulating drug delivery systems with stable drug retention (Gabizon, 1995), turns them into an attractive probe to test the feasibility of ligand-mediated liposome targeting to tumor cells. We have examined the likelihood that such a strategy could succeed in a human tumor model using immunoliposomes to erbB-2 oncoprotein, a receptor that is overexpressed by certain carcinomas. The antibody was coupled to the liposome surface by reacting a hydrazide group on the outer end of PEG with the carbohydrate moiety of the immunoglobulin Fc. Despite optimal preparation of antibody-conjugated liposomes with high *in vitro* affinity to tumor cells and reduced RES uptake (Goren et al., 1996), no advantage in tumor targeting with immunoliposomes over plain liposome was gained and biodistribution pattern of plain and immunoliposomes was quite similar (Goren et al., 1996). Therapeutic experiments correlated with pharmacokinetic studies, i.e., no enhancement of therapeutic efficacy was achieved with antibody-targeted liposomes (Goren et al., 1996). These studies suggest that the rate limiting factor of liposome accumulation in tumors is the liposome extravasation process, irrespective of liposome affinity or targeting to tumor cells. Conceivably, antibody-targeted liposomes have no advantage over non-targeted liposomes in the extravasation step, which is an essential prerequisite for interaction between the liposome ligand and the tumor cell target antigen.

We are currently examining the concept of tumor targeting by coupling a low molecular weight ligand, such as folic acid to the liposome surface (Lee and Low, 1994). Folate is attached to the distal end of a DSPE-conjugated PEG molecule on the liposome surface, at the γ-carboxy position (Goren et al., 1997). As a small, non-bulky ligand, folate should not hinder the process of liposome extravasation. This would enable liposomes to bind specifically to the folate receptor which is over-expressed in several carcinomas. Folate is a non-immunogenic ligand as opposed to immunoglobulins. Moreover, rapid removal of liposomes from circulation through the macrophage Fc-receptor should not be a problem with this approach. The reports from Lee and Low (1994) and our preliminary findings (Goren et al., 1997) suggesting that folate mediated targeting can overcome some of the hurdles obstructing the traditional antibody based approach. However, an *in vivo* phamacologic advantage over non-targeted stealth liposomes in tumor localization and therapeutic effect remains to be shown.

REFERENCES

Allen, T.M., Hansen, C.B., and Guo, L.S.S., 1993, Subcutaneous administration of liposomes: a comparison with the intravenous and intraperitoneal routes of injection. *Biochim.Biophys. Acta*, 1150:9.

Cabanes, A., Tzemach, D., Goren, D., Horowitz, A.T., and Gabizon A., 1998, Comparative study of the antitumor activity of free doxorubicin and polyethylene glycol-coated liposomal doxorubicin in a mouse lymphoma model, *Clin. Cancer Res.* 4:499.

Forssen, E.A., and Tokes, Z.A., 1981, Use of anionic liposomes for the reduction of chronic doxorubicin-induced cardiotoxicity, *Proc Natl Acad Sci* 78:1873.

Forssen, E.A., and Tokes, Z.A., 1983, Improved therapeutic benefits of doxorubicin by entrapment in anionic liposomes, *Cancer Res.* 43:546.

Gabizon, A., 1992, Selective tumor localization and improved therapeutic index of anthracyclines encapsulated in long-circulating liposomes, *Cancer Res.* 52:891.

Gabizon, A., 1995, Liposome circulation time and tumor targeting :implications for cancer chemotherapy, *Adv. Drug DelivRev.* 16:285.

Gabizon, A., Meshorer, A., and Barenholz, 1986b, Y., Comparative long term study of the toxicities of free and liposomes associated doxorubicin in mice after intravenous administration, *J.Natl. Cancer Inst.* 77:459.

Gabizon, A., Goren, D., and Barenholz, Y., 1988a, Investigation on the antitumor efficacy of liposome-associated doxorubicin in murine tumor models, *Isr.J.Med.Sci.* 24:517.

Gabizon, A., Barenholz, Y., and Bialer, M., 1993a, Prolongation of the circulation time of doxorubicin encapsulated in liposomes containing a polyethylene glycol-derivatized phospholipid: Pharmacokinetic studies in rodents and dogs, *Pharm.Res.* 10:703.

Gabizon, A., Goren, D., Ramu, A., and Barenholz, Y., 1986a, Design, characterization and anti-tumor acticity of adriamycin containing phospholipid vesicles, In: *Targeting of Drugs with Synthetic Systems*, G. Gregoriadis, J. Senior, G. Poste eds., Plenum, London.

Gabizon, A., Goren, D., Fuks, Z., Meshorer, A., and Y., Barenholz, 1985, Superior therapeutic activity of liposome associated adriamycin in a murine metastatic tumor model, *Br.J.Cancer,* 51:681.

Gabizon, G., Goren, D., Fuks, Z., Barenholz, Y., Dagan, A., and Meshorer, A., 1983, Enhancement of adriamycin delivery to liver metastatic cells with increased tumoricidal effect using liposomes as drug carriers, *Cancer Res.* 43:4730.

Gabizon, A., Dagan, A., Goren, D., Barenholz, Y., and Fuks, Z., 1982, Liposomes as *in vivo* carriers of adriamycin: Reduced cardiac uptake and preserved antitumor activity in mice. *Cancer Res.* 42:4734.

Gabizon, A., Pappo, O., Goren, D., Chemla, M., Tzemach, D., and Horowitz, A.T., 1993b, Preclinical studies with doxorubicin encapsulated in polyethylene glycol-coated liposomes, *J. Liposome Res.* 3:517.

Gabizon, G., Chemla, M., Tzemach, D., Horowitz, A.T., and Goren, D., 1996, Liposome longevity and stability in circulation: effects on the *in vivo* delivery to tumors and therapeutic efficacy of encapsulated anthracyclines, *J. Drug Targeting*, 3:391.

Gabizon, A., Goren, D., Horowitz, A.T., Tzemach, D., Lossos, A., and Siegal, T., 1997, Long-circulating liposomes for drug delivery in cancer therapy : a review of biodistribution studies in tumor-bearing animals, *Adv. Drug Deliv.Rev.* 24:337.

Gabizon, A., Catane, R., Uziely, B., Kaufman, B., and Barenholz, Y., 1994, Prolonged circulation time and enhanced accumulation in malignant exudates of doxorubicin encapsulated in polyethylene-glycol coated liposomes, *Cancer Res.* 54:987.

Goren, D., Horowitz, A.T., Zalipsky, S., Woodle, M.C., Yarden, Y., and Gabizon, A., 1996, Targeting of Stealth liposomes to erbB2 (Her/2) receptor: in vitro and *in vivo* studies. *Br.J. Cancer*, 74:1749.

Goren, D., Horowitz, A.T., Mandelbaum-Shavit, F., Tzemach, D., Zalipsky, S., and Gabizon, A., 1997, In vitro and *in vivo* studies of folate-targeted liposomes, *Proc.Cont.Rel. Soc.* 24:865.

Gregoriadis, G., ed., 1988, *Liposomes as Drug Carriers: Recent Trends and Progress*, Wiley, London.

Haran, G., Cohen, R., Bar., L.K. and Barenholz, Y., 1993, Transmembrane ammonium sulfate gradients in liposomes produce efficient and stable entrapment of amphipathic weak bases, *Biochim.Biophys. Acta*, 1151:201.

Harrison, D., Tomlinson, D., and Stewart, S., 1995, Liposomal entrapped doxorubicin:an active agent in aids-related Kaposi`s sarcoma, *J.Clin.Oncol.* 13:914.

Van Hoesel, Q.G., Steerenberg, P.A., Crommelin, D.J., Van Dijk, A., Van Oort, W., Klein, S., Douze, J.M., de Wildt, D.J., and Hillen, F.C., 1984, Reduced cardiotoxicity and nephrotoxicity with preservation of anti-tumor activity of doxorubicin entrapped in stable liposomes in the Lou/M Wsl Rat, *Cancer Res.* 44:3698.

Lasic, D., and Martin, F., eds., 1995, *Stealth Liposomes,* Pharmacology and Toxicology series, CRC Press, Boca Raton, FL

Lasic, D.D., Frederik, P.M., Stuart, M.C., Barenholz, Y., and McIntosh, T.J., 1992, Gelation of liposome interior a novel method for drug encapsulation, *FEBS Lett.* 312:255.

Lee, R.J., and Low, P.S., 1994, Delivery of liposomes into cultured KB cells via folate receptor-mediated endocytosis, *J.Biol.Chem.* 269:3198.

Legha, S.S., Benjamin, R.S., Mackay, B., et al., 1982, Reduction of doxorubicin cardiotoxicity by prolonged continuous infusion, *Ann Intern. Med* 96:133.

Mayer, L.D., Tai, L.C., Bally, M.B., Mitilens, G.N., Ginsberg, R.S., and Cullis, P.R., 1990, Characterization of liposomal systems containing doxorubicin entrapped in response to pH gardients, *Biochim.Biophys. Acta*, 1025:143.

Mayhew, E., and Papahadjopoulos, D., 1983, Therapeutic application of liposomes. In: *Liposomes*, M.J.Ostro, ed., Marcel Dekker, New-York.

Minow, R.A., Benjamin, R.S., and Gottlieb, J.A., 1975, Adriamycin cardiomyopathy; an overview with determination of risk factors, *Cancer Chemother. Rep.* 6:195.

Muggia, F.M., Hainsworth, J.D., Jeffers, S., Miller, P., et al., and Liang.-Jung, L.J., 1997, Phase II study of liposomal doxorubicin in refractory ovarian carcinoma: anti-tumor activity and toxicity modification by liposomal encapsulation, *J.Clin.Oncol.* 15:987.

Olson, F., Mayhew, E., Maslow, D., Rustum, Y., and Szoka, F., 1982, Characterization, toxicity and therapeutic efficacy of adriamycin encapsulated in liposomes, *Eur.J. Cancer Clin.Oncol.* 18:167.

Rahman, A., White, G., More, N., and Schein, P.S., 1985, Pharmacological, toxocological and therapeutic evaluation in mice of doxorubicin entrapped in cardiolipin liposomes, *Cancer Res.* 45:769.

Rahman, A., Fumagali, A., Barbieri, B., Schein, P.S., and Casazza, A.M., 1986, Anti-tumor and toxicity evaluation of free doxorubicin and doxorubicin entrapped in cardiolipin liposomes, *Cancer Chemother.Pharmacol.* 16:22.

Ranson, M.R., Carmichael, J., O`Byrne, K., Stewart, S., Smith, D., and Howell, A., 1997, Treatment of advanced breast cancer with sterically stabilized liposomal doxorubicin: results of a multicenter phase II trial. *J.Clin.Oncol.* 15:3185.

Torchilin, V.P., 1995, Long circulating drug delivery systems, *Adv. Drug Deliv.Rev.* 16:125.

Unezaki, S., Maruyama, K., Ishido, O., Suginaka, A., Hosoda, J., and Iwatsuru, M., 1995, Enhanced tumor targeting and improved antitumor activity of doxorubicin by long-circulating liposomes containing amphipathic poly(ethyleneglycol), *Int.J.Pharm.* 126:41.

Uziely, B., Jeffers, S., Isacson, R., Kutsch, K., Wei-Tsao, D., Yehoshua, Z., Muggia, F.M., and Gabizon, A., 1995, Liposomal doxorubicin: anti-tumor activity and unique toxicities during two complementary phase I studies, *J.Clin.Oncol.* 13:1777.

Young, R.C., Ozols, R.F., and Myers, C.E., 1981, The anthracycline anti-neoplastic drugs, *N.Engl.J.Med.* 305:139.

LIPOSOME-MEDIATED DELIVERY OF RETINOIDS

Kapil Mehta[1], Elihu Estey[2], Gabriel Lopez-Berestein[1]

[1]Immunobiology and Drug Carriers Section, Department of Bioimmunotherapy and [2]Department of Hematology, The University of Texas M.D. Anderson Cancer Center, Houston, TX 77030

INTRODUCTION

The term retinoid is used to describe the natural and synthetic derivatives of vitamin A. Based on their growth inhibition and maturation-inducing properties *in vitro*, retinoids have been used and shown to be clinically active in such diverse premalignant and malignant conditions as, cutaneous T-cell lymphomas, leukoplakia, squamous cell carcinomas of the skin, and basal cell carcinomas (Lippman et al., 1987; Smith et al., 1992). Recent clinical data, demonstrating the high response rates using retinoids in patients with cervical cancer (Lippman et al., 1992a) and metastatic squamous cancer of the skin (Lippman et al., 1992b), have kindled further interest in this group of compounds as therapeutic agents.

Recently retinoids have also been used in the treatment of preleukemic and leukemic conditions. Thus, hematological improvement was observed in patients treated with 13-*cis*-retinoic acid (isotretinoin). Gold and co-workers (Gold et al., 1983) described hematological responses in 5 of the 15 patients with myelodysplastic syndrome. Responses required long-term therapy, usually for 3-4 weeks. More recently, several investigators have documented dramatic improvement in patients with acute promyelocytic leukemia (APL) treated with isotretinoin and, even more so with all-*trans*-retinoic acid (ATRA) (Flynn et al., 1983; Meng-er et al., 1988; Castaigne et al., 1990; Chomienne et al., 1989; Warell et al., 1991; Warell 1991). APL is a subtype of acute myelogenous leukemia, representing a clonal expansion of immature promyelocytes carrying a t (15;17) translocation that fuses the promelocytic leukemia gene (PML) on chromosome 15 to the retinoic acid receptor alpha gene (RAR(alpha)) on chromosome 17 (Early and Dmitrovsky, 1995). As a result of this fusion, the chimeric gene encodes the PML-RAR(alpha) fusion protein that is thought to play a crucial role in the pathogenesis of APL by interfering with factors essential for the differentiation of myloid precursors. ATRA causes APL blasts to differentiate into polymorphonuclear leukocytes *in vitro*. Based on this principle, Meng-er and coworkers (1988), treated 24 APL patients with ATRA at doses of 45-100 mg/m^2/day. Twenty three patients achieved a complete remission without intervening bone marrow hypoplasia. These results were independently confirmed by

Targeting of Drugs 6: Strategies for Stealth Therapeutic Systems
Edited by Gregoriadis and McCormack, Plenum Press, New York, 1998

Castaigne et al (Castaigne et al., 1990) in France, who initially used a 13-*cis* isomer, but later changed over to ATRA which had been shown to be 10-fold more potent *in vitro*. Their study revealed that all *trans*-retinoic acid induced a complete hematological and cytogenetic remission in 64% of the patients and partial remission in 18%, for overall objective response rate of 82%. Similarly, (Warrell et al., 1991) reported the results of their study in which they treated 11 APL patients with ATRA (45 mg/m^2/day for 30 days). Of the 11 patients treated with all-*trans*-retinoic acid, 9 entered complete remission. Cellular morphological features, cell surface immunophenotypic analysis, and fluorescence *in situ* hybridization with a chromosome 17 probe revealed that clinical response was associated with maturation of the leukemic clone. All the patients showed an aberrant expression of retinoic acid receptor-alpha (RAR-alpha). However, as the patients entered complete remission, the expression of the abnormal RAR-alpha message decreased.

In view of these results, ATRA warrants further investigation to delineate optimal strategies for treatment of other leukemic states. Although the doses used (45-100 mg/m^2/day) in these studies were effective for treating APL, these doses may not be adequate for treating other types of malignancies. Moreover, the mean peak plasma levels in patients receiving 45 mg/m^2 of ATRA are 1 μM and the drug is rapidly eliminated from the plasma with a $t_{1/2}$ of less than 1 hr (Warrell, Jr, 1991). Furthermore, chronic administration of ATRA in mice, monkeys and some patients seems to be associated with lower peak concentrations and total systemic exposure (AUC) (Creech-Kraft et al., 1991; Kalin and Starling, 1981; Muindi et al., 1992a,b). The mechanism(s) responsible for this phenomenon are currently unkown, but could include such possibilities as progressive malabsorption, accelerated metabolism (Muindi et al., 1992; Roberts et al., 1979; Leo et al., 1984) or elevation of tissue retinoic acid binding protein (CRABP) resulting in an increased drug retention by non-target tissues (Boylan and Guda, 1991; Hirshel et al., 1989). Progressive diminution of ATRA levels in plasma with chronic administration may represent an important mechanism of acquired resistance to ATRA therapy in patients with APL (Muindi et al., 1992a,b).

LIPOSOMAL DELIVERY CAN CIRCUMVENT THE PROBLEMS ASSOCIATED WITH ORAL ATRA

The rationale for using liposomes as a drug delivery system is to improve the performance of all-*trans*-retinoic acid by its delivery to the specific tissue sites from which the malignant cells are believed to be seeded (e.g. bone marrow, liver and spleen) (Gregoriadis, 1976). Since liposomal-ATRA (L-ATRA) is an intravenous formulation, it is possible that its use may prove useful in surmounting the problem of lower plasma drug concentrations seen during chronic drug administration.

Furthermore, administration of ATRA at therapeutic dose levels is associated with acute and chronic toxicities. The side effects of oral ATRA therapy are similar to effects seen with vitamin A. In a study by Stuttgen (1975), 30 patients given oral ATRA at doses of 50-100 mg/day for 4 weeks showed toxicities similar to those for vitamin A: headache, other central nervous system problem (lethargy, visual disturbances, dizziness etc), gastrointestinal effects, and mucocutaneous problems (e.g. dryness of mucosal tissues, erythema and desquamation of skin, and cheilitis). In another study, of the 27 patients receiving 70-100 mg/day ATRA, 50% experienced headaches and/or dizziness, with 11% requiring interruption of therapy because of these symptoms (Koch, 1978). Mucocutaneous toxicities are common with ATRA administration. Ocular effects (particularly, dryness of the eyes and blepharoconjunctivitis) and corneal erosions have also been reported. Additionally, myalgias and arthralgias, as well as chronic retinoid use, can lead to more significant musculoskeletal toxicity, including premature epiphyseal closure in children, spinal ligament calcification, and osteoporosis (Kilcoyne, 1988;

Digiovanna et al, 1986; McGuire et al., 1985). Laboratory abnormalities associated with ATRA treatment include elevated serum liver enzymes as a manifestation of reversible hepatotoxicity as well as hypertriglyceridemia and hypercholesterolemia. A considerable effort has been devoted in an attempt to develop vitamin A derivatives with improved therapeutic/toxicity ratio, but to-date these efforts have been unsuccessful.

Use of a drug carrier system is an alternate method for reducing the toxicity associated with retinoid administration. Liposomes offer a good choice as a drug delivery system, as the drug can be easily intercalated into their phospholipid bilayers. Liposomes are useful for controlling the topography of drug distribution *in vivo*. This in essence involves attaining higher concentration of drug and /or long duration of drug action at a target site where beneficial effects may occur, while maintaining a low concentration and/or reduced exposure at other sites where adverse side effects may occur. For the past several years we have studied retinoid-regulated modulation of cell growth and differentiation functions in myeloid cells (Mehta et al., 1985; Mehta and Berestein, 1986; Mehta et al., 1987a; Lee et al., 1987; Mehta et al., 1987b; Murray et al., 1988; Turpin et al., 1990; Sanduja et al., 1991; Mehta, 1989). When incorporated in lipid particles (liposomes), ATRA-associated toxicity is markedly reduced whereas anti-tumor properties (growth inhibition and differentiation induction) of ATRA are maintained or even enhanced (Mehta, 1989; Sacks et al., 1992; Parthasarathy et al., 1994; McCarthy et al., 1992; Nastruzzi et al., 1990). Hence, liposome-mediated delivery of retinoids could lead to their wider clinical use. Furthermore, we and others have previously shown that liposomes target preferentially to liver, spleen, and bone marrow (Gray and Morgan, 1991; Berestein et al., 1987; Kasi et al., 1984), where most of the hematopoietic activity is present. Liposomal formulation of ATRA may potentially lead to enhanced targeting and decreased toxicity. An intravenous formulation, such as liposomal-ATRA, may represent a potentially useful formulation to overcome acquired ATRA-resistance seen in APL patients during ATRA therapy. Since rapid clearance from plasma seems to be the principal mechanism of acquired ATRA-resistance (Muindi et al., 1992a,b), liposomal-ATRA may have the potential to sustain therapeutically effective drug concentrations due to altered pharmacokinetics and reduced systemic toxicity (Gray and Morgan, 1991). Furthermore, an intravenous formulation will be particularly advantageous in patients with other malignancies since this way compliance will be ascertained and its use in acutely ill patients will be facilitated.

ANTI-PROLIFERATIVE AND DIFFERENTIATION-INDUCING PROPERTIES OF LIPOSOMAL-ATRA

Fortunately, the reduced toxicity of liposomal ATRA (L-ATRA) was not associated with a parallel reduction in its anti-proliferative or differentiation-inducing properties. The three human myeloid leukemia cell lines, HL-60, KG-1 and THP-1 exhibited similar morphological features of differentiated cells when cultured in the presence of equimolar concentrations of free or L-ATRA (Drach et al., 1993). The extent of free- or L-ATRA-induced differentiation in three leukemia cell lines was also similar when assessed by differential counts. HL-60 and KG-1 cells were predominantly at the stage of metamyelocytes and band cells (Table 1). A decreased nucleus/cytoplasm ratio and indented nuclei indicated monocytic differentiation of THP-1 cells. Further evidence for L-ATRA-induced differentiation became evident from analysis of cell surface markers. CD11b is a surface marker that is expressed on differentiated cells (myelocytes or monocytes) but not on the precursor promyelocytes or myeloblasts. Staining with a fluorescent monoclonal antibody directed against CD11b is thus indicative of cell differentiation. CD33 is a surface protein present on both undifferentiated and differentiated cells, and serves as a normalization factor. Both HL-60 and KG-1 cells normally express the CD33 $^+$/CD11b-phenotype.

Table 1. Morphological evaluation of leukemia cells treated with L-ATRA (6 days, 2 μM).

	Control	Free-ATRA	L-ATRA
HL-60			
Myeloblasts	0%	0%	0%
Promyelocytes	96%	2%	0%
Myelocytes	3%	15%	9%
Metamyelocytes	1%	43%	59%
Band forms	0%	36%	33%
Segmented neutrophils	0%	4%	1%
KG-1			
Myeloblasts	97%	0%	0%
Promyelocytes	2%	5%	2%
Myelocytes	1%	24%	16%
Metamyelocytes	0%	49%	54%
Band forms	0%	22%	28%
Segmented neutrophils	0%	0%	0%
THP-1			
Monoblast+Promonocytes	95%	8%	7%
Monocytes	5%	92%	93%

Following treatment with L-ATRA for 7 days, 89.4% of KG-1 cells acquired the $CD33^+/CD11b^+$ phenotype. HL-60 cells had a significant increase in $CD33^+$/CD11b cells after 7 days of culture (Drach et al., 1993).

Furthermore, acridine orange flow cytometry technique also confirmed the equipotent effects of free- and L-ATRA on cell growth of leukemia cells. In culture, proliferating cells are normally distributed in G_1, S, G^2 or M phase, with few in G_0. Commitment of these cells to differentiate in response to ATRA or other differentiating agent, is associated with a cell growth arrest and an increase the proportion of cells in G_0. After six days of culture in the presence of L-ATRA, 53.5% of HL-60 cells were in G_0, as opposed to 2.5% of untreated cells. Similarly, after seven days of culture in the presence of L-ATRA, 63.1% of KG-1 cells were in G_0, as opposed to 7% of controls. 48.8% of THP-1 cells were in G_0 after three days of culture, as opposed to 13.9% of controls (Drach et al., 1993). Similar immunophenotypic changes were observed in leukemia cells collected from bone marrow aspirates of patients following their incubation with L-ATRA. These data suggested that L-ATRA is effective at inducing the differentiation of the promyelocytes into metamyelocytes and neutrophils.

CIRCUMVENTION OF 'RETINOID RESISTANCE' BY L-ATRA

The ability of isolated liver microsomes from rats that have been consecutively treated for fifteen days with either free or oral ATRA or intravenous L-ATRA was compared for their ability to metabolize ^{14}C-labelled ATRA *in vitro* (Mehta et al., 1994). Three important observations emerged from these experiments: a) significantly higher plasma levels of intact ATRA can be achieved by injecting L-ATRA than by feeding same amount of ATRA orally; b) intravenous administration of L-ATRA can circumvent the altered pharmacokinetic behavior of the drug; c) decreased plasma levels of tretinoin associated with chronic oral administration of the drug are, at least in part, due to accelerated metabolism of tretinoin in liver microsomes. We also took advantage of the fact that in F9 teratocarcinoma cells, p450-dependent metabolism of retionic acid can be induced following short term exposure of these cells to ATRA. Exposure of F9 cells to ATRA in free- or liposome-encapsulated forms further confirmed that in liposomal form, ATRA-inducible metabolism is significantly lower than when the drug is presented in free form.

Taken together, these studies provided a strong rationale for developing an intravenous formulation of tretinoin in a lipid carrier. Since the limitation of tretinoin in APL appears to be due to access of the drug to the malignant cells and not resistance at the cellular level, an intravenous formulation like L-ATRA would avoid adaptations in drug absorbance or clearance that may, in part, underlay the observed differences in plasma levels after prolonged administration. The phase I and II clinical studies, summarized below, indeed confirmed this contention; the plasma drug levels were maintained at high concentrations even after prolonged treatment of patients with L-ATRA.

CLINICAL STUDIES

A phase I study of L-ATRA in hematologic malignancies at M.D. Anderson Cancer Center was completed in 1996. L-ATRA was administered by intravenous infusion over 0.5 hr every other day for 28 days (14 doses). Patients who responded or in whom the diasese became stable were continued on treatment until progression in disease or toxicity was observed. At least 3 patients were studied at each dose level and evaluated for at least 4 weeks before starting additional patients on an escalated schedule. Thirty-seven patients were entered in the study. These included eighteen patients with lymphoma and six patients with APL. Escalations proceeded from the starting dose of 15 mg/m^2 to 170 mg/m^2.

Toxicities observed to date in patients treated with L-ATRA have been considered as minor and consisted mainly of transient headaches. About 90% patients reported minor headaches that were easily treatable (Estey et al., 1996). Severe toxicity developed in 3 of the eight patients treated at 175 mg/m^2 dose (joint pains in 2, skin toxicity in 1). The maximum tolerated dose (MTD) for L-ATRA was declared to be 140 mg/m^2, the dose at which no grade 3 or 4 toxicities were observed.

Six patients with APL were entered in phase I study. Three of them did not respond, two had CR, and one patient who is in her fourth week of therapy had a reduction of bone marrow promyelocytes from 88% to 0.8 percent. One patient in her first relapse 10 months after oral ATRA and chemotherapy was treated with L-ATRA at 90 mg/m^2 dose. At the initiation of therapy, she had 90% blasts in the marrow. After 7 doses of L-ATRA, the number of blasts in the marrow went down to 30% and after 28 days to 8%. Three weeks into treatment the patient's white count had risen and there was a evidence of differentiation in the blood and in the marrow. The three patients that did not respond were treated with doses lower than 90 mg/m^2 while the three responders were treated with doses of L-ATRA at 90mg/m^2 (2 patients) or at 110 mg/m^2 (1 patient).

Three of five patients with T cell cutaneous lymphoma also attained minor responses. Although this disease is retinoid-sensitive, the patient was considered resistant to oral ATRA when treatment with L-ATRA began. The patient was treated at 30 mg/m^2 dose of L-ATRA and a minor response was noted during the first cycle. One patient with T-cell lymphoma with lung involvement when treated with 90 mg/m^2 L-ATRA dose, showed a partial response in the lung (by x-ray examination).

Based on the the lack of serious toxicity, the absence of the retinoid acid syndrome and the activity observed in APL patients during phase I study of L-ATRA, a phase II clinical study was initiated (Estey et al., 1997). Also, to assess whether the difference in pharmacokinetics between oral ATRA and L-ATRA is clinically relevant, a multi-institutional phase II study of L-ATRA in recurrent APL patients is being conducted. Disease is characterized as either (1) in first relapse with patients never having received oral ATRA, or having last received ATRA 1 year from recurrence (group 1) or (2) in subsequent relapse, or in first relapse with patients having received oral ATRA within 1 year of relapse date (group 2). Since the historical data in APL patients (including oral ATRA-treated patients) indicate that patients in group 1 have a CR rate of 80-90% but only a 40% probability of remaining in CR at 6 months, the objective in group 1 is to improve the 6 month remission duration rate. In group 2 the objective is to improve the CR rate from the historical 13%. Patients in both groups are given L-ATRA 90 mg/m^2 QOD for 56 days or until CR whichever comes first. If not in CR after 56 days, they are considered failures. In CR group 1 patients alternate courses of chemotherapy (mitoxantrone + etoposide cycles 1, 2 idarubicin cycles 2, 4) with courses of L-ATRA 90 mg/m^2 three-times weekly for 4 weeks, while group 2 patients receive only L-ATRA 90 mg/m^2 twice a week for 6 months. Initial results from this study indicate recurrence (6 months post CR) in 1 of the 8 CR patients in group 1. Three of the remaining 7 received an allogeneic transplant after 2, 3 and 11 months in CR while the other 4 remain in CR after 4-14 months follow-up. The one CR in group 2 occurred in a patient with disease in first relapse who had last received oral ATRA 12 months before. Of the 7 failures in group 2, 4 were in first, 2 in second and 1 in third relapse. Five of the 7 died within 3 weeks of starting L-ATRA treatment while disease did not enter CR despite 7-8 weeks of L-ATRA in the remaining two patients. These initial observations suggest that L-ATRA has activity in recurrent APL but its efficacy with oral ATRA cannot yet be assessed.

Pharmacokinetics of L-ATRA in Patients

The concentration of ATRA was measured in patients' whole blood using an HPLC assay. Blood samples (5 ml total) were obtained prior to infusion (baseline), at the infusion midpoint, at the end of infusion (t=0), at 0, 1, 3, 5, 10, 15, 30 minutes and at 2, 4, 8, 12, and 24 hours after L-ATRA administration. Adequate blood concentrations to derive pharmacokinetics parameters were obtained in patients treated with doses of 30 mg/m^2 and above. Blood concentrations of L-ATRA in blood were subjected to analysis for pharmacokinetics parameters using nonlinear regression analysis. Patients were monitored for pharmacokinetics on day 1 and day 15 of therapy. The data obtained revealed that the clearance of drug from blood closely fit ($r^2 > 0.9$) a two compartment mathematical model. Clearance of drug showed a rapid alpha phase with a mean of 0.7 min at the 30 mg/m^2, 2.8 min at the 60 mg/m^2, 3.5 min at 75 mg/m^2, and 2.4 min at 90 mg/m ^{2}dose level (Estey et al., 1996). The beta phase half life was approximately 58 min at the 30 mg dose, 87.8 min at the 60 mg dose, 116 min at 90 mg, and 121 min at 90 mg/m^2 dose level. There did not appear to be a difference at either dose level between day 1 and day 15 pharmacokinetics parameters suggesting that repetitive dosing with this agent does not saturate clearance mechanisms nor does repeated dosing evoke an increased clearance or metabolism of the agent as observed with oral dosing with retinoic acid. At the 15 mg/m^2 dose level, the drug cleared with a terminal phase half life of about 56 min. The immediate apparent volume of distribution (V_d) of L-ATRA approximated the total body water (25 ± 2.1) suggesting extensive distribution of the drug outside the vasculature. At 15 mg/m^2 dose level , the mean V_d increased to 36.4 ± 4.7 on day 15 however when V_d was adjusted for body weight, there appeared to be no statistically significant differences in the V_d between day 1 and day 15. Increases in the dose of drug from 15 mg to 90 mg/m^2 resulted in no significant changes in their terminal phase half-life or the volume of distribution. In addition, we noted no significant differences in pharmacokinetics parameters on day 15 or day 85 compared to the values on day 1.

These results suggested that plasma levels with L-ATRA, contrary to what is observed with oral ATRA, are maintained even after chronic administration of L-ATRA. These results further strengthen the use of L-ATRA in patients with APL. Thus, the hypothesis proposed to use L-ATRA in order to circumvent the problems associated with chronic administration of oral ATRA in the Phase I study has been confirmed in humans and possibly other malignant diseases.

CONCLUSION

Standard therapy for treatment of human leukemias includes chemotherapy and irradiation that results in cell-growth arrest and destruction of malignant cells. Although these approaches are effective to an extent; some patients relapse or even fail to respond to these therapies. Therefore, the search for alternative approaches to cure leukemias has continued. One such approach is the differentiation therapy. Since the differentiated mature cells stop to devide, differentiation therapy should result in mature non-proliferating leukemic cells that will eventually die via apoptosis. The strategy of inducing differentiation in cancerous cells has been tested *in vitro* and in anaimal models for more than two decades. Retinoids represent one class of such compounds that have been found to inhibit the growth and induce differentiation of a number of leukemic cells and to inhibit the proliferation of a wide variety of normal and malignant cell types. Based on these observations, ATRA has been successfully used for the treatment of APL. In general, the use of retinoids is safe and induces complete remission in 80-90% APL patients. However, chronic oral administration results in reduced plasma levels that is associated with the disease relapse in the majority of patients. Our preclinical data suggest that this problem of pharmacological adaptation can be circumvented by using liposome-ATRA and thus this formulation may be extremely useful in accomplishing long-term remissions.

REFERENCES

Berestein, G.L., Bodey, G.P., Frankel, L.S., and Mehta, K., 1987, Treatment of hepato-splenic candidiasis with liposomal-amphotericin B. *J Clin Oncol* 5:310.

Boylan, J., and Guda, L.,1991, Over expression of CRABP-I results in reduction in differentiation-specific gene expression in F9 teratocarcinoma cells. *J Cell Biol* 112:965.

Castaigne, S., Chomienne, C., Daniel, M. T., Ballerini, P., Berger, R., Fanaux, P., and Degos, L., 1990, All trans retinoic acid as a differentiation therapy for acute promyelocytic leukemia I. Clincal results. *Blood* 76:1704.

Chomienne, C., Ballerini, P., Balitrans, N., Amar, M., Bernard, J.F., Biovin, P., Daniel, M.T., Berger, R., Castaigen, S., and Degos, L., 1989, Retinoic acid therapy for promyelocytic leukemia. *Lancet* 2:746.

Creech-Kraft, J., Slikker, W., Bailey, J., Roberts, L.G., Fischer, B., Wittfoht, W., and Nau, H., 1991, Plasma pharmacokinetics and metabolism of 13-cis and all-trans retinoic acid in cynomolgus monkey and the identification of 13-cis and all-trans retinoyl beta glucuronides. *Drug Metab Dispos* 19:317.

Digiovanna, J., Helffgott, R., Gerber, L., and Peck, G.L., 1986, Extraspinal tendon and ligament calcification associated with long term therapy with etretinate. *New Eng J Med* 315:1177.

Drach, J., Lopez-Berestein, G., Andreeff, M., and Mehta, K., 1993, Induction of differentiation in myeloid leukemia cell lines and promyelocytic cells by liposomal retinoic acid. Cancer Res., 53:2100.

Early, E., and Dmitrovsky, E., 1995, Acute promyelocytic leukemia: retinoic acid response and resistance. *J. Invest. Med.* 43:337.

Estey, E.,Thall, P.F., Mehta, K., Rosenblum, M.G., Brewer, T., Simmons, V., Cabanillas, F., Kurzrock, R., and Berestein, G.L., 1996, Alterations in tretinoin pharmacokinetics following administration of liposomal all-trans retinoic acid. *Blood* 87:3650.

Estey, E., Douer, D., Wiernik, P., Reed, J., Thompson, J., and Gordon, D., 1997, Treatment of recurrent acute promyelocytic leukemia (APL) with liposomal ALL-trans retinoic acid (L-ATRA) Blood 90:1479 (Abstr).

Flynn, P.J., Miller, W.J., Weisdrof, D.J., Arthur, D.C., Brunning, R., and Branda, R.F., 1983, Retinoic acid treatment of acute promyelocytic leukemia: *in vitro* observations. *Blood* 62:1211.

Gold, E., Mertelsmann, R., Itri, L.M., Gee, T., Arlin, Z., Kempin, S., Clarkson, B., and Moore, M.A., 1983, Phase I clinical trial of 13-cis-retinoic acid in myelodysplastic syndrome. *Cancer Treat Rep.* 67: 981.

Gray, A., and Morgan, J., 1991, Liposomes in hematology. *Blood Rev* 5:258.

Gregoriadis, G., 1976, The carrier potential of liposomes in biology and medicine. *New Eng J Med* 295:704.

Hirshel, S., Seigenthaler, G., and Saurat, J. H., 1989, Ligand specific and non specific in vivo modulation of human epidermal CRABP. *Europ J Clin Invest* 19:220.

Kalin, J., Starling, M., and Hill, D., 1981, Disposition of ATRA in mice following oral doses. *Drug Metab Dispos* 9:196.

Kasi, L., Berestein, G.L., Mehta, K., Juliano, R.L., Rosenblum, M.B., Glen, H.J., Hynie, T.P., Mavligit, G.M., and Hersh, E.M., 1984, Distribution and pharmacology of intravenous ^{99m}Tc-labelled multilameller liposomes in rats and mice. *Int J Nucl Med Biol* 11:35.

Kilcoyne, R., 1988, Effects of retinoids in bone. *J Am Acad Dermatol* 19:212.
Koch, H., 1978, Biochemical treatment of precancerous oral lesions: The effectiveness of various analogues of retinoic acid. *J Max-fac Surg* 6:59.
Lee, J., Mehta, K., Blick, M., and Lopez-Berestein, G., 1987, Expression of c-fos, c-myb, c-myc in human monocytes: Correlation with monocytic differentiation. *Blood* 69:1542.
Leo, M., Iida, S., and Lieber, C., 1984, Retinoic acid metabolism by a system reconstituted with cytochrome P-450. *Arch Biochem Biophys* 234:305.
Lippman, S. M., Kessler, J. F., and Meyskens, F. L., 1987, Retinoids as preventive and therapeutic anticancer agents (part II) *Cancer Treat Rep* 17:493
Lippman, S., Kavanagh, J., Espinoza, M.P., Madrueno, F.D., Cassilas, P.P., Hong, W.K., Holdner, D., and Krakoff, I.W., 1992a, Highgly active systemic therapy for squamous cells carcinoma of the cervix, *J.Natl.Cancer Inst.,* 84:241.
Lippman, S., Parkinson, D., Itri, L., Weber, R.S., Schantz, S.P., Mota, D., Schusterman, M.A., Krakoff, I.W., Gutterman, J.U., and Hong, W.K.,1992b, 13-cis retinoic acid and interferon-alpha 2a: effective combination therapy for advanced squamous cell carcinoma of the skin. *J Natl Cancer Inst* 84:235.
McCarthy, D.J., Dollar, G.R., and Hill, D.L., 1992, Toxicity and antitumor activity of liposome entrapped retinoid Ro13-7410. *Selective Cancer Therap* 7:151.
McGuire, J., Milstone, L., and Lawson, J., 1985, Isotretinoin administration alters juvenile and adult bone. Basel, Switzerland, Karger, pp419-439.
Mehta, K., and Berestein, G.L., 1986, Expression of transglutaminase in cultured monocytic leukemia cells during differentiation. *Cancer Res* 46:1388.
Mehta, K., Turpin, J., and Berestein, G.L., 1987a, Induction of transglutaminase in human blood monocytes by intracellular delivery of retinoids. *J Leukocyte Biol* 41:341.
Mehta, K., Claringbold, P., and Berestein, G.L., 1987b, Suppression of macrophage cytostatic activation by serum retinoids. *J Immunol* 138:3902.
Mehta, K., Sadeghi, T., McQueen, T., and Lopez-Berestein, G., 1994, Liposome encapsulation circumvents the hepatic clearance mechanisms of all *trans*-retinoic acid. *Leukemia Res.*, 18:587.
Mehta, K., 1989, Interaction of liposome-encapsulated retinoids with normal and leukemic cells. In: Pharmacology and the Skin (Shroot B, Schaefer H eds), Karger, Basel, pp74-80.
Mehta, K., Berestein, G.L., Moore, W.T. , and Davies, P.J., 1985, Interferon-gamma requires serum retinoids to promote the expression of transglutaminase in cultured human blood monocytes. *J Immunol* 134:2053.
Meng-er, H., Yu-chen, Y., Shu-rong, C., Jin-ren, C., Jia-ziang, L., Lim, Z., Ling-jun, G., and Zhen-yi, W., 1988, Use of all *trans*-retinoic acid in the treatment of acute promyelocytic leukemia. *Blood* 72:567.
Muindi, J., Frankel, S., Miller, W., Jakubowski, A., Scheinberg, D.A., Young, C.W., Dmistrovsky, E., and Warrell, D.P., Jr, 1992a, Cutaneous treatment with ATRA causes a progressive reduction in plasma drug concentrations: Implication for relapse and retinoid resistance in patients with APL. *Blood* 79:299.
Muindi, J., Frankel, S.R, Huselton C., DeGrazia, F., Garland, W.A., Young, C.W., and Warrell, R.P., Jr. 1992b Clinical pharmacology of oral ATRA in patients with acute promyelocytic leukemia. *Cancer Res* 52:2138.
Murray, L., Mehta, K., Berestein, G.L., 1988, Induction of ADA and 5' nucleotidase activity in cultured human blood monocytes and monocytic leukemia cells by differentiating agents. *J Leukocyte Biol* 44:205.
Nastruzzi, C., Walde, P., Menegatti, E., and Gambari, R., 1990, Liposome associated retinoic acid: Increased in vitro antiproliferative effects on neoplastic cells.
Parthasarathy, R., Sacks, P. G., Harris, D., and Mehta, K., 1994, Interaction of liposomal-associated all trans retinoic acid with squamous carcinoma cells. *Cancer Chemother. Pharmacol.* 34:527.
Roberts A, Frolik, C.A., Nichols M, and Newton, D., 1979, In vitro metabolism of retinoic acid in hamster intestine and liver. *J Biol Chem* 245:6296.
Sacks, P,, Oke,V., and Mehta, K., 1992, Antiproliferative effects of free and liposome-encapsulated retinoic acid in a squamous carcinoma model: monolayer cells and multicellular tumor spheroids. *Cancer Res Clin Oncol* 118:490
Sanduja, S., Mehta, K., Xu, X.M., Hsu, S.M., Sanduja, R., and Wu, K.K., 1991, Differentiation associated expression of prostaglandin H and thromboxane A synthases in monocytoid leukemia cell lines. *Blood* 78: 3178.
Smith, M. A., Parkinson , D.R., Cheson, B.D., and Friedman, M.A., 1992, Retinoids in cancer therapy. J Clin Oncol active systemic therapy for squamous cell carcinoma of the cervix. N NAtl *Cancer Inst* 84:241.
Stuttgen, G., 1975, Oral vitamin A acid therapy. *Acta Dermatovener* (Stockholm) 74:174.
Tafuri, A., Jakubowski, A., Gabrilove, J., Gordon, M.S., and Demitrovski, E., 1991, Differentiation therapy of acute promyelocytic leukemia with tretinoin. *New Engl J Med* 324:1385.
Turpin, J., Mehta, K., Blick, M., and Lopez-Berestein, G., 1990, Effect of retinoids on the release and gene expression of TNF-alpha in human blood monocytes. *J Leukocyte Biol* 48:444.
Warell, R.P. Jr., Frankel, S.R., Miller, W.H. Jr., Scheinberg, D., Itri, L.M., Hittelman, W., Vyas, R., Andreeff, M.,
Warrell, R.P. Jr., 1991, Retinoic acid and acute promyelocytic leukemia. *Biol Ther Cancer Updates* 3:1.
Warrell, R.P. Jr., 1994, Application for retinoids in cancer therapy. *Sem. Hematol.,* 31:1.

APPLICATIONS OF LIPOSOME TECHNOLOGY TO OVERCOME MULTIDRUG RESISTANCE IN SOLID TUMORS

Rajesh Krishna and Lawrence D. Mayer

Department of Advanced Therapeutics, BC Cancer Agency,
600 West 10 Avenue, Vancouver, BC, V5Z 4E6, Canada;
Faculty of Pharmaceutical Sciences, University of British
Columbia, 2146 East Mall, Vancouver, BC, V6T 1Z3, Canada.

INTRODUCTION

A significant obstacle to chemotherapy of many human malignancies is the development of drug resistance. Multidrug resistance (MDR) is defined as the ability of tumor cells to develop resistance to the cytotoxic effects of a variety of chemically unrelated chemotherapeutic agents. Several mechanisms have been proposed to explain this phenomenon; however, the P-glycoprotein (PGP) based MDR phenotype has received the most attention and has been correlated with poor patient outcome for a number of tumor types. Conventional strategies employed for overcoming MDR involve the use of a PGP inhibitor coadministered with the anticancer agent. However, progress in this area has been hindered by the relatively low specificity of PGP modulators for tumor tissue. This has resulted in problems associated with inherent modulator toxicity as well as modulator induced changes in anticancer drug pharmacokinetics. Several avenues have been pursued using liposome technology to overcome these difficulties. This review summarizes some of the work done in this area and how long circulating non-leaky liposomes may be applied to circumvent adverse drug-drug interactions between MDR modulators and anticancer drugs, resulting in effective therapy of MDR in solid tumors.

Multi-Drug Resistance (MDR)

The phenomenon of drug resistance can be classified into two major categories: cellular and non-cellular. Cellular mechanisms (defined in terms of alterations in the biochemistry of cancer cells) include increased drug inactivation, altered DNA repair, altered intracellular drug transport, decreased therapeutic targets, and altered cell cycle control mechanisms. These MDR mechanisms (Table 1) are characterized by cross resistance to several chemically different antineoplastic agents. Resistance may also occur as a consequence of non-cellular mechanisms

Targeting of Drugs 6: Strategies for Stealth Therapeutic Systems
Edited by Gregoriadis and McCormack, Plenum Press, New York, 1998

Table 1. Classification of MDR mechanisms.

Mechanism	Specific enzyme/transporter
1. Decreased therapeutic target levels	(a) Topo I (b) Topo II (Giaccone et al., 1992)
2. Altered drug metabolism	(a) GSH and related enzymes (Batist et al., 1986)
3. Altered cell death (apoptosis)	(a) mutant p53 (b) increased bcl-2, bcl-xl (c) decreased bax, bcl-xs
4. Increased DNA repair enzymes	(a) alkylators
5. Altered drug transport	
Increased efflux	(a) P-Glycoprotein (Ling and Thompson, 1974; Juliano and Ling, 1976; Riordan and Ling, 1979; Kartner et al., 1983) (b) Multidrug Resistance Protein (Cole et al., 1992; Krishnamachary and Center, 1992; Grant et al., 1994; Almquist et al., 1995)
Decreased influx	(a) drug specific transport (methotrexate)
6. Intercellular/tissue factors	(a) hypoxia (b) altered growth fraction (c) poor blood flow

NB: (a), (b), or (c) are sub-classifications.

arising from solid tumor growth *in vivo*. These processes can reduce drug tumor-availability due to poor tumor vascularisation, or reduce the fraction of vulnerable cells due to presence of non-cycling tumor cells resistant to proliferation dependent drugs.

Clinical Significance of MDR

Of the 1,300,000 new cases of cancer diagnosed in North America each year, approximately 500,000 will exhibit properties consistent with inherent or acquired MDR. Perhaps the most widely studied MDR mechanism is the PGP phenotype. Studies on patient biopsy samples have revealed a strong correlation between PGP expression and outcome of therapy (Chan et al., 1990; Sugawara, 1990; Chan et al., 1993). Further, chemosensitization of fresh human tumors with cytotoxic drugs and MDR modulators have been shown (Ross et al., 1993; Ludescher et al., 1993). However, other studies have failed to corroborate this correlation (Haak et al., 1993; Marie et al., 1993; Vergier et al., 1993) and have shown the presence of functional MDR in the absence of PGP expression (Versantvoort et al., 1992; Haber et al., 1989) which has been, in some instances, shown to be due to MRP overexpression (Cole et al., 1992; Krishnamachary and Center, 1992). There are cases where other mechanisms have been shown to result in ineffective therapy, where tumors exhibit resistance that is linked to GST, topoisomerase II, or MRP in addition to PGP (Batist et al., 1986; Giacconne et al., 1992). Given the fact that PGP is essentially a substrate for several anticancer drugs, it is likely that the ability to block the activity of PGP will be an important component of therapeutic approaches for chemosensitizing solid tumors and constitutes an important target for chemosensitizer-based therapeutic strategies.

Table 2. Diversity of MDR reversing agents

Category	Examples
1. Calcium channel blockers	Verapamil, nifedipine, diltiazem, beperidil
2. Calmodulin inhibitors	Trifluperazine, prochlorperazine, transflupenthixol
3. Coronory vasodilators	Dipyridamole, amiodarone
4. Indole alkaloids	Vindoline, reserpine
5. Quinolines	Chloroquine, quinine
6. Lysomotropic agents	Nigericin, monexin
7. Hormones and antihormones	Progesterone, tamoxifen
8. Cephalosporins	Cefoperazone, ceftriaxone
9. Anticancer drug analogs	N-Acetyldaunorubicin, c-20'-vinblastine
10. Cyclosporins	Cyclosporine A, PSC 833
11. Protein kinase inhibitors	Staurosporine, H-87
12. Surfactants and lipids	Cremophor-EL, Tween 80, PS
13. Antibodies	MRK16, UIC-2
14. New agents	Dexniguldipine, BF 120918, VX-710

CHEMOSENSITIZATION OF MDR TUMORS

Of the several approaches investigated for circumventing MDR, chemosensitization has been widely studied as a strategy for reversal of drug resistance. Several compounds have been shown to bind directly to PGP and block its transport function or impair its phosphorylation, resulting in modulation of MDR by restoring the ability of the anticancer drug to accumulate in the cancer cells. As seen in Table 2, a wide range of previously approved pharmaceutical agents and investigational drugs have been evaluated as MDR reversing agents *in vitro*. First generation sensitizers were identified from studies that investigated the ability of certain drugs from various pharmacological classes to sensitize MDR tumor cells to anticancer drugs in culture. These include the calcium channel blockers such as verapamil and immunosuppressants such as cyclosporin A. Between 10-20 fold increases in intracellular doxorubicin (DOX) and vincristine concentrations can be obtained when these agents are used to reverse drug resistance (Tsuruo et al., 1981, 1982, 1983; Slater et al., 1986). However, these agents exhibit serious problems of inherent pharmacological activity that lead to toxicities at doses required for MDR reversal *in vivo*.

Early clinical trials with first generation MDR modulators made it apparent that agents such as verapamil were too toxic at higher MDR reversing concentrations, and were therefore of limited use as PGP modulating agents. Consequently, newer analogs of such first generation modulators with MDR reversal properties and low inherent toxicity as well as completely new chemical entities with selective action in blocking PGP were identified. Perhaps the most widely studied agent in this group is PSC 833, a non-immunosuppressive analog of cyclosporin A, which is highly potent in reversing MDR (Keller et al., 1992; Watanabe et al., 1995) and possesses latent modulating activity (Krishna et al., 1997). Further, PSC 833 is non-toxic at typical MDR reversing concentrations. Other second generation modulators include novel tiapamil analogs such as Ro11-2933 as well as dexniguldipine, GW918, and VX710 (for a detailed review see Fan et al., 1994). While *in vitro* conditions can often be constructed under which significant chemosensitization of MDR tumor cells can be achieved, the ultimate utility of a MDR modulator *in vivo* will depend on some or all of the characteristics outlined in Table

Table 3. Characteristics of an optimized MDR modulator

1.	Able to completely reverse resistance with negligible residual resistance
2.	Able to modulate MDR at low concentrations
3.	Low inherent pharmacological activity
4.	Exhibits high potency/toxicity ratio
5.	Exhibits latent or sustained modulating activity
6.	Easy to formulate; exhibits acceptable bioavailability
7.	Tumor selective; does not affect normal tissues

3. Many of these features reflect the complexity for interdependent combined dosing of a modulator and one or more anticancer drugs.

We have performed a comprehensive evaluation of the comparative pharmacology of MDR modulators in two PGP positive cell lines, the P388/ADR murine lymphocytic leukemia and the MCF7/ADR human breast carcinoma (R. Krishna et al., unpublished observations) in order to select a MDR modulator for applications of liposome technology. Some of the results of the evaluation are summarized in Table 4 which includes first generation modulators (verapamil, phenothiazines) as well as second generation MDR reversing agents (PSC 833, dexniguldipine, and Ro11-2933). This comparison is based on two key properties used to define the extent of MDR reversal, namely fold reversal and residual resistance factor. While the former is widely used in the literature to define the extent of reversal, we believe that MDR modulators exhibiting high fold reversal may still exhibit significant residual resistance factors, resulting in modest modulation efficiencies for *in vivo* applications.

Table 4. Comparison of the MDR reversing ability of chemosensitizers[1]

Modulator	**Conc. (μM)**	**P388 ADR**		**MCF7 ADR**	
		FR[1]	**RRF[2]**	**FR[1]**	**RRF[2]**
Verapamil	10	23.6	7.2*	1.417	132.8*
Dexniguldipine	5	312.8#	0.582	1952	0.194
	2.5	1.741	110.1*	ND	ND
PSC 833	1	75.4	2.259	263.4	0.714
Ro11-2933	10	95.4	1.787	516.6	0.734
Prochlorperazine	10	3.156	85.4*	1.221	258.3*
Quinidine	10	2.947	91.4*	1.459	216.1*
Chlorpromazine	10	1.707	157.8*	1.116	282.5*
Tamoxifen	10	1.613	167.1*	ND	ND

1: P388/ADR or MCF7/ADR cells were preincubated for 30 min with the chemosensitizer and plated in 96-well culture plates. Serial DOX concentrations were added and incubated for 72h, after which the MTT assay was performed. IC50s were calculated using the sigmoidal E_{max} model. # 30% viability; * difference (IC50) relative to sensitive WT control is significant at $p<0.05$

[1] Fold Reversal (FR) =

$$= \frac{\text{IC50 (DOX - Modulator) P388/ADR or MCF7/ADR}}{\text{IC50 (DOX + Modulator) P388/ADR or MCF7/ADR}}$$

[2] Residual Resistance Factor (RRF) =

$$= \frac{\text{IC50 (DOX + Modulator) P388/ADR or MCF7/ADR}}{\text{IC50 (DOX - Modulator) P388/WT or MCF7/WT}}$$

In addition to the potency of MDR reversal and residual resistance, the ability of an MDR modulator to demonstrate sustained or latent MDR modulating activity is also important. This is due to the complexities in synchronizing therapeutic levels of modulator and anticancer drug at the site of tumor progression. We have demonstrated that of the compounds studied (Table 4), only PSC 833 exhibits significant latent modulating activity (Krishna et al., 1997). Based on the potency and extent of reversal criteria outlined in Table 3, only a few modulators (PSC 833, dexniguldipine, and Ro11-2933) evaluated in Table 4 would be expected to provide any significant improvements for *in vivo* MDR reversal. In this case, PSC 833 was observed to be the most potent modulator able to reverse MDR in a concentration 5-fold lower than dexniguldipine and 10-fold lower than Ro11-2933. Further, in view of latency considerations, PSC 833 would appear one of the most suited MDR reversing agents available.

While second generation MDR modulators such as PSC 833 have several improvements such as low inherent toxicity and increased potency, the adverse pharmacokinetic interaction between these MDR modulators and the coadministered anticancer drug have created new complexities (Pourtier-Mazanedo et al., 1995; Gonzalez et al., 1995; Erlichman et al., 1994). Two consequences of this pharmacokinetic interaction are: (i) increased toxicity due to altered pharmacokinetics from modulator-induced changes in biodistribution properties of the anticancer drug; (ii) problems with interpreting preclinical and clinical data with respect to: are therapeutic improvements due to altered pharmacokinetics or PGP modulation? does decreasing anticancer drug dose to that equitoxic in the absence of modulator potentially compromise tumor therapy?

Although many of the difficulties associated with coadministration of MDR modulators and anticancer drugs are manifested by toxicity effects, it is ultimately the ability to obtain effective antitumor activity against resistant tumors that will determine the utility of chemosensitization approaches. Liposomes appear to be well suited to solve many of the problems noted above associated with conventional anticancer drugs and MDR modulators. We believe that inadequate tumor selectivity of MDR reversing agents and anticancer drugs is primarily responsible for the attenuated therapy of extravascular MDR solid tumors. Studies in our laboratory have confirmed this speculation indicating that liposomes not only provide a versatile delivery system for anticancer drugs, but also circumvent free drug - modulator adverse pharmacokinetic interactions resulting in effective therapy of MDR solid tumors. In the following sections we review the use of liposomes to overcome various problems facing conventional tumor chemosensitization strategies.

APPLICATIONS OF LIPOSOME TECHNOLOGY FOR MDR REVERSAL

Many properties of liposomal delivery systems appear well suited to solving the problems associated with therapeutic strategies utilizing conventional anticancer drugs and MDR modulators. Liposomes have been applied to the treatment of MDR tumors in three basic approaches. These are: 1) the use of liposomes alone as carriers for anticancer drugs to provide increased dose intensity and intracellular delivery, 2) inclusion of PGP modulators in the lipid bilayer or entrapped inside liposomes for delivery to MDR tumors, and 3) combining potent conventional MDR modulators with long circulating liposomal anticancer drugs.

Liposomal anticancer drugs alone

Liposomes have been demonstrated to confer increased delivery of the encapsulated anticancer agents to sites of tumor growth. Improvements in liposome technology such as the production of target specific systems and extended circulation life times have led to a greater optimization of drug therapy, particularly in cancer (Yaun et al., 1994; Huang et al., 1992;

Williams et al., 1993). In addition to this, the advantages of small (100 nm vesicles) liposomes to passively extravasate in tumor tissues have resulted in the ability of liposomes to selectively localize the anti-cancer drugs at the tumor site, markedly reduced toxicity as well as improved therapeutic activity due to higher drug levels being delivered to the tumor (Gabizon, 1992; Gabizon and Papahadjopoulos, 1988; Mayer et al., 1989). This passive targeting of liposomes to tumors has been shown to be due to increased permeability of tumor neovasculature and poor lymphatic flow contributing to enhanced extravasation and retention of liposomes in many solid tumors (Gabizon and Papahadjopoulos, 1988; Yuan et al., 1994). Sterically stabilized liposomes have been shown to provide enhanced therapeutic activity compared to short circulating leakier liposome systems in a number of animal models (Gabizon and Papahadjopoulos, 1988; Ahmad et al., 1993; Mayhew et al., 1992; Gabizon, 1992; Williams et al., 1993). These stealth liposomes possess extended circulation life times (Allen and Hansen, 1991; Gabizon et al., 1993), exhibit dose-independent pharmacokinetics (Allan and Hansen, 1991), and demonstrate increased solid tumor uptake properties (Gabizon and Papahadjopoulos, 1988; Gabizon, 1992). This increased delivery of anticancer drugs has the potential of overcoming MDR based solely on mass action, provided that the level of drug resistance is of a magnitude comparable to the increase in tumor drug levels.

One of the early indications that liposomes can be beneficial for MDR reversal was observed in a study that demonstrated enhanced activity of a long circulating liposomal formulation incorporating either the ganglioside GM1 or PEG-polymerized DSPE for doxorubicin and epirubicin against the murine C26 colon carcinoma (Huang et al., 1992; Mayhew et al., 1992; Papahadjopoulos et al., 1991). These C26 colon carcinoma cells are known to express moderate levels of PGP and doses of free doxorubicin up to 10 mg/kg are rendered ineffective (Huang et al., 1992) *in vivo*. Results indicated that mice receiving stealth formulations of DOX or epirubicin resulted in tumor regression to nonmeasurable sizes, with 90 and 100% long term 120 day survivors in groups that were treated with stealth liposomal epirubicin and DOX, respectively. In comparison, free drugs did not have any effect on delaying tumor growth (Huang et al., 1992). The authors attributed this enhanced activity in a relatively resistant tumor model to be due to enhanced localization of the drug in tumor tissue, drug release from liposomes into the extravascular spaces and uptake of the released contents by the tumor (Huang et al., 1992). The therapeutic efficacy of free epirubicin, stealth liposomal epirubicin (PEG-DSPE) and conventional liposomal epirubicin was also evaluated in the C26 colon carcinoma model by Mayhew et al., (1992). As observed with the study of Huang et al., (1992), liposomal epirubicin exhibited enhanced activity in this anthracycline resistant murine colon carcinoma model.

This ability of liposomes to inherently overcome a certain degree of MDR, stems from the fact that liposomes can markedly enhance (between 3-10 fold) the amount of drug that can be delivered to tumors compared to free drug. For example, if the tumor exhibits a resistance factor of 5-fold, this could theoretically be overcome by increasing anticancer drug dose intensity through liposomal delivery. The fact that stealth liposomes were able to provide tumor regression in C26 colon carcinoma cells (Huang et al., 1992; Mayhew et al., 1992) by delivering higher amounts of the drug supports this argument. However, for tumors exhibiting higher resistance levels, liposomes by themselves may be unable to circumvent MDR significantly as demonstrated in a rat glioblastoma tumor model (Hu et al., 1995). In further support of this argument, DSPC/Chol liposomal DOX alone at a dose of 10 mg/kg was unable to circumvent PGP mediated MDR in a murine P388/ADR solid tumor model, where resistance factors are in the order of 50-100 fold (see section below).

Antibody-coated liposomes (also called immunoliposomes) have the ability to provide targeting and selective toxicity towards tumor cells and this has been demonstrated in KLN-205 squamous cell carcinoma *in vivo* (Ahmad et al., 1993). By targeting tumor cells expressing a specific internalizable surface epitope, these immunoliposomes provide a unique approach of providing a intracellular tumor delivery of the drug (Marjan et al., 1996). This is exemplified by a recent *in vitro* study which reported that DOX resistance was modulated by an immunoliposome targeting transferring receptor in MDR human leukemic cells (Suzuki et al., 1997). However, site-directed targeting may not result in enhanced antitumor activity unless targeting is directed to surface markers that can be endocytosed since most tumor cells do not actively internalize non-targeted liposomes. Further, studies have shown that the PEG on the surface of the sterically stabilized liposome causes a steric hindrance which may interfere with antibody when directly conjugated to the lipid bilayer (Lee and Low, 1995; Storm et al., 1994; Hansen et al., 1995). In addition, given the penetration difficulties experienced for immunoliposomes in larger solid tumors, it is likely that their application to circumvent MDR may be limited to circulating hematopoietic malignancies as has been suggested by Allen and co-workers (Lopes de Menezes et al., 1995).

Some of the alternative approaches using liposomal anticancer drugs include the use of thermosensitive liposome-encapsulated DOX in conjunction with hyperthermia which resulted in increased activity against MDR MCF7 cells (Merlin et al., 1993), and the use of folate-mediated tumor cell targeting of liposomal DOX in tumor cells that overexpress folate binding proteins such as ovarian carinomas (Lee and Low, 1995). However, these novel approaches have limited applicability *in vivo* given the fact that the conditions for hyperthermia are unlikely to be useful for visceral or widespread systemic malignancies. Use of folate-PEG-liposomes may increase the specificity for tumor cells overexpressing the folate receptor, however, this approach relies on effective internalization of the liposomes into the vast majority of tumor cells which may be difficult in view of the problems experienced with tumor penetration for immunoliposomes.

Use of liposomes for delivery of MDR modulators

Studies with liposomes composed of certain acidic phospholipid-based systems such as phosphatidylserine (Fan et al., 1990) or cardiolipin liposomes (Rahman et al., 1992) have shown that these lipids are able to increase the cytotoxicity of encapsulated or complexed anticancer drugs against MDR cells. MDR reversal using such phospholipids has been related to increased intracellular delivery of anticancer agents as well as a direct PGP blocking effect (Thierry et al., 1992). Several *in vitro* investigations with acidic phospholipids have been described. *In vitro* work in colon cancer cell lines (Oudard et al., 1991), human ovarian carcinoma cells (Thierry et al., 1992), human leukemia cells, and human breast carcinoma cells (Thierry et al., 1994) suggested that liposomal encapsulation may have beneficial effects in MDR cells. In some systems, liposomes can increase intracellular drug accumulation and work by Thierry et al., (1992) suggested that cardiolipin containing liposomes directly alter PGP function as illustrated by the fact that empty liposomes inhibited specific [^{3}H]-vincristine binding to PGP-enriched membranes. Cardiolipin liposomes complexed with DOX have been shown to reverse MDR in human breast carcinoma MCF7/ADR cells *in vitro* (Thierry et al., 1994). While these results provided encouraging data implicating certain lipids as PGP modulating agents, acidic lipids such as cardiolipin and phosphatidylserine are readily recognized by phagocytic cells of the reticulo-endothelial system and liposomes containing these lipids are cleared within minutes from the circulation offering little systemic availability *in vivo*. Consequently, whereas cardiolipin based liposomal encapsulation of DOX has been shown to have enhanced efficacy in an *in vivo* murine ascitic L1210 leukemia model compared to free drug (Gokhale et al., 1996), this model does not reflect conditions for systemic administration of therapeutic agents to treat a distal extravascular disease site. Given the rapid clearance of negatively charged liposomes from the plasma, it is yet to be demonstrated that such liposome systems will be of use in treating solid tumors.

Since liposomal formulations offer considerable advantages over free drugs for many biological agents, novel strategies involving liposomes have been increasing. As an example, antisense oligonucleotides against mdr-1 mRNA exhibits greater therapy when encapsulated in liposomes and reduces PGP synthesis (Thierry et al., 1993). It was been recently demonstrated (Alahari et al., 1996) that phosphorothioate antisense oligonucleotides reduced levels of the mdr1 message inhibiting the expression of PGP as well as affecting drug uptake. However, effective reduction in mdr1 mRNA and protein levels was only obtained when used in conjunction with cationic liposomes (Alahari et al., 1996). Currently, the therapeutic utility of such cationic liposomes are limited to local delivery since these systems are also cleared very rapidly after systemic administration and are unlikely to provide any systemic delivery. A related study reported the liposome-mediated transfer of MDR-1 ribozymes into MDR human pleural mesothelioma cells which restored sensitivity towards anticancer drugs (Kiehntopf et al., 1994).

Studies have also been performed whereby MDR modulators were encapsulated in liposomes. However, to date the MDR modulators are either not amenable for liposomal encapsulation or exhibit high leakage rates when encapsulated agents are exposed to circulating components after i.v. injection. It has been shown that modulators such as verapamil, prochlorperazine (Webb et al., 1995), and cyclosporin A (Ouyang et al., 1995; Choice et al., 1995) could be effectively entrapped within liposomes. However, their very high membrane permeabilities resulted in rapid leakage from the liposomes on i.v. adminstration (Choice et al., 1995; L.D. Mayer, unpublished observations) which compromises any potential advantages of such liposomes to provide improved selectivity of tumor delivery for these agents through the use of long circulating liposomes. It is speculated that stable liposomal modulator formulations will increase the selectivity of their reversal properties in solid tumors. However, in order for such liposomes to be formulated, the modulator must possess either higher affinity for the liposomal membranes or significantly reduced membrane permeability when entrapped in the liposome interior.

Use of liposomal anticancer drugs in combination with second generation MDR modulators

Conventional methods to circumvent multidrug resistance (MDR) often utilize co-administration of chemosensitizers and anticancer drugs. The cyclosporin analog PSC 833 has been shown to possess powerful chemosensitization properties *in vitro* besides being intrinsically non-toxic. However, co-administration with anticancer drugs has resulted in exacerbated toxicity of the anticancer drugs, due to altered anticancer drug pharmacokinetics (Keller et al., 1992; Scheithauer et al., 1993), necessitating anticancer drug dose reduction by at least 2-fold in humans (Boote et al., 1996). Liposomal carriers have been demonstrated to provide tumor specific delivery of anticancer agents as well as to circumvent many toxicities associated with these agents by altering the pharmacodistribution properties of encapsulated drugs (Gabizon and Papahadjopolous, 1988; Gabizon, 1992; Mayer et al., 1989; Mayer et al., 1990; Mayer et al., 1995; Forssen et al., 1996). Given the pharmacokinetic changes induced by PSC 833 on conventional (free) DOX, we postulated that liposomes may limit these effects by virtue of their ability to reduce the exposure of encapsulated DOX to the kidneys and alter clearance of DOX in the liver (Van Hossel et al., 1984; Mayer et al., 1989). These tissues appear to be key factors involved in PSC 833-induced DOX pharmacokinetic changes (Colombo et al., 1996). Given the ability of small liposomes to passively extravasate in tumors (Gabizon, 1992; Gabizon and Papahadjopolous, 1988), we have shown that liposomes avoid these adverse pharmacokinetic interactions with PSC 833 leading to substantially increased selectivity of MDR modulation at the tumor site and result in improved therapy of MDR solid tumors (Krishna and Mayer, 1997).

Our toxicity results indicated that oral administration of PSC 833 at a dose of 100 mg/kg reduced the maximum tolerated dose (MTD) of free drug administered intravenously by 2.5-3 fold. In contrast, PSC 833 administration resulted in only a 20% reduction of the MTD for DOX encapsulated in 100 nm distearoylphosphatidylcholine (DSPC)/cholesterol (Chol) liposomes (55:45 molar lipid ratio). Similar results were obtained for single dose and multiple dose injections. Further, alterations in toxicity induced by PSC 833 correlated well with changes in plasma DOX elimination properties and the increased toxicity for free DOX caused by PSC 833 was accompanied by increased DOX levels in susceptible normal tissues, whereas the DOX pharmacokinetics for the liposomal formulation was minimally affected by PSC 833 (Krishna and Mayer, 1997).

Modest modulation of PGP-mediated MDR was observed in the murine P388/ADR solid tumor model when PSC 833 was administered with free DOX at the MTD. Specifically, our results demonstrated that free DOX at 7.5 mg/kg (MTD) conferred negligible therapeutic activity in the MDR solid tumor model. In the presence of PSC 833, however, there is transient inhibition of tumor growth until day 9 after which the growth characteristics are similar to the untreated P388/ADR control tumors (Fig. 1). Furthermore, the dose of the free DOX had to be reduced from 7.5 mg/kg to 2 mg/kg when combined with PSC 833. Tissue distribution studies indicated that this dose reduction decreased DOX tumor levels which may be inadequate to provide sustained antitumor activity.

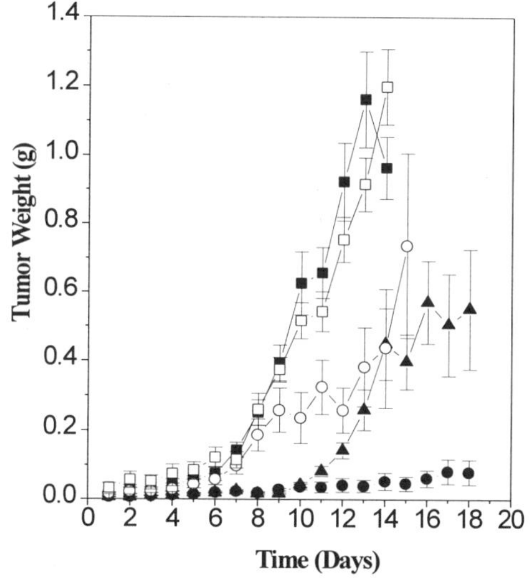

Figure 1: Liposomes increase the therapy of DOX against MDR solid tumors when combined with PSC 833. P388/ADR solid tumors were grown sub-cutaneously in BDF1 mice. Oral PSC 833 (100 mg/kg) and i.v. DOX treatments were initiated once tumors were established (20-100 mg) and were given on days 1, 5, and 9 at the indicated doses of free and liposomal DOX. PSC 833 was administered 4 hours prior to DOX injection. P388/ADR control untreated tumors (solid squares), P388/ADR tumors treated with free DOX 7.5 mg/kg (open squares), P388/ADR tumors treated with lipo DOX 10 mg/kg (open circles), P388/ADR tumors treated with free DOX 2.0 mg/kg in conjunction with 100 mg/kg p.o. PSC 833 (solid triangles), and P388/ADR tumors treated with liposomal DOX 10 mg/kg in conjunction with 100 mg/kg p.o. PSC 833 (solid circles). (Source: with permission from Cancer Research, Krishna and Mayer, 1997).

Liposomal DOX in the absence of PSC 833 resulted in limited antitumor activity as manifested by a small delay in tumor growth. In contrast, liposomal DOX combined with PSC 833 resulted in tumor growth inhibition comparable to that observed for drug sensitive P388/WT tumors (Krishna and Mayer, 1997, see Figure 1). The fact that only the combination of PSC 833 with the liposomal DOX formulation provided complete chemosensitization strongly supports the direct involvement of PGP blockade in MDR solid tumor therapy. This direct role of PSC 833 in the therapy of PGP overexpressing tumors is corroborated by the fact that DOX pharmacokinetic properties were not significantly altered by PSC 833 when combined with the liposomal formulation. Pharmacokinetic data demonstrated that DOX encapsulated in DSPC/Chol liposomes exhibited comparable tumor accumulation levels in the presence or absence of PSC 833.

The results above indicate that improved tumor selectivity of anticancer drugs administered in liposomal formulations will avoid complications experienced with free drugs when combined with MDR reversing agents and enhance the therapy of MDR tumors. Tumor DOX levels following liposomal delivery were significantly higher than that observed for the free drug administration. Most importantly, the tumor associated DOX levels were unaffected by PSC 833 when liposomes were used, unlike the observations seen with free DOX. These studies suggest that the significant improvements in MDR therapy obtained using conventional modulators and long circulating liposomal anticancer drugs arise from an increased dose intensity of DOX at the tumor site in addition to effective PGP blockade induced by PSC 833. This is consistent with results from an *in vitro* study using human colon cancer cells which suggested that the activity of liposome encapsulated vincristine was potentiated when combined with MRK-16, a monoclonal antibody (Sela et al., 1995).

SUMMARY

Multidrug resistance is a continuing therapeutic obstacle in the chemotherapy of many human malignancies. The ability of long circulating liposomes to permit higher amounts of drug to be effectively delivered to the tumors to address the issue of anticancer dose intensity is an attractive advantage. Further, these carriers are tumor-selective with little toxicity to susceptible healthy tissues. Our results utilizing liposomal DOX in combination with a potent MDR modulator such as PSC 833 show that liposomes can circumvent the modulator-induced adverse anticancer drug pharmacokinetic alterations. Further improvements in the target specific delivery of anticancer drugs with MDR modulators as well as the development of liposome based MDR modulators will likely reveal additional therapeutic strategies for treating MDR solid tumors.

Acknowledgements

This work has been supported by the National Cancer Institute of Canada with funds from the Canadian Cancer Society. RK is a BC Medical Services Foundation Pre-Doctoral Fellow.

REFERENCES

Ahmad, I., Longenecker, M., Samuel, J., and Allen, T. M., 1993, Antibody-targeted delivery of doxorubicin entrapped in sterically stabilzed liposomes can eradicate lung cancer in mice, *Cancer Res.*, 53:1484.

Alahari, S.K., Dean, N. M., Fisher, M. H., Delong, R., Manoharan, M., Tivel, K. L., and Juliano, R. L., 1996, Inhibition of expression of the multidrug resistance associated P-glycoprotein by phosphorothioate and 5' cholesterol-conjugated phosphorothioate antisense oligonucleotides, *Mol. Pharmacol.*, 50:808.

Allen, T. M., and Hansen, C., 1991, Pharmacokinetics of stealth versus conventional liposomes: effect of dose, *Biochim. Biophys. Acta*, 1068:133.

Almquist, K. C., Loe, D. W., Hipfner, D. R., Mackie, J. E., Cole, S. P., and Deeley, R. G., 1995, Characterization of the M(r) 190,000 multidrug resistance protein (MRP) in drug selected and transfected human tumor cell, *Cancer Res.*, 55:102.

Batist, G., Tulpule, A., Sinha, B. K., Kakti, A. G., Myers, C. E., and Cowan, K. H., 1986, Overexpression of a novel anionic glutathione transferase in multidrug resistant human breast cancer cells, *J Biol Chem.*, 33:15544.

Boote D.J., Dennis I.F., Twentyman P.R., Osborne R.J., Laburte C., Hensel S., *et al.*, 1996, Phase I study of etoposide with PSC-833 as a modulator of multidrug resistance in patients with cancer, *J. Clin. Oncol.*, 14:610.

Chan, H. S., Thorner, P. S., Haddad, G., DeBoer, G., Gallie, B. L., and Ling, V., 1993, Multidrug resistance in cancers of childhood: clinical relevance and circumvention, *Cancer Res.*, 41:1967.

Chan, H. S., Thorner, P., Haddad, G., and Ling, V., 1990, Immunohistochemical detection of P-glycoprotein: prognostic correlation in soft tissue sarcoma of childhood, *J. Clin. Oncol.*, 8:689.

Choice, E., Masin, D., Bally, M. B., Meloche, M., and Madden, T. D., 1995, Liposomal cyclosporine. Comparison of drug and lipid carrier pharmacokinetics and biodistribution, *Transplantation*, 60:1006.

Cole, S.P., Bhardwaj, G., Gerlach, J. H., Mackie, J. E., Grant, C. E., Almquist, K. C., Stewart, A. J., Kurz, E. U., Duncan, A. M., and Deeley, R. G., 1992, Overexpression of a transporter gene in a multidrug resistant human lung cancer cell line, *Science*, 258:1650.

Colombo T., Paz O. G., and D'Incalci M., 1996, Distribution and activity of doxorubicin combined with SDZ PSC 833 in mice with P388 and P388/DOX leukemia, *Br. J. Cancer*, 73:866.

Erlichman C., Moore M., Thiessen J., De Angelis C., Goodman P., and Manzo J., 1994, A Phase I trial of doxorubicin (DOX) and PSC 833, a modulator of multidrug resistance (MDR), *Anti-Cancer Drugs*, 5:42.

Fan, D., Bucana, C.D., O'Brian, C. A., Zwelling, L.A., Seid, C., and Fidler, I.J., 1990, Enhancement of murine tumor cell sensitivity to adriamycin by presentation of the drug in phosphatidylcholine-phosphatidylserine liposomes, *Cancer Res.*, 50:3619.

Fan D., Beltran P.J., and O'Brien C.A., 1994, Reversal of multidrug resistance, in: "Reversal of Multidrug Resistance in Cancer", J. A. Kellen (Ed.), CRC Press, Boca Raton.

Forssen, E. A., Male-Brune, R., Adler-Moore, J. P., Lee, M. J., Schmidt, P. G., Krasieva, T. B., Shimizu, S., and Tromberg, B. J., 1996, Fluorescence imaging studies for the disposition of daunorubicin liposomes (DaunoXome) within tumor tissue, *Cancer Res.*, 56:2066.

Gabizon, A., 1992, Selective tumor localization and improved therapeutic index of anthracyclines encapsulated in long circulating liposomes, *Cancer Res.*, 52:891.

Gabizon, A., and Papahadjopolous, D., 1988, Liposome formulations with prolonged circulation time in blood and enhanced uptake by tumors, *Proc. Natl. Acad. Sci. USA*, 85:6949.

Gabizon, A., Barenholz, Y., and Bialer, M., 1993, Prolongation of the circulation time of doxorubicin encapsulated in liposomes containing a polyethylene glycol-derivatized phospholipid: pharmacokinetic studies in rodents and dogs, *Pharm. Res.*, 10:703.

Giaccone, G., Gazdar, A. F., Beck, H., Zunino, F., and Capranico, G., 1992, Multidrug sensitivity phenotype of human lung cancer cells associated with topoisomerase II expression, *Cancer Res.*, 52:1666.

Gokhale, P. C., Radhakrishnan, B., Husain, S. R., Abernethy, D. R., Sacher, R., Dritschilo, A., and Rahman, A., 1996, An improved method of encapsulation of doxorubicin in liposomes: pharmacological, toxicological, and therapeutic evaluation, *Br. J. Cancer*, 74:43.

Gonzalez O., Colombo T., De Fusco M., Imperatori L., Zucchetti M., and D'Incalci M., 1995, Changes in doxorubicin distribution and toxicity in mice pretreated with the cyclosporin analogue SDZ PSC 833, *Cancer Chemother. Pharmacol.*, 36:335.

Grant, C. E., Valdimarsson, G., Hipfner, D. R., Almquist, K. C., Cole, S. P., and Deeley, R. G., 1994, Overexpression of multidrug resistance-associated protein increases resistance to natural product drugs, *Cancer Res.*, 54:357.

Haak, H. R., van Seters, A. P., Moolenaar, A. J., and Fleuren, G. J., 1993, Expression of P-glycoprotein in relation to clinical manifestation, treatment and prognosis of adrenocortical cancer, *Eur. J. Cancer*, 29A:1036.

Haber, M., Norris, M. D., Kavallaris, M., Bell, D. R., Davey, R. A., White, L., and Stewart, B. W., 1989, Atypical multidrug resistance in a therapy-induced drug resistant human leukemia cell line (LALW-2): resistance to vinca alkaloids independent of P-glycoprotein, *Cancer Res.*, 49:5281.

Hansen, C. B., Kao, G. Y., Moase, E. H., Zalipsky, S., and Allen, T. M., 1995, Attachment of antibodies to sterically stabilized liposomes: evaluation, comparison, and optimization of coupling procedures, *Biochim. Biophys. Acta*, 1239:133.

Hu, Y. P., Henry-Toulme, N., and Robert, J., 1995, Failure of liposome encapsulation of doxorubicin to circumvent multidrug resistance in an *in vitro* model of rat glioblastoma cells, *Eur. J. Cancer*,

31A:389.
Huang, S. K., Mayhew, E., Gilani, S., Lasic, D. D., Martin, F. J., and Papahadjopoulos, D., 1992, Pharmacokinetics and therapeutics of sterically stabilized liposomes in mice bearing C-26 colon carcinoma, *Cancer Res.*, 52:6744.
Juliano, R. L., and Ling, V., 1976, A surface glycoprotein modulating drug permeability in chinese hamster ovary cell mutants, *Biochem Biophys Acta.*, 455:152.
Kartner, N., Riordan, J. R., and Ling, V., 1983, Cell surface P-glycoprotein associated with multidrug resistance in mammalian cell lines, *Science*, 221:1285.
Keller R.P., Altermatt H.J., Donatsch P., Zihlmann H., Laissue J.A., and Hiestand P.C., 1992, Pharmacologic interactions between the resistance-modifying cyclosporine SDZ PSC 833 and etoposide (VP 16-213) enhance *in vivo* cytostatic activity and toxicity, *Int. J. Cancer*, 51:433.
Keller R.P., Altermatt H.J., Nooter K., Poschmann G., Laissue J.A., Bollinger P., *et al.*, 1992, SDZ PSC 833, a non-immunosuppressive cyclosporine: Its potency in overcoming P-glycoprotein mediated multidrug resistance of murine leukemia, *Int. J. Cancer*, 50:593.
Kiehntopf, M., Brach, M.A., Licht, T., Petschauer, S., Karawajew, L., Krischning, C., and Herrmann, F., 1994, Ribozyme-mediated cleavage of the MDR-1 transcript restores chemosensitivity in previously resistant cancer cells, *EMBO J.*, 13:4645.
Krishna, R., de Jong, G., and Mayer, L. D., 1997, Pulsed exposure of SDZ PSC 833 to multidrug resistant P388/ADR and MCF7/ADR cells in the absence of anticancer drugs can fully restore sensitivity to doxorubicin, *Anticancer Res.*, 17:3329.
Krishna, R., and Mayer, L.D., 1997, Liposomal doxorubicin circumvents PSC 833-free drug interactions, resulting in effective therapy of multidrug resistant solid tumors, *Cancer Res.*, 57:5246.
Krishnamachary, N., and Center, M. S., 1992, Detection and characterization of membrane protein changes in multidrug resistant HL-60 cells, *Oncology Res.*, 4:23.
Lee, R.J., and Low, P.S., 1995, Folate-mediated tumor cell targeting of liposome-entrapped doxoubicin *in vitro*, *Biochim. Biophys. Acta*, 1233:134.
Ling, V., and Thompson, L. H., 1974, Reduced permeability in CHO cells as a mechanism of resistance to colchicine, *J Cell Physiol.*, 83:103.
Lopes de Menezes, D. E., Pilarski, L. M., and Allen, T. M., 1995, Selective cytotoxicity of immunoliposomal doxorubicin to B lympocytes, *Proc. AACR*, 36:A1825.
Ludescher, C., Hilbe, W., Eisterer, W., Preuss, E., Huber, C., Gotwald, M., Hofmann, J., and Thaler, J., 1993, Activity of P-Glycoprotein in B-cell chronic lymphocytic leukemia determined by a flow cytometry assay, *J. Natl. Cancer Inst.*, 85:1751.
Marie, J., Faussat-Suberville, A., Zhou, D., and Zittoun, R., 1993, Daunorubicin uptake by leukemic cells: correlations with treatment outcome and mdr1 expression, *Leukemia*, 7:825.
Marjan, J., Charrios, G., Lopes de Menezes, D., and Allen, T. M., 1996, Antibody-mediated targeting of liposomal doxorubicin to lymphoblastic cells can reverse multidrug resistance, *Proc. AACR*, 37:A2103.
Mayer L.D., Bally M.B., Cullis P.R., Wilson S.L., and Emerman J.T., 1990, Comparison of free and liposomal encapsulated doxorubicin tumor drug uptake and antitumor efficacy in the SC115 murine mammary tumor, *Cancer Lett.*, 53:183.
Mayer L.D., Tai L.C.L., Ko D.S.C., Masin D., Ginsberg R.S., Cullis P.R., *et al.*, 1989, Influence of vesicle size, lipid composition, and drug-to-lipid ratio on the biological activity of liposomal doxorubicin in mice, *Cancer Res.*, 49:5922.
Mayer, L. D., Masin, D., Nayar, R., Boman, N. L., and Bally, M. B., 1995, Pharmacology of liposomal vincristine in mice bearing L1210 ascitic and B16/BL6 solid tumors, *Br. J. Cancer*, 71:482.
Mayhew, E.G., Lasic, D., Babbar, S., and Martin, F.J., 1992, Pharmacokinetics and antitumor activity of epirubicin encapsulated in long-circulating liposomes incorporating a polyethylene glycol-derivatized phospholipid, *Int. J. Cancer*, 51:302.
Merlin, J. L., Marchal, S., Ramacci, C., Notter, D., and Vigneron, C., 1993, Antiproliferative activity of thermosensitive liposome-encapsulated doxorubicin combined with 43 degrees C hyperthermia in sensitive and multidrug resistant MCF7 cells, *Eur. J. Cancer*, 29A:2264.
Oudard, S., Thierry, A., Jorgensen, T. J., and Rahman, A., 1991, Sensitization of multidrug resistant colon cancer cells to doxorubicin encapsulated in liposomes, *Cancer Chemother. Pharmacol.*, 28:259.
Ouyang, C., Choice, E., Holland, J., Meloche, M., and Madden, T. M., 1995, Liposomal cyclosporine. Characterization of drug incorporation and interbilayer exchange, *Transplantation*, 60:999.
Papahadjopoulos, D., Allen, T. M., Gabizon, A., Mayhew, E., Mathay, K., Huang, S. L., Lee, K.-D., Woodle, M. C., Lasic, D. D., Redemann, C., and Martin, F. J., 1991, Sterically stabilized liposomes: improvements in pharmacokinetics and antitumor therapeutic efficacy, *Proc. Natl. Acad. Sci. USA*, 88:11460.
Pourtier-Manzanedo A., Didier A., Froidevaux S., and Loor F., 1995, Lymphotoxicity and myelotoxicity of doxorubicin and SDZ PSC 833 combined chemotherapies for normal mice, *Toxicology*, 99:207.

Rahman, A., Husain, S. R., Siddiqui, J., Verma, M., Agresti, M., Center, M., Safa, A. R., and Glazer, R. I., 1992, Liposome-mediated modulation of multidrug resistance in human HL-60 leukemia cells, *J. Natl. Cancer Inst.*, 84:1909.

Riordan, J. R., and Ling, V., 1979, Purification of P-glycoprotein from plasma membrane vesicles of chinese hamster ovary cell mutants with reduced colchicine permeability, *J Biol Chem.*, 254:12701.

Ross, D. D., Wooten, P. J., Sridhara, R., Ordonez, J. V., Lee, E. J., and Schiffer, C. A., 1993, Enhancement of daunorubicin accumulation, retention, and cytotoxicity by verapamil or cyclosporin A in blast cells from patients with previously untreated acute myeloid leukemia, *Blood*, 82:1288.

Scheithauer, W., Schenk, T., and Czejka, M., 1993, Pharmacokinetic interaction between epirubicin and the multidrug resistance reverting agent D-verapamil, *Br. J. Cancer*, 68:8.

Sela, S., Husain, S. R., Pearson, J. W., Longo, D. L., and Rahman, A., 1995, Reversal of multidrug resistance in human colon cancer cells expressing the human MDR1 gene by liposomes in combination with monoclonal antibody or verapamil, *J. Natl. Cancer Inst.*, 87:123.

Slater, L. M., Murray, S. L., Wetzel, M. W., Sweet, P., and Stupeck, M., 1986, Verapamil potentiation of VP-16-213 in acute lymphatic leukemia and reversal of pleiotropic drug resistance, *Cancer Chemother. Pharmacol.*, 16:50.

Storm, G., Bakker-Woudenberg, I. A., Woodle, M. C., Blume, G., Nassander, U. K., Vingerhoeds, M. H., Haisma, H., and Crommelin, D. J. A., 1994, Liposomal drug delivery: possibilities for manipulation, in *Targeting of drugs 4: Advances in System Constructs*, G. Gregoriadis, B. McCormack and G. Poste (Eds.), Plenum Press, New York.

Sugawara, I., 1990, Expression and functions of P-glycoprotein (mdr1 gene product) in normal and malignant tissues, *Acta Pathol. Jap.*, 40:545.

Suzuki, S., Inoue, K., Hongoh, A., Hashimoto, Y., and Yamazoe, Y., 1997, Modulation of doxorubicin resistance in a doxorubicin-resistant human leukemia cell by an immunoliposome targeting transferring receptor, *Br. J. Cancer*, 76:83.

Thierry, A. R., Dritschilo, A., and Rahman, A., 1992, Effect of liposomes on P-glycoprotein function in multidrug resistant cells, *Biochem. Biophys. Res. Comm.*, 187:1098.

Thierry, A. R., Rahman, A., and Dritschilo, A., 1993, Overcoming multidrug resistance in human tumor cells using free and liposomally encapsulated antisense oligonucleotides, *Biochem. Biophys. Res. Comm.*, 190:952.

Thierry, A. R., Rahman, A., and Dritschilo, A., 1994, A new procedure for the preparation of liposomal doxorubicin: biological activity in multidrug-resistant tumor cells. *Cancer Chemother. Pharmacol.*, 35:84.

Tsuruo, T., Iida, H., Kitatani, Y., Yokota, K., Tsukagoshi, S., and Sakurai, Y., 1982, Enhancement of vincristine- and adriamycin-induced cytotoxicity by verapamil in P388 leukemia and its sublines resistant to vincristine and adriamycin, *Biochem Pharmacol.*, 31:3138.

Tsuruo, T., Iida, H., Naganuma, K., Tsukagoshi, S., and Sakurai, Y., 1983, Promotion of verapamil of vincristine responsiveness in tumor cell lines inherently resistant to the drug, *Cancer Res.*, 43:808.

Tsuruo, T., Iida, H., Tsukagoshi, S., and Sakurai, Y., 1981, Overcoming of vincristine resistance in P388 leukemia *in vivo* and *in vitro* through enhanced cytotoxicity of vincristine and vinblastine by verapamil, *Cancer Res.*, 41:1967.

Van Hossel Q.G.C.M., Steerenberg P.A., Crommelin D.J.A., van Dijk A., van Oort W., Klein S., *et al.*, 1984, Reduced cardiotoxicity and nephrotoxicity with preservation of antitumor activity of DOX entrapped in stable liposomes in the LOU/M Ws1 rat, *Cancer Res.*, 44:3698.

Vergier, B., Cany, L., Bonnet, F., Robert, J., de Mascarel, A., and Coindre, J. M., 1993, Expression of MDR1/P-glycoprotein in human sarcomas, *Br. J. Cancer*, 68:1221.

Versantoort, C. H. M., Broxterman, H. J., Pinedo, H. M., de Vries, E. G. E., Feller, N., Kupier, C. M., and Lankelma, J., 1992, Energy-dependent processes involved in reduced drug accumulation in multidrug resistant human lung cancer cell lines without P-glycoprotein expression, *Cancer Res.*, 52:17.

Watanabe T., Tsuge H., Oh-hara T., Naito M., and Tsuruo T., 1995, Comparative study on reversal efficacy of SDZ PSC 833, cyclosproin A and verapamil on multidrug resistance *in vitro* and *in vivo*, *Acta Oncologica*, 34:235.

Webb, M. S., Wheeler, I. J., Bally, M. B., and Mayer, L. D., 1995, The cationic lipid stearylamine reduces the permeability of the cationic drugs verapamil and prochlorperazine to lipid bilayers: implications for drug delivery, *Biochim. Biophys. Acta*, 1238:147.

Williams, S.S., Alosco, T. R., Mayhew, E., Lasic, D. D., Martin, F. J., and Bankert, R. B., 1993, Arrest of human lung tumor xenograft growth in severe combined immunodeficient mice using doxorubicin encapsulated in sterically stabilized liposomes, *Cancer Res.*, 53:3964.

Yuan, F., Leunig, M., Huang, S. K., Berk, D. A., Papahadjopoulos, D., and Jain, R. K., 1994, Microvascular permeability and interstitial penetration of sterically stabilized (stealth) liposomes in a human tumor xenograft, *Cancer Res.*, 54:3352.

USE OF RADIOLABELED LIPOSOMES FOR PEG-LIPOSOME-BASED DRUG TARGETING AND DIAGNOSTIC IMAGING APPLICATIONS

William T. Phillips

Department of Radiology, University of Texas Health Science Center at San Antonio, San Antonio, Texas 78284, USA

INTRODUCTION

The development of new targeted diagnostic and therapeutic imaging agents is at the cutting edge of diagnostic imaging (Torchilin,1995). When a specific targeted agent accumulates in pathological tissue, the diagnostic imaging physician is able to determine simultaneously that a disease process is present and its location in the body. Usually, the uptake of a targeted agent is due to alteration of a physiological process, and hence a disease process may be detected prior to manifestation of any anatomical changes. Targeted diagnostic imaging, therefore, has the ability to be more sensitive for the detection of some disease processes in early stages. Although targeted diagnostic imaging agents are currently being developed for use in imaging modalities such as computed tomography (CT), magnetic resonance imaging (MRI) and ultrasound, nuclear medicine is the branch of diagnostic medicine with the longest history and the largest number of diagnostic and therapeutic targeted agents. This success in the field of nuclear medicine with targeted diagnostic imaging agents is probably due to the high signal response generated from an atomic decay which requires localization of only a small number of atoms at a particular body location in order for the photon emissions to be detected. While CT requires 10 to 100 mg of contrast agent per study and MRI requires 10-1 to 10-3, nuclear medicine imaging requires only 10-7 to 10-9 mg for diagnostic imaging (Wolf,1995). Targeted agents developed for nuclear scintigraphy can be used in virtually all the organs of the body including bones, heart, liver, gallbladder, kidney, brain, soft tissues and tumors.

Scintigraphic nuclear imaging as a modality for the development of targeted imaging agents offers several advantages over higher resolution cross sectional imaging modalities such as CT and MRI. One advantages is its rapid ability to image and screen the whole body for sites of abnormal uptake as can be seen in on the bone scan (figure 1) showing focal areas of increased uptake secondary to metatastatic prostate cancer at multiple sites over the body.

Another significant advantage of scintigraphic imaging is the ability to quantitate non-invasively the amount of activity injected that reaches a particular location in the body. This quantitation requires: 1) correction for the attenuation that occurs when parts of the body

Targeting of Drugs 6: Strategies for Stealth Therapeutic Systems
Edited by Gregoriadis and McCormack, Plenum Press, New York, 1998

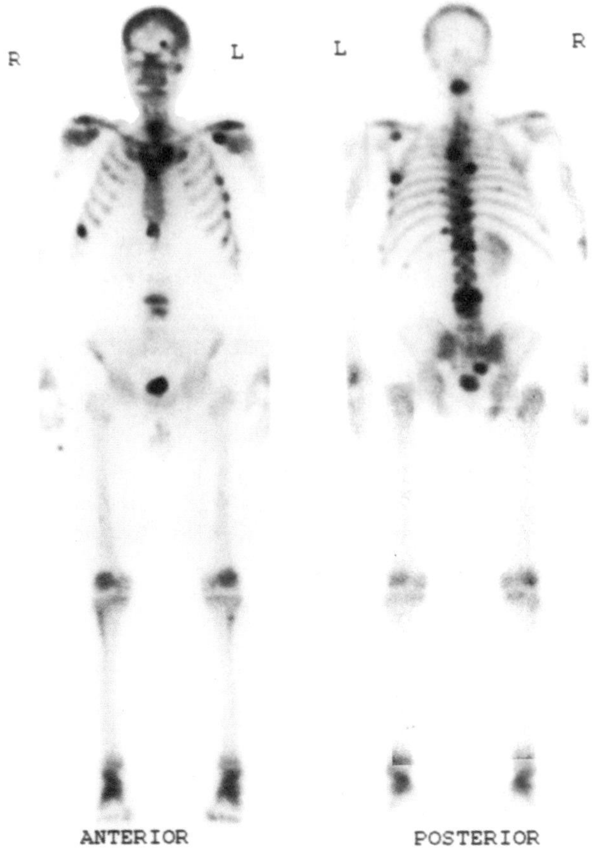

Figure 1. Anterior view and posterior view of a whole body bone scan of a patient with metastatic breast cancer. The hot spots are areas of increased bone growth in response to cancer metastasis. The time required to complete this study was 15 minutes.

absorb the radiation before it can be emitted from the body; and 2) subtraction of the amount of activity in the blood contained by a particular part or organ of the body.

TECHNETIUM-99M

The most commonly used isotope in nuclear medicine is technetium-99m (^{99m}Tc). ^{99m}Tc is readily available, inexpensive and has ideal imaging and dosimetry characteristics. Its energy of 140 keV is high enough for emitted photons to escape the body without absorption of overlying body tissues, but low enough to be readily collimated by lead and absorbed by the sodium iodide scintillation crystal. With higher energy photon emission of other radioisotopes, the difficulty in collimating the energy with lead increases and the efficiency of absorption by the scintillation crystal decreases. Both of these factors result in decreased image resolution. On the other hand, photons of less than 100 keV have more attenuation by overlying body tissues resulting in a lower amount of photons escaping from the body and imaging artifacts created by the irregular distribution of overlying body tissue.

The most common clinically used nuclear scinitigraphic agents are targeted agents based on their chemical structure. The bone scanning agent, ^{99m}Tc-methylene diphosphonate

(^{99m}Tc-MDP), accumulates at sites of active bone growth due to its chemical similarity to calcium phosphate. Radioactive iodine (^{131}I for therapy and ^{123}I for diagnosis) targets normal thyroid and thyroid metastasis due to the normal uptake of iodine by thyroid follicular cells for the manufacture of thyroid hormone.

SPECIFIC RECEPTOR TARGETING IN NUCLEAR MEDICINE

For the specific targeting of tumors, researchers in nuclear medicine have pursued radioactively tagged monoclonal antibodies for specific targeting of tumors. Several imaging agents based on monoclonal antibodies have been approved for the diagnosis of colon and ovarian cancer. The most recent interest in the field of nuclear medicine is the use of radioactively tagged receptor peptides (Reubi,1997). The quality of the images appears to be improved compared to those produced by monoclonal antibodies and the regulatory approval process has been faster compared to monoclonal antibody imaging agents. The rapid approval process was facilitated in part by the easier preparation of peptide imaging agents with less possibility of contamination as compared with hybridoma cell culture for monoclonal antibody production (Hoffman,1990; Kuus-Reichel et al,1994). Peptide agents also do not produce human antimouse antibodies (HAMA) which occurs in 26% of patients receiving a monoclonal antibody agent for the detection of colon cancer (Winzelberg et al,1992).

A recently approved receptor peptide imaging agent, ^{111}In-pentetreotide, is a radiolabeled peptide analogue of somatostatin (Olsen et al,1995). ^{111}In-pentetreotide is proving very useful in detecting tumors derived from neuroendocrine tissue such as carcinoid tumors. More peptide based receptor imaging agents are under development for the detection of colon cancer, sites of infection and thrombus imaging.

LIPOSOMES AS IMAGING AGENTS

Research involving liposomes as carriers for targeted diagnostic imaging applications has shown much promise (Phillips and Goins,1995). Liposome possess the ability to carry contrast agents for almost all imaging modalities including iodine for X-ray and CT contrast (Seltzer,1989), gadolinium and iron for MRI contrast (Pauser et al,1997; Trubetskoy et al,1995), gas for ultrasound contrast (Unger et al,1992) and ^{99m}Tc, indium-111 (^{111}In) and gallium-67 (^{67}Ga) for scintigraphic nuclear imaging (Gabizon et al,1989; Phillips et al,1992; Presant et al,1990; Woodle,1993).

HMPAO-GLUTATHIONE METHOD FOR LABELING LIPOSOMES WITH ^{99m}Tc

In our laboratory, we have developed a method of labeling preformed liposomes with ^{99m}Tc which is convenient, stable in vivo and highly efficient (Phillips et al,1992). With this method, technetium-99m pertechnetate (^{99m}TcO4) (the commercially available form of ^{99m}Tc) in saline is added to lyophilized hexamethylpropyleneamine oxime (HMPAO), a clinically approved and commercially available chelator of ^{99m}Tc used for brain imaging. During incubation for 5 minutes, the HMPAO chelates the ^{99m}Tc into lipophilic ^{99m}Tc-HMPAO. The ^{99m}Tc-HMPAO is then added to previously manufactured liposomes that encapsulate glutathione. It is generally believed that lipophilic HMPAO carries the ^{99m}Tc into the liposomes where it interacts with the encapsulated glutathione, resulting in its conversion to

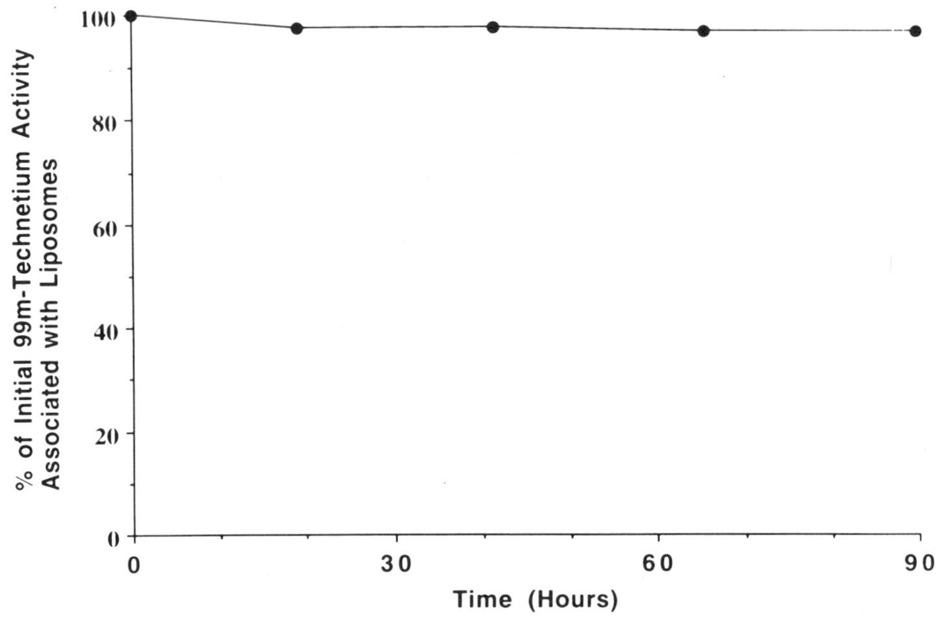

Figure 2. In vitro incubation of ^{99m}Tc-liposomes with plasma at 37°C over a 90 hour period. Liposome were labeled using the glutathione-HMPAO method. ^{99m}Tc does not dissociate from the liposomes over this period.

hydrophilic ^{99m}Tc-HMPAO. The hydrophilic ^{99m}Tc-HMPAO is irreversibly trapped in the aqueous phase of the liposome because it is unable to cross the lipid membrane. A similar mechanism has been proposed to explain the process whereby ^{99m}Tc-HMPAO becomes trapped in brain cells for use as a brain imaging agent (Neirinckx et al,1988). The liposome labeling efficiency using the method developed in our laboratory ranges from 85 to 98%; the label being very stable with virtually no dissociation of the ^{99m}Tc from the liposomes. This stability is well demonstrated in figure 2 with an in vitro study in which with ^{99m}Tc-liposomes were incubated with plasma at 37°C.

The labeling method described here is now being used to label liposomes for a variety of diagnostic imaging applications in many locations of the world (Boerman et al,1997a; Boerman et al,1997b; Ogihara-Umeda et al,1996; Ogihara-Umeda et al,1997; Oyen et al,1996; Tilcock et al,1994). Other proteins such as hemoglobin can also be co-encapsulated with glutathione without disturbing the ^{99m}Tc labeling process (Rudolph et al,1991). Furthermore, all types of liposomes can be labeled with ^{99m}Tc with equal label stability including multilamellar and unilamellar liposomes composed of a variety of lipids. Surface modification of the liposomes with polyethylene glycol (PEG) does not interfere with the labeling of liposomes with ^{99m}Tc (Awasthi et al,1998; Boerman et al,1997a; Tilcock et al,1994). It is also possible to encapsulate the metal chelator desferoxamine in liposomes along with glutathione so that liposomes can be simultaneously labeled with ^{99m}Tc and radioisotopes with longer half-lives such as indium-111(^{111}In) and gallium-67 (^{67}Ga) (Awasthi et al,1998). Using desferoxamime allows initial imaging to be performed with the gamma camera window on 140 keV for ^{99m}Tc, thereby taking advantage of the high count rate and higher resolution of ^{99m}Tc during imaging for the first 24 hours, while the isotopes ^{111}In-(half life of 67 hours) and ^{67}Ga-(half life of 5.1 days) can be used to image the distribution of liposomes up to 5 days. This technique may be useful for chronic processes in which the liposomes accumulate at a slow rate.

POTENTIAL DIAGNOSTIC IMAGING USES OF ^{99m}Tc-LABELED LIPOSOMES

Infection Imaging

There are promising applications of ^{99m}Tc-labeled liposomes for the detection of hidden sites of infection in the body which have significant advantages over the currently available agents used to detect occult infection (^{111}In and ^{99}mTc-labeled leukocytes and ^{67}Ga-citrate). Labeling of leukocytes requires a labor intensive 3-4 hour procedure with extensive blood handling and thus potential exposure to infectious agents. Moreover, gallium-67 has poor imaging energy characteristics with a high dose delivered to the patient and a normal path of excretion into the bowel that can obscure occult infections in the abdomen. In contrast, ^{99m}Tc-labeled liposomes are convenient to prepare and use for infection imaging. Prior research using ^{99m}Tc liposomes labeled by an earlier less stable method showed some promise for detection of infection, but these efforts were hampered by the instability of the ^{99m}Tc label in vivo. ^{99m}Tc activity peaked at the site of infection at 30 minutes and decreased thereafter (Morgan et al,1981).

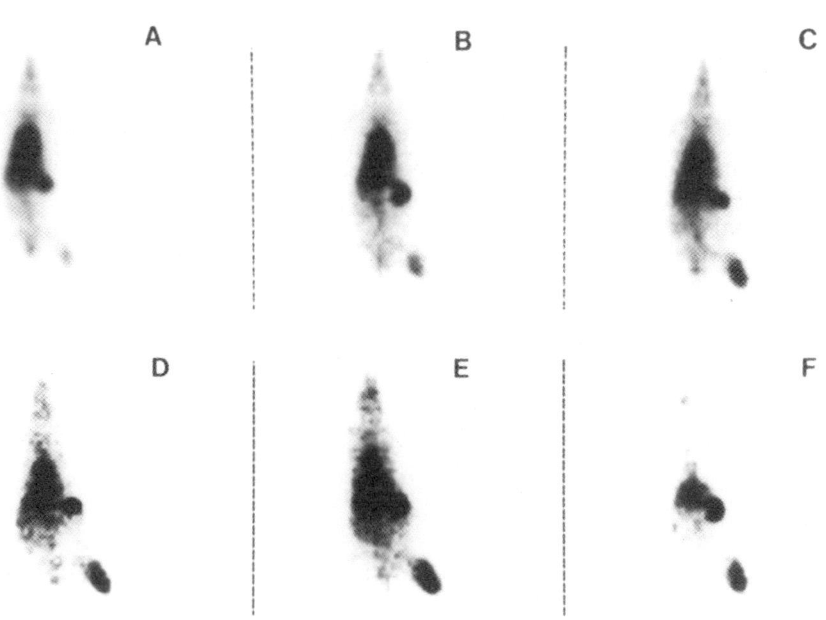

Figure 3. Scintigraphic images of rats with S. Aureus infection in the right thigh at various time points after injection with dual radiolabeled liposomes. (A)2h; (B) 4 h; (C) 8 h (all ^{99m}Tc window); (D) 8 h; (E) 24 h; (F) 48 h (all 111In window) Reprinted with permission from Nuclear Medicine and Biology 25:155-160, 1997)

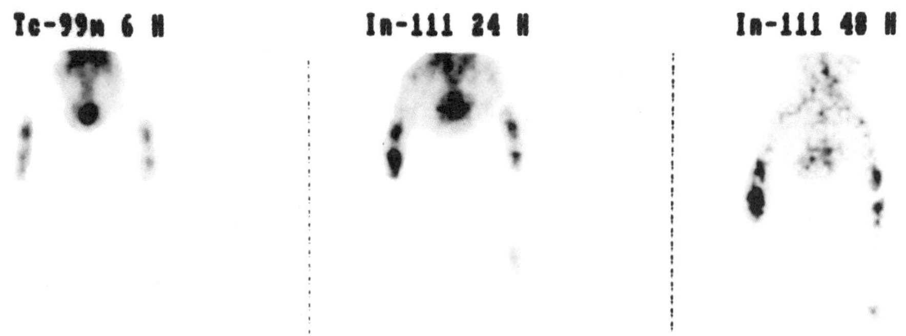

Figure 4. Radiolabeled ^{99m}Tc-PEG-liposome scan in rabbit with right tibial infection which was found to be negative in an early (6h) ^{99m}Tc image but became positive in the delayed ^{111}In images at 24 and 48 hours. Reprinted with permission from the Journal of Nuclear Medicine, 1998 (In Press).

In our laboratory, Goins et al (1993) first demonstrated the potential of ^{99m}Tc liposomes for imaging infection using a rat thigh abscess model. These images of infection were of high quality, and the activity at the infection site continued to rise throughout the study. The continued accumulation after injection is well demonstrated in figure 3 in the rat thigh abscess model. This rat thigh abscess was injected with liposomes labeled with both ^{99m}Tc and ^{111}In and continues to accumulate liposomes for 48 hours.

It is thought that the mechanism by which the liposomes accumulate in the infection is through nonspecific leakage out of the vasculature at the site of inflammation due to increased permeability of capillaries within inflamed tissue. Retention at the site of rupture could be due to the phagocytosis of liposomes by phagocytic cells that are present at the inflamed site.

Oyen, et al. (1996)have demonstrated that PEG liposomes labeled with ^{99m}Tc using the HMPAO glutathione method may also be effective for detecting sites of infection in both a rat thigh abscess model and in a lung infection model. We have also shown that ^{99m}Tc-PEG-liposomes are useful for detecting osteomyelitis, as demonstrated in figure 4. In this work we used liposomes labeled with both ^{99m}Tc and ^{111}In. The dual labeled liposomes produced good early images yet permitted prolonged imaging. Using both, long circulating PEG- liposomes and an isotope of ^{111}In with a long half-life, was helpful for imaging the slower, more chronic osteomyelitis infection which appears to require a longer period for liposomes to accumulate.

Inflammation

^{99m}Tc-liposomes were shown to be useful in determining localization and extent of inflammation. In a rabbit model of colitis, ^{99m}Tc-PEG-liposomes (2.5 mole% PEG-lipid) have visualized the segments of inflamed colon as well as or better than the current gold standard, ^{99m}Tc labeled white blood cells (Oyen et al,1997). The degree of localization correlated with the amount of inflammation in each segment of colon varying from normal to inflamed to ulcerated. ^{99m}Tc PEG-liposomes have also been shown to be useful for imaging arthritis

(Boerman et al,1997b). Liposomes effectively accumulated in an experimental model of arthritis in rats induced by *Mycobacterium butyricum.*

Blood Pool Imaging

^{99m}Tc-PEG-liposomes have been used as blood imaging agents (Goins et al,1996; Tilcock et al,1994). ^{99m}Tc labeled red cells are commonly used as blood-pool imaging agents to study the motion of the heart and determine the percentage of blood ejected. In our laboratory, we studied several different mole percentages of PEG-lipid and determined that 5 mole percent of PEG-lipid resulted in a blood-pool with the longest half-life (36 hours). The ^{99m}Tc-PEG-liposomes gave exceedingly high quality images of the heart that currently cannot be obtained without handling red blood cells (Goins et al,1996); obviously, the handling of blood is a great inconvenience and always carries the risk of contamination. Because of this risk, a lower quality technique is most commonly used to label red cells *in vivo* for cardiac nuclear imaging . Often, the quality of the images suffers as a significant number of patients are taking drugs that interfere with the labeling of red blood cells with ^{99m}Tc.

In addition to heart imaging, ^{99m}Tc labeled PEG liposomes appear to have great advantages for detecting sites of gastrointestinal bleeding. Currently, labeled red cells are used for this purpose, however, the ^{99m}Tc is not strongly bound to the red cell and dissociates resulting in clearance of the free ^{99m}Tc through the kidney and into the bladder. This disassociation greatly interferes with the detection of sites of bleeding in the lower abdomen. Because ^{99m}Tc PEG liposomes labeled with the HMPAO glutathione technique are very stable *in vivo* with virtually no dissociation of ^{99m}Tc, these ^{99m}Tc liposomes would appear to have much promise for the detection of gastrointestinal bleeding.

We have used the information gained from the development of this long circulating blood pool imaging agent and applied it to the development of a long circulating liposome encapsulated hemoglobin (LEH) red cell substitute. The same 10 mole percent PEG-lipid

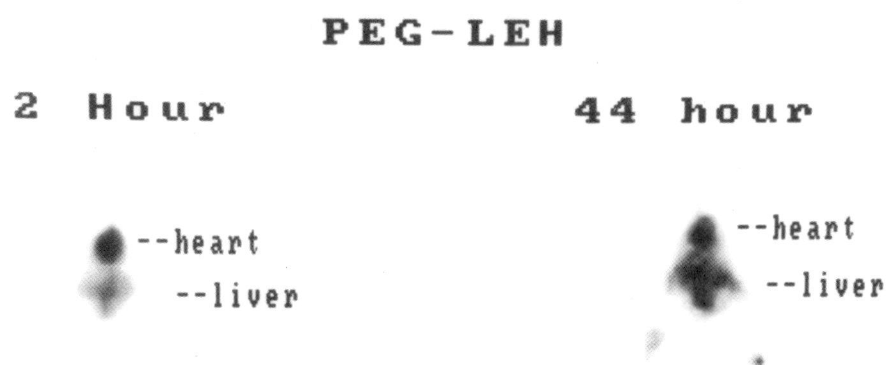

Figure 5. Images at 1 hour (left) and 44 hours (right) after injection of a 25% topload of ^{99m}Tc-PEG-Liposome encapsulated hemoglobin. The blood pool activity continues to be well visualized in the heart at 44 hours, indicating the long circulation time of this PEG-liposome based blood substitute.

formulation was incorporated into LEH, which resulted in a PEG-LEH formulation that had a circulation time that was greater than 3 times longer than the previous LEH formulation (65 hours half-life in circulation for PEG-LEH vs 18 hours for unmodified LEH). Retention of activity in the heart (which is the visual representative of the blood pool) can be clearly seen at 48 hours in figure 5. Development of longer circulating red cell substitutes has been a long sought-after goal in blood substitute research.

Tumor Imaging

The first imaging studies that were performed with liposomes described their potential as diagnostic agents for detecting tumors (Richardson et al,1978). In clinical studies, tumors were detected with ^{111}In labeled liposomes in 22 out of 24 participants (Presant et al,1990). These studies were generally performed with conventional liposomes. The use of ^{99m}Tc-PEG liposomes, however, appears to have some significant advantages for the detection of tumors. We have found excellent images with delivery of > 15% of the dose of 99mTc-liposomes to a rat model of prostate cancer induced with met-Ly-Leu prostate tumor cells. The high uptake in the tumor can be seen in the thigh in figure 6; greater than 15% of the infected dose accumulates in the tumor with the remaining portion accumulating in the liver and spleen. These studies are consistent with previous reports PEG-lipsomes having greatly increased drug delivery compared to unmodified liposomes (Gabizon et al,1990).

Lymphoscintigraphy

Liposomes are promising agents for the delivery of imaging agents and drugs to lymph nodes one of the first reports involving the use of ^{99m}Tc liposomes for lymphoscintigraphy, due to instability of the ^{99m}Tc-labeling method, the distribution of the ^{99m}Tc failed to represent intact

24 Hours Post-Injection

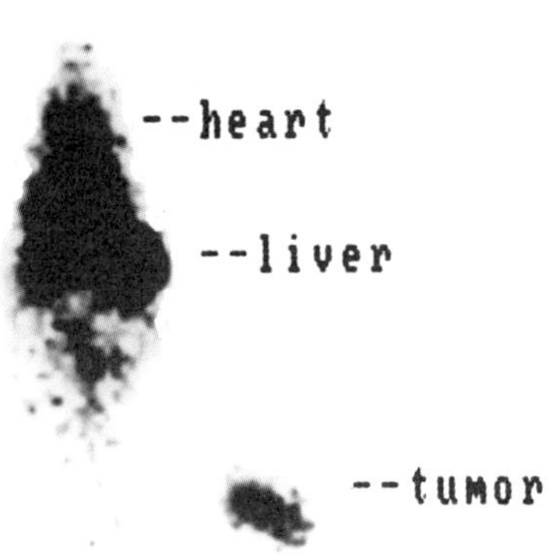

Figure 6. Images taken 24 hour after injection of a rat with implanted prostate tumor. Greater than 10% of the injected dose is in the tumor.

liposomes in lymph nodes (Patel et al,1984). More recent research involving PEG-liposomes has found that PEG-liposomes do not accumulate in lymph node at increased levels compared to unmodified liposomes (Oussoren and Storm,1997; Velinova et al,1996).

We have also used ^{99m}Tc-liposomes labeled by the HMPAO glutathione technique to determine the effect of PEG on lymphoscintigraphy. Our research demonstrates that ^{99m}Tc-PEG-liposomes had a slightly greater movement from the initial site of injection as compared to unmodified neutral liposomes. Eventually greater than 50% of the ^{99m}Tc-PEG liposomes left the subcutaneous site of injection in the foot. The addition of PEG, however, reduced the retention of liposomes in the initially encountered lymph node. From the point of view of diagnostic imaging, this decreased retention in the first lymph node has advantages and disadvantages; it is an advantage for determining if the lymphatic system is functioning normally and not obstructed, it is a disadvantage for determining the location of the first lymph node encountered near a tumor. This first lymph node is known as the sentinel lymph node and it has important diagnostic and therapeutic aspects as it is the usual location of the first metastasis from a cancer. This lack of retention of ^{99m}Tc-PEG-liposomes can be seen in figure 7. It is apparent that the ^{99m}Tc-PEG-liposomes which initially accumulate in this first popliteal lymph node at 30 minutes are able to pass through this node with passive leg movement administered after 30 minutes. Neutral liposomes, not modified with PEG, had somewhat improved retention in this initial popliteal lymph node. This retention of neutral liposomes is probably related to increased recognition and uptake by phagocytic cells in the lymph node.

^{99m}Tc-Liposomes as Tools For Determination of the Distribution of Liposome Encapsulated Drugs

Development of drugs: As can be seen from the previous studies, ^{99m}Tc-labeled liposomes can be useful tools for the determination of liposome drug delivery and in the development of liposome surface modification techniques to increase targeting to a particular site in the body. The same experimental animal can be studied at multiple time points so that more information is gained from each study. A good estimate can be determined for organ uptake in the liver, spleen, kidneys and muscle and brain after correcting for photon attenuation and the amount of activity remaining in the blood pool. The correction of blood pool activity is best done by correcting activity removed from the blood samples and counting them in a scintillation well counter.

Diagnostic predictor of the effectiveness of therapy: The use of ^{99m}Tc-liposomes has much potential for use in patient management. The non-invasive imaging prior to liposomal drug will allow for a very accurate estimate of the delivery of a liposome encapsulated drug to a particular targeted region such as infection or tumor prior to the commencement of liposomal drug therapy. This feature could be very useful when considering the ease of performing whole body imaging with ^{99m}Tc and the possibility of delivering therapeutic liposomes to an undesirable location in the body. This is particularly important considering the previously demonstrated affinity of liposomes for inflammatory sites in the body. It may be important to know that a high amount of potentially toxic anticancer drug will not be delivered to a normal, albeit, inflamed location of the body. Imaging with ^{99m}Tc-liposomes prior to therapy would prevent this occurrence.

FUTURE DEVELOPMENTS

^{99m}Tc-labeled liposomes can be an important tool in the development of more specifically targeted liposomal drugs in humans. The modification of liposomes with monoclonal antibody

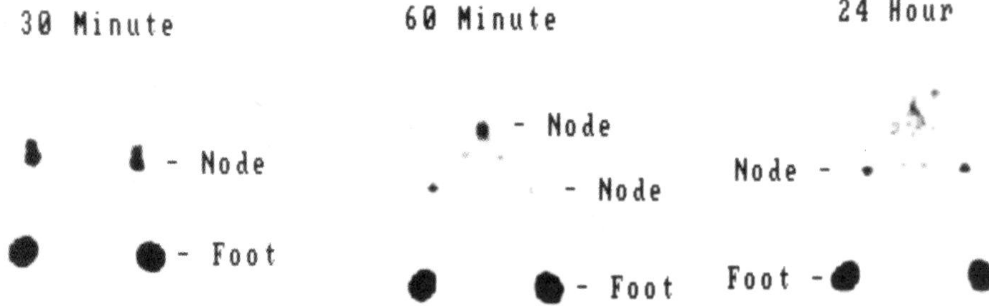

Figure 7. Lymphoscintigraphic images in a rabbit taken 30 minutes after subcutaneous injection in the foot with ^{99m}Tc labeled PEG-liposomes (left image); at 1 hour after passive leg movement for 3 minutes between the 30 minutes and 1 hour (center image) and at 20 hours (right image). Although the ^{99m}Tc-PEG-liposomes had good initial distribution to the popliteal lymph node at 30 minutes, most of the liposomes had passed through the popliteal lymph node after passive leg movement as visualized on the 1 and 20 hour images.

fragments or peptide receptors would appear to have great potential especially when the target is a site of the body for which liposomes have ready access. These readily accessible targets would include those targets in the vasculative which could provide important diagnostic information on potentially valuable therapeutic targets. Examples of these would be myocardial or cerebral infarcts detection for diagnosis or therapy. Radiolabeled antimyosin monoclonal antibodies have already been developed for detecting myocardial infarction. However, the time required for clearance of the background in the blood requires 24 hours after injection before the study can be acquired and interpreted.

Liposomes with targeting receptors would have ready access to these infarcted areas and the circulation kinetics of the liposome can be manipulated in order to achieve rapid detection of infarcted areas. Torchilin, et al, (1996) have already demonstrated accumulation of antibody and PEGcoated liposomes in infarcted myocardium. The PEG modification appears to be important in order to reduce rapid recognition by the RES.

Another interesting and readily accessible liposome target for both diagnosis and therapy is that of vascular embolus and thrombus. These disease processes are important diagnostic and therapeutic targets to which liposomes may prove useful. Improved methods of delivering drugs to these intravascular sites of thrombus and embolus would appear to have great therapeutic and diagnostic possibilities.

The use of peptides as targeting moieties on the surface would appear to offer much potential as surface receptors for use in liposome targeting. Peptides do not appear to be as immunogenic as monoclonal antibodies and regulatory approval may be easier to obtain. Scintigraphic liposome imaging can be an important tool in the development of these targeted liposomal agents.

Another system to which liposomes have ready access is the lymphatic system. Lymphatic channels are the major route of tumor cell migration and liposomes, when injected subcutaneously, follow these same pathways. Liposomes modified with specific tumor targeting receptors could deliver therapeutic drugs to micrometastasis within the lymphatic system, potentially destroying tumor cells before they grow large enough to be detected. This type of targeting has been previously suggested by Moghimi et al (Moghimi and Rajabi-Siahboomi,1996).

SUMMARY

Scintigraphic imaging of liposomes has much potential for use in a multitude of diagnostic imaging applications and in the development of liposomes for targeted drug delivery.

Acknowledgments

The author would like to thank the Naval Medical Research and Development Command and the Heart, Lung and Blood Institute of the National Institute for Health for the support of this research.

REFERENCES

Awasthi, V. D., Goins, B., Klipper, R., and Phillips, W. T., 1998. Dual radiolabeled liposomes: Biodistribution studies and localization of focal sites of infection in rats. *Nuclear Medicine and Biology*. 25:155.

Boerman, O. C., Oyen, W. J. G., Bloois, L. v., Koenders, E. B., Meer, J. W. M. v. d., and Corstens, F. H. M., 1997a. Optimization of technetium-99m-labeled PEG liposomes to image focal infection: Effects of particle size and circulation time. *Journal of Nuclear Medicine*. 38:489.

Boerman, O. C., Oyen, W. J. G., Storm, G., Corvo, M. L., Boois, L. v., Meer, J. W. M. v. d., and Corstens, F. H. M., 1997b. Technetium-99m labelled liposomes to image experimental arthritis. *Ann. Rheum. Dis.* 56:369.

Gabizon, A., Huberty, J., Straubinger, R. M., Price, D. C., and Papahadjpoulos, D., 1989. An improved method for in vivo tracing and imaging of liposomes using gallium-67-desferoxamine complex. *J. Liposome Res.* 1:123.

Gabizon, A., Price, D. C., Huberty, J., Bresalier, R. S., and Papahadjopoulos, D., 1990. Effect of liposome composition and other factors on the targeting of liposomes to experimental tumors: biodistribution and imaging studies. *Cancer Research*. 50:6371.

Goins, B., Klipper, R., Cliff, R. O., Blumhardt, R., and Phillips, W. T., 1993. Biodistribution and imaging studies of technetium-99m-labeled liposomes in rats with focal infection. *Journal of Nuclear Medicine*. 34:2160.

Goins, B., Phillips, and W. T., Klipper, R., 1996. Blood-pool imaging using technetium-99m-labeled liposomes. *Journal of Nuclear Medicine*. 37:1374.

Hoffman, T., 1990. Anticipating, recognizing, and preventing hazards associated with *in vivo* use of monoclonal antibodies: special considerations related to human anti-mouse antibodies. *Cancer Research*. 50:1049S.

Kuus-Reichel, K., Grauer, L. S., Karavodin, L. M., Knott, C., Drusemeier, M., and Kay, N. E., 1994. Will immunogenicity limit the use, efficacy, and future development of therapeutic monoclonal antibodies. *Clin Diagn Lab Immunol*. 1:365.

Moghimi, S. M., and Rajabi-Siahboomi, A. R., 1996. Advanced colloid-based systems for efficient delivery of drugs and diagnostic agenst to the lymphatic tissues. *Prog Biophys Molecular Biology*. 65:221.

Morgan, J. R., Williams, K. E., Davies, R. L., Leach, K., Thomason, M., and Williams, L. A. P., 1981. Localisation of experimental staphylococcal abscesses by ^{99m}Tc-technetium-labelled liposomes. *Journal of Medical Microbiology*. 14:213.

Neirinckx, R. D., Burke, J. F., Harrison, R. C., Foster, A. M., Andersen, A. R., and Lassen, N. A., 1988. The retention mechanism of technetium-99m-HM-PAO: Intracellular reaction with glutathione. *J. Cerebral Blood Flow and Metabolism*. *: S4-S12.

Ogihara-Umeda, I., Sasaki, t., Kojima, S., and Nishigori, H., 1996. Optimal radiolabeled liposomes for tumor imaging. *Journal of Nuclear Medicine*. 37:326.

Ogihara-Umeda, I., Sasaki, T., Toyama, H., Oda, K., Senda, M., and Nishigori, H., 1997. Rapid diagnostic imaging of cancer using radiolabeled liposomes. *Cancer Detection and Prevention*. 21:490.

Olsen, J. O., Pozderac, R. V., Hinkle, G., Hill, T., O'Dorisio, T. M., Schirmer, W. J., Ellison, E. C., and O'Dorisio, M. S., 1995. Somatostatin receptor imaging of neuroendocrine tumors with indium-111 pentetreotide (Octreoscan). *Seminars in Nuclear Medicine*. 3:251.

Oussoren, C., and Storm, G., 1997. Lymphatic uptake and biodistribution of liposomes after subcutaneous injection: III. Influence of surface modification with poly(ethyleneglycol). *Pharmacology Research*. 14:1479.

Oyen, W. J. G., Boerman, O. C., Dams, E. T. M., Storm, G., Bloois, L. v., Koenders, E. B., Haelst, U. J. G. M.

v., Meer, J. W. M. v. d., and Corstens, F. H. M., 1997. Scintigraphic evaluation of experimental colitis in rabbits. *Journal of Nuclear Medicine.* :72P.

Oyen, W. J. G., Boerman, O. C., Storm, G., Bloois, L. V., Koenders, E. B., Crommelin, D. J. A., Meer, J. W. M. v. d., and Corstens, F. H. M., 1996. Labelled Stealth liposomes in experimental infection: An alternative to leukocyte scintigraphy? *Nuclear Medicine Communications.* 17:742.

Patel, H. M., Boodle, K. M., and Vaughan-Jones, R., 1984. Assessment of the potential uses of liposomes for lymphoscintigraphy and lymphatic drug delivery. Failure of 99m-Technetium to represent intact liposomes in lymph nodes. *Biochimica et Biophysica Acta.* 801:76.

Pauser, S., Reszka, R., Wagner, S., Wolf, K. J., Buhr, H. J., and Berger, G., 1997. Liposomes-encapsulated superparamagnetic iron oxide particles as markers in an MRI-guided search for tumor-specific drug carriers. *Anticancer Drug Des.* 12:125.

Phillips, W., and Goins, B., 1995. Targeted delivery of imaging agents by liposomes. In: Handbook of Targeted Delivery of Imaging Agents. V.P. Torchilin, ed., CRC Press, Boca Raton.

Phillips, W. T., Rudolph, A. S., Goins, B., Timmons, J. H., Klipper, R., and Blumhardt, R., 1992. A simple method for producing a technetium-99m-labeled liposome which is stable in vivo. *Nucl. Med. Biol.* 19:539.

Presant, C. A., Ksionski, G., and Crossley, R., 1990. ^{111}In-labeled liposomes for tumor imaging: clinical results of the international liposome imaging study. *J. of Liposome Research.* 1:431.

Reubi, J. C., 1997. Regulatory peptide receptors as molecular targets for cancer diagnosis and therapy. *Q Journal Nuclear Medicine.* 41:63.

Richardson, V. J., Jeyasingh, K., Jewkes, R. F., Ryman, B. E., and Tattersall, M. H. N., 1978. Possible tumor localization of Tc-99m-labeled liposomes: effects of lipid composition, charge, and liposome size. *Journal of Nuclear Medicine.* 19:1049.

Rudolph, A. S., Klipper, R. W., Goins, B., and Phillips, W. T., 1991. In vivo biodistribution of a radiolabeled blood substitute: ^{99m}Tc-labeled liposome encapsulated hemoglobin in an anesthetized rabbit. *Proc Natl Acad Sci USA.* 88:10976.

Seltzer, S. E., 1989. The role of liposomes in diagnostic imaging. *Radiology.* 171:19.

Tilcock, C., Yap, M., Szucs, M., Utkhede, D., 1994. PEG-coated lipid vesicles with encapsulated technetium-99mas blood pool agents for nuclear medicine. *Nuclear Medicine and Biology.* 21:165.

Torchilin, V. P., 1995. Handbook of Targeted Delivery of Imaging Agents. CRC Press, Boca Raton

Torchilin, V. P., Narula, J., Halpern, E., and Khaw, B. A., 1996. Poly(ethylene glycol)-coated anti-cardiac myosin immunoliposomes: Factors influencing targeted accumulation in the infarcted myocardium. *Biochim Biophys Acta.* 1279:75.

Trubetskoy, V. S., Cannillo, J. A., Milshtein, A., Wolf, G. L., and Torchilin, V. P., 1995. Controlled delivery of Gd-containing liposomes to lymph nodes: surface modification may enhance MRI contrast properties. *Magnetic Resonance Imaging.* 13:31.

Unger, E. C., Lund, P. J., Shen, D. K., Fritz, T. A., Yellowhair, D., and New, T. E., 1992. Nitrogen-filled liposomes as a vascular US contrast agent: preliminary evaluation. *Radiology.* 185:453.

Velinova, M., Read, N., Kirby, C., and Gregoriadis, G., 1996. Morphological observations on the fate of liposomes in the regional lymph nodes after footpad injection into rats. *Biochim Biophys Acta.* 1299:207.

Winzelberg, G. G., Grossman, S. J., Rizk, S., Joyce, J. M., Hill, J. B., Atkinson, D. P., Sudina, K., Anderson, K., McElwain, D., and Jones, A. M., 1992. Indium-111 monoclonal antibody B72.3 scintigraphy in colorectal cancer. Correlation with computed tomography, surgery, histopathology, immunohistology, and human immune response. 69:1656.

Wolf, G. L., 1995. Targeted delivery of imaging agents: An overview. In: Handbook of Targeted Delivery of Imaging Agents. V.P.Torchilin, ed., CRC Press, Boca Raton.

Woodle, M. C., 1993. ^{67}Gallium-labeled liposomes with prolonged circulation: preparation and potential as nuclear imaging agents. Nuclear Medicine and Biology. 20: 149-155.

DIAGNOSTIC AND THERAPEUTIC TARGETING OF INFECTIOUS AND INFLAMMATORY DISEASES USING STERICALLY STABILIZED LIPOSOMES

G. Storm[1], I.A.J.M. Bakker-Woudenberg[2], R.M. Schiffelers[1,2],
W.J.G. Oyen[3], D.J.A. Crommelin[1], F.H.M. Corstens[3], O.C. Boerman[3]

[1]Dept of Pharmaceutics, Utrecht Inst. for Pharmaceutical Sciences,
Utrecht University, Utrecht, The Netherlands
[2]Dept of Clinical Microbiology, Erasmus University, Rotterdam, The Netherlands
[3]Dept of Nuclear Medicine, University Hospital Nijmegen, The Netherlands

INTRODUCTION

The ability to selectively target molecules to specific sites within the body has been one of the most coveted goals in experimental and clinical therapeutics. The performance of many diagnostic and therapeutic agents, both existing and promising new ones (like biomacromolecules such as proteins and DNA), would benefit greatly from targeted delivery strategies. Perhaps the major reason for the resurgence of interest in liposomal delivery systems for targeted delivery relates to the development of long-circulating "stealth" liposomes. As discussed elsewhere in this volume, the principal results have been obtained by modification of the liposomal surface through the use of the hydrophilic polymer polyethylene glycol conjugated to the lipid phosphatidylethanolamine (PEG-PE). This contribution will discuss the utility of PEG-coated liposomes (often referred to as sterically stabilized liposomes) for the imaging and therapy of infectious and inflammatory diseases.

IMAGING OF INFECTION AND INFLAMMATION

Accurate and rapid detection of infection and inflammatory lesions facilitates both elucidation of the course of the disease as well as installation of a tailored therapeutic regimen. Imaging modalities include ultrasonography, X-ray computed tomography, magnetic resonance imaging, and scintigraphy. However, using the former three techniques, it is often not possible to differentiate between active infection or inflammation and residual structural abnormalities persisting after cure or resulting from local, postoperative changes. Scintigraphic techniques do not depend upon the local anatomic situation but are based on local accumulation of radiolabeled compounds resulting from local 'physiochemical' processes and changes therein. The ability to perform whole body imaging without inconvenience to the patient is an important additional value of nuclear medicine techniques. When the site and source of suspected infection or inflammation are unknown or the presence of additional foci is to be verified, whole body imaging is particularly advantageous. However, all currently available and commonly used scintigraphic agents have various limitations (Boerman et al., 1997). Therefore, the development of radiopharmaceuticals for the noninvasive detection of infectious and inflammatory sites remains an active field of research.

During the past two decades several radiopharmaceuticals have been developed for infection and inflammation imaging (Gallium-67 citrate, radiolabeled leukocytes, ^{111}Indium-IgG). Each of the currently available scintigraphic agents for infection and inflammation imaging has its particular disadvantages, stimulating the search for new radiopharmaceuticals. Gallium-67-citrate has been used ever since its discovery in 1971, especially for evaluation of abnormalities of the lungs and the motoric system (Lavender et al., 1971; Ito et al., 1971). Scanning of the abdomen is difficult due to the high physiological uptake in the GI and urinary tract, specificity is low and its radiation burden is relatively high. In addition, due to its slow background clearance in most cases, a final diagnosis can only be made based on the images made 48 h after injection of ^{67}Ga-citrate. The use of radiolabeled autologous leukocytes for imaging infection and inflammation has been a major step forward (McAfee and Thakur, 1976; Peters et al., 1986). This approach exploits the ability of leukocytes to migrate towards the site of infection and/or inflammation (chemotaxis). In acute infection, high quality images of infectious foci can be obtained (Froelich and Swanson, 1984; Baba et al., 1990). In chronic infection, the sensitivity of labeled leukocyte scintigraphy appears to be limited probably due to the relatively low influx of granulocytes in this type of infection (Coleman, 1982). In addition, this technique is laborious, expensive and requires manipulation of potentially (HBV- or HIV-) contaminated blood.

Recently, ^{111}In-labeled human polyclonal immunoglobulin G (^{111}In-IgG) has been introduced as an infection and inflammation imaging agent. ^{111}In-IgG accumulates nonspecifically in infectious and inflammatory foci and can be administered safely without any side-effects (Rubin et al., 1989). Good results have been obtained in several groups of patients, including patients with infections in the locomotor system, patients with abdominal infections, patients with diabetic foot infections or inflammatory bowel disease, febrile granulocytopenic patients, and patients with fever of unknown origin (Oyen et al., 1991,1992a-d). Limitations of ^{111}In-IgG are (i) inability to image vascular lesions due to its slow blood clearance, (ii) moderate sensitivity in particular patient groups, (iii) relatively long time span between injection and final diagnosis (± 48 h) and (iv) the use of ^{111}Indium as a gamma emitter with its relatively high radiation burden. Recently, human polyclonal IgG labeled with ^{99m}Tc has been developed for infection and inflammation imaging (Blok et al., 1990; Buscombe et al., 1992). However, stable labeling of IgG with ^{99m}Tc is difficult to achieve, resulting in loss of the radiolabel *in vivo* and subsequent accumulation of the released label in nontarget organs especially the kidneys (Claessens et al., 1996).

PEG-Liposomes labeled with ^{111}Indium

Our group evaluated radiolabeled PEG-liposomes for infection imaging. By encapsulating desferal in PEG-liposomes, these liposomes could be remotely labeled with ^{111}In-oxine with high efficiency (>90%). The biodistribution of ^{111}In-labeled PEG-liposomes (mean diameter approximately 100 nm) was studied in rats with focal *Staphylococcus aureus* infection using ^{111}In-IgG as a reference agent (Boerman et al., 1995). The circulatory half-life of the ^{111}In-labeled PEG-liposomes in this model ($t1/2 = 20$ h) was similar to that of ^{111}In-IgG. However, the uptake of the ^{111}In-labeled liposomes in the abscess was more than twice as high as the abscess uptake obtained with ^{111}In-IgG (2.7 %ID/g vs. 1.1 %ID/g at 48 h post injection). Besides high uptake in the abscess, marked localization was also observed in the spleen (16 %ID/g at 48 h post injection). Uptake in

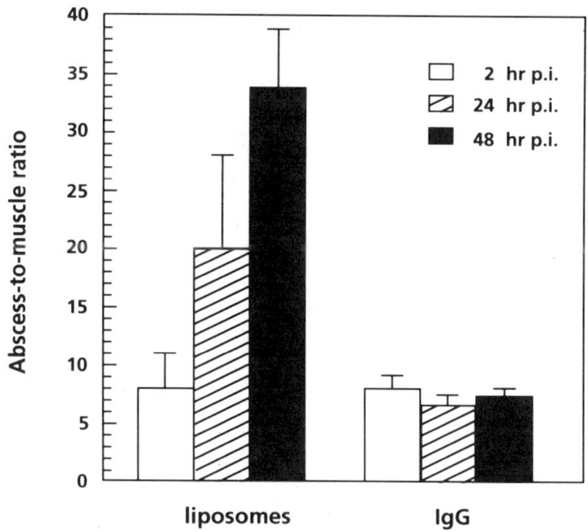

Figure 1. Abscess-to-muscle ratios at 2, 24 and 48 h post injection for ^{111}In-labeled PEG-liposomes and ^{111}In-IgG in rats with *S. aureus* infection. The biodistribution data of five rats per time point were used. Error bars represent standard deviations.

the other nontarget organs was relatively low and decreased with time. Consequently, target-to-background ratios steadily increased during the 48 h following i.v. administration. Based on tissue counting, abscess-to-muscle ratios were approximately four times as high as those obtained with ^{111}In-IgG in this model (34 versus 8 at 48 h post injection) (Fig. 1).

The distribution of ^{111}In-labeled PEG-liposomes was also monitored by gamma camera imaging. With ^{111}In-liposomes, the abscess was well delineated as early as 1 h post injection. In terms of contrast between the infection and the background, the images improved with time. From the biodistribution data based on gamma counting of dissected tissues as well as from imaging the rats on the gamma camera, it was clear that the ^{111}In-PEG-liposomes had favorable imaging characteristics as compared to ^{111}In-IgG. One of the most striking beneficial characteristics of the ^{111}In-labeled PEG-liposomes in this model was the almost complete absence of kidney uptake (< 1.0 % of injected dose/g). With ^{111}In-IgG, kidney activity exceeds 4.0 % of injected dose/g throughout the 48 h time interval. In patients, abdominal abscess localization with ^{111} In-IgG is hampered by the relatively high uptake in the kidneys (Oyen et al., 1991). The abscess-to-blood ratios obtained with ^{111} In-liposomes were also much higher than those obtained with ^{111}In-IgG in this model (3.3 vs. 1.0 at 48 h). Based on this favorable characteristic one might anticipate that PEG-liposomes (as with the radiolabeled autologous leukocytes) may be used to image vascular lesions.

PEG-Liposomes labeled with Technetium-99m

Because of its low costs, good availability, attractive physical characteristics and low radiation burden, technetium-99m is by far the most ideal radionuclide for scintigraphy. Several methods have been described to label liposomes with ^{99m}Tc. Phillips et al. (1992) have labeled liposomes with encapsulated glutathione similarly to the widely applied leukocyte labeling method, using the lipophilic chelator HMPAO for the transport of the ^{99m}Tc through the lipid bilayer. The ^{99m}Tc-HMPAO complex is reduced by glutathione to its more hydrophilic form and thus trapped within the liposome. We have studied the imaging potential of ^{99m}Tc-labeled PEG-liposomes (100 nm) in several models of infection and inflammation (Oyen et al., 1996). The images obtained in rats with *S. aureus* infection are shown in Fig. 2.

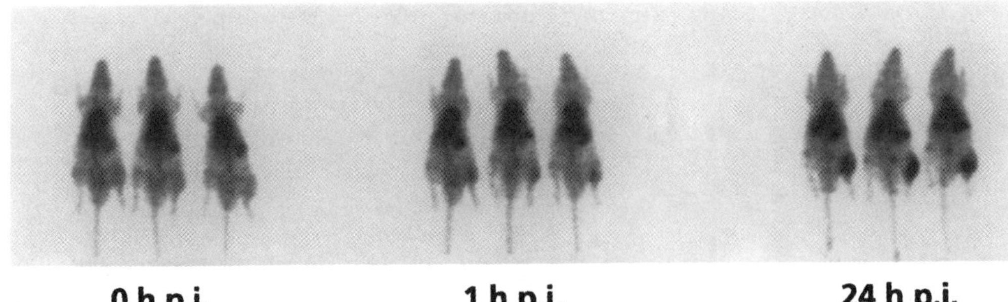

Figure 2. Scintigrams of rats with unilateral *S. aureus* infection in calf muscle imaged before and 1 and 24 h after injection of ^{99m}Tc-labeled PEG-liposomes.

High absolute abscess uptake (up to 1.0 % of injected dose/g) and abscess-to-muscle ratios (up to 42, at 24 h post injection) were observed. Uptake in background tissues was relatively low, except for the spleen (approximately 12.5 % of injected dose/g). Excretion of radioactivity from the body was much higher than that observed with the ^{111}In-labeled PEG-liposomes (<5 % of injected dose for ^{111}In-liposomes versus 23 % of injected dose for ^{99m}Tc-liposomes over 24 hours). The enhanced whole-body clearance of the ^{99m}Tc label may be either due to the reduced stability of the ^{99m}Tc-labeled liposomes *in vivo*, or to association of the ^{111}In-label with metal binding proteins following its release from the liposomes.

The ^{99m}Tc-labeled PEG-liposomes were also tested in a rat model of sterile inflammation. An inflammatory focus in the left calf muscle was induced by intramuscular injection of 0.1 ml turpentine (Fig. 3).This type of inflammatory focus was also clearly visualized, indicating that the presence of microorganisms is not required for accumulation of the liposomes in the inflammatory focus.

Target uptake and retention of the ^{99m}Tc-labeled PEG-liposomes appeared to correlate with the intensity of inflammatory response. *E. coli* infection (with a milder inflammatory response at visual inspection and palpation compared to *S. aureus* infection) showed abscess uptake of 0.6

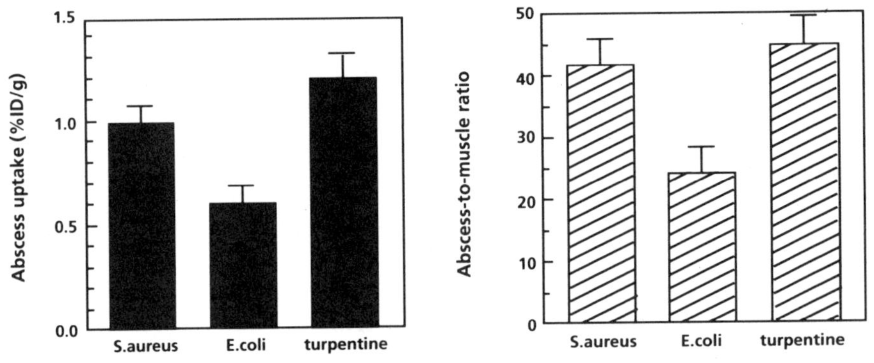

Figure 3. Abscess uptake (left panel) and abscess-to-muscle ratios (right panel) at 24 h post injection of ^{99m}Tc-labeled PEG-liposomes. The biodistribution data of five rats per time point were used. Error bars represent standard errors of the mean.

% of injected dose/g and abscess-to-muscle ratios of 25 at 24 h post injection. In the case of the more severe *S. aureus* infection abscess uptake amounted to 1.0 % of injected dose/g and abscess-to-muscle ratios exceeded 40 (24 h post injection). Uptake of PEG-liposomes in turpentine-induced inflammation (1.2 % of injected dose/g) was even higher than the abscess uptake observed in the *S. aureus* infection (Fig. 3).

^{99m}Tc-liposomes compared favorably with ^{99m}Tc-IgG, a commercially available radiopharmaceutical for imaging infection and inflammation in humans (Blok et al., 1990; Buscombe et al., 1990). Both abscess uptake and abscess-to-muscle ratios were significantly higher for the labeled PEG-liposomes in a variety of animal models.

The success of scintigraphic imaging of infectious or inflammatory foci, apart from considerations such as availability, costs, toxicity and ease of preparation, largely depends upon two factors. First, there needs to be sufficient absolute uptake in the target (the signal) and secondly, there should be a good target to background ratio (contrast). In a subsequent study we investigated whether the formulation of the PEG-liposomes could be tailored for this application (Boerman et al., 1997). The mean diameter of the liposomes had a marked effect on their *in vivo* distribution upon i.v. administration. PEG-liposomes of 90 nm in diameter combined the two favorable characteristics for imaging: optimal uptake in the infectious focus and minimal splenic accumulation. This preparation not only revealed high absolute uptake of the radiolabel in the target (1.6 ± 0.4 % of injected dose/g, 24 h post injection), but also high abscess-to-contralateral muscle ratios (34.8 ± 8.4). Most importantly, the 90 nm liposomes showed marked reduced splenic uptake (6.9 ± 0.7 % of injected dose/g) as compared to the larger liposomes (Fig. 4).

It has been reported that somewhat larger long-circulating liposomes tend to accumulate in the spleen (Storm et al., 1995; Boerman et al., 1995; Oyen et al., 1996). The relatively high and rapid splenic uptake of larger PEG-liposomes is most likely due to physical filtration rather than phagocytosis by spleen macrophages. The effect of reduced circulation time of PEG-liposomes was investigated by including increasing amounts of phosphatidylserine (PS) in the lipid bilayer (Boerman et al., 1997). In the rat model, liposomes with 1.0 mole% PS in the bilayer showed a

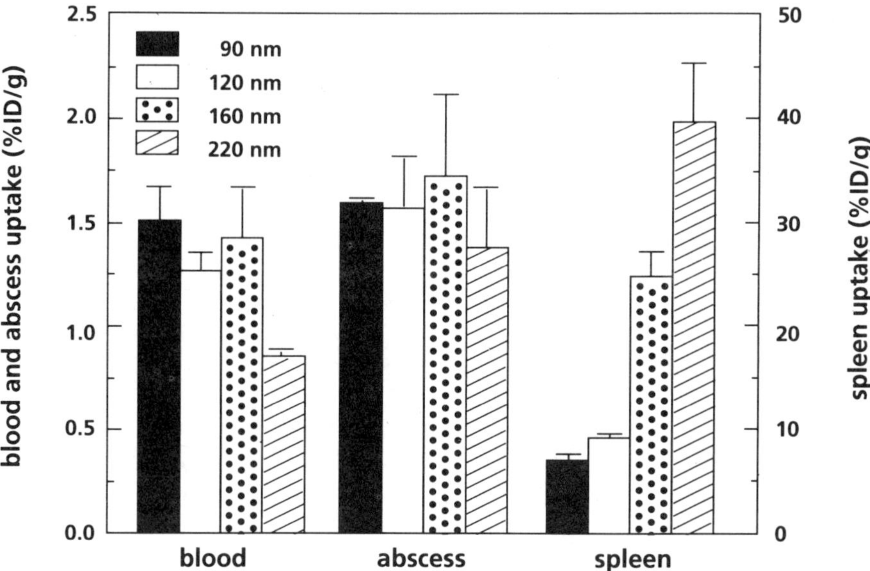

Figure 4. Effect of liposome diameter on the blood level, abscess uptake and spleen uptake of ^{99m}Tc-labeled PEG-liposomes in rats with *S. aureus* infection of the left calf muscle 24 h post injection. Five rats per group were used. Error bars represent standard deviations.

three-fold higher abscess-to-blood ratio. However, while improved abscess-to-blood ratios were obtained with these PS-containing liposomes, abscess-to-background ratios for liver and spleen decreased. Overall, these studies showed that the *in vivo* behavior of PEG-liposomes can be modulated to optimize their characteristics for infection imaging; in particular, by carefully tuning their size the localization of long-circulating liposomes in the spleen could be reduced.

In terms of diagnostic accuracy, radiolabeled leukocytes are currently the best radiopharmaceutical available for imaging infection and inflammation. However, as indicated above their preparation bears some important disadvantages (blood handling, lack of quality control method, laborious and time-consuming preparation). A liposome-based radiopharmaceutical would have several advantages over radiolabeled leukocytes: (i) continuous availability, (ii) easy labeling procedure, (iii) no need to handle blood. In patient care, labeled liposomes would then be an attractive agent to replace labeled leukocytes, especially in acute infectious disease. In order to compare the *in vivo* imaging characteristics of radiolabeled liposomes with those of radiolabeled leukocytes, a comparative study was carried out in a model of infection. Because radiolabeled leukocytes cannot be used in rats, we compared the imaging potential of radiolabeled leukocytes and liposomes in rabbits with focal *S. aureus* infection (Oyen et al., 1996). The results pointed to similar *in vivo* imaging characteristics of labeled liposomes and labeled leukocytes. Therefore, in patient care, labeled liposomes may be an attractive replacement for labeled leukocytes for the reasons outlined above.

In conclusion, the promising preclinical work along with other attractive features of the PEG-liposome formulation (in terms of availability and ease of preparation) indicates that radiolabeled PEG-coated liposomes have important advantages for imaging infectious and inflammatory foci. Further validation is awaiting the results of ongoing clinical studies.

ANTIBACTERIAL THERAPY

Many of the antibiotics in use today were discovered more than 25 years ago. Finding or designing new antibiotics is a laborious process involving the screening of thousands of soil samples for new antibiotic-producing organisms or the testing of hundreds of analogs of known antibiotics for new properties. Although efforts in these directions continue, alternative approaches are also being developed. One alternative is to use PEG-liposomes as carriers of existing antibiotics to obtain targeting to sites of infection beyond the mononuclear phagocyte system (MPS).

Clinical experience shows that bacterial infections frequently occur in immunocompromised patients (due to e.g. malignant disease, chemotherapy, organ transplantation, and AIDS) and constitute a major cause of morbidity and mortality. Failure of antibiotic treatment of these infections occurs despite the availability of new and potent antibiotics. Two major factors are contributing to the lack of success of antibiotic treatment in these patients:

1. One factor is an only moderate antibiotic susceptibility of the micro-organism. In such a case relatively high antibiotic concentrations at the site of infection at the earliest time possible are needed. Relevant in this respect is the gram-negative bacterium *Pseudomonas aeruginosa* causing lung infection which proceeds to dissemination of the infection resulting in sepsis due to insufficient antibiotic treatment. Mortality associated with sepsis is high (40%). Among patients who develop the complications of 'septic shock' induced by the triggering of the production of mediators by bacterial endotoxin, mortality increases up to 80-90% (Bone, 1993). The failure of the antibiotic therapy to prevent septic shock as seen for *Pseudomonas aeruginosa* is due to the moderate antibiotic susceptibility of this bacterium along with the inability to achieve sufficient antibiotic concentrations at the infected site.

2. Another factor responsible for antibiotic treatment failure is that successful antibiotic treatment is hampered by the intracellular location of the micro-organism. Relevant in this respect is the intracellular bacterium *Mycobacterium tuberculosis* that shows a low multiplication rate

intracellularly. Eradication of the mycobacteria in the dormant state requires treatment for prolonged periods of time (1/2 - 1 year) with high doses of antibiotic to obtain a bactericidal effect. To prevent the development of antibiotic resistance which can be expected when bacteria are exposed to high antibiotic concentration for a long period, combinations of antibiotics (three or four) are applied. It is clear that antibiotic treatment of mycobacterial infections is demanding and almost unmanageable for the patient. Therapy failure due to inappropriate treatment results in outbreaks of tuberculosis disease caused by multidrug-resistant strains of *Mycobacterium tuberculosis* (Ellner et al., 1993), the 'return of an old enemy'.

Both examples of antibiotic treatment failure underline the need to achieve enhanced delivery of antibiotics to infections localized in non-MPS tissues. Below, the utility of PEG-liposomes in this regard is discussed.

We have recently demonstrated in a rat model of *Klebsiella pneumoniae* infection that intravenously injected PEG-liposomes show prolonged circulation and enhanced localization in infected lung tissue when compared to liposomes lacking PEG (Bakker-Woudenberg et al., 1992,1993). In this model of unilateral pneumonia, bacterial inoculation in the left lung results in a progressive infection of the left lung, whereas the right lung of the same animal remains uninfected. This experimental lung infection is a clinically relevant model as evidenced by the many characteristics of the infection which are also observed in human acute bacterial lung infections, such as the temperature course (subnormal values after an initial rise), infectious complications (septicemia, pleuritis), and the histological features of the pneumonic lesion. The *in vivo* fate of PEG-liposomes was followed using the ^{67}Ga-Desferal complex as aqueous liposomal marker. It was found that the circulation half-life of 80 nm PEG-liposomes was about 20 h in uninfected rats. The prolonged circulation time was accompanied by a relatively low hepatosplenic uptake. The degree of localization of the PEG-liposomes in the infected left lung was significantly higher compared with the localization of a 'conventional' liposome type which exhibited a relatively short blood circulation time. At 40 h after i.v. administration, i.e. at the time when about 10% of the PEG-liposomes were still circulating in the bloodstream, up to 9% of the injected dose of PEG-liposomes could be recovered from the infected lung tissue. The increased localization of liposomes in the infected left lung is related to the local infectious process as in the uninfected right lung of the same rat, the degree of localization of liposomes was not increased compared to the localization observed in uninfected rats. The increased localization of PEG-liposomes in the infected lung appeared to be highly correlated with the intensity of the infection, and was dependent on the liposomal lipid dose administered.

To study the effect of particle size on the degree of liposome localization in the infected lung, three different PEG-liposome dispersions each with a different mean size (i.e. 100, 280, and 360 nm) were prepared and injected i.v. at 40 h after bacterial inoculation at a lipid dose of 75 μmol/kg. The 280 nm and 360 nm liposome preparations gave similar values for the area under the blood-concentration time curve ($AUC_{0\text{-}24h}$) but these values were substantially lower (about 1.8-fold) as compared to the $AUC_{0\text{-}24h}$ obtained after administration of the 100 nm liposomes. The reduced circulation time of the two preparations with larger mean size was parallelled by an increased MPS uptake, in particular by the spleen. Furthermore, at 24 h after injection, the 100 nm liposomes localized in the infected lung to an approximately 2.3-fold higher degree as compared both preparations of larger mean size which showed a similar degree of localization. In an additional experiment the $AUC_{0\text{-}24h}$ of the 100 nm liposomes was reduced by about 2-fold by a 2-fold reduction of the lipid dose (i.e., the lipid dose was 37.5 mmol/kg). This resulted in a 2-fold decreased target localization as compared with that obtained at a lipid dose of 75 mmol/kg but similar to those obtained after injection of the 280 nm and 360 nm liposomes at the 75 μmol/kg dose. Altogether, these findings indicate the AUC rather than the absolute particle size is the major factor determining the degree of accumulation in the infected lung.

Table 1. Gentamicin-containing PEG-liposomes in *Klebsiella pneumoniae* lung infection

Treatment	*Day 2*		*Day 4*		*Day 7*	
	Surv (%)	Log no. bacteria in left lung of surviving rats (n=10)	Surv (%)	Log no. bacteria in left lung of surviving rats (n=10)	Surv (%)	Log no. bacteria in left lung of surviving rats (n=15)
PBS	100	ND	30	ND	0	ND
Free gentamicin	100	8.3±0.2	90	8.7±0.5	47	9.5±0.3
Free gentamicin mixed with empty PEG-liposomes	100	ND	93	ND	47	9.9±0.3
Gentamicin in PEG-liposomes	100	8.6±0.2	100	7.6±0.2	100	5.6±0.4

Rats were injected i.v. with a single dose of 5 mg gentamicin/kg 26 hr after inoculation of the left lung (on day 0) with 10^6 cfu *K. pneumoniae* in the logarithmic phase of growth. PEG-liposomes were composed of PEG-DSPE distearoylphosphatidylethanolamine derivatized with polyethylene glycol), PHEPC (partially hydrogenated phosphatidylcholine), and cholesterol at a molar ratio of 0.15:1.85:1. Mean size was about 100 nm. Surv = survival, ND = not determined

The antibacterial activity of PEG-liposomes containing antibiotics was studied in the same pneumonia model (Bakker-Woudenberg et al., 1995). Results on therapeutic activity of PEG-liposomes loaded with the aminoglycoside gentamicin are shown in Table 1.

In control (PBS) rats, inoculation of the left lung with 10^6 cfu of *K. pneumoniae* resulted in a progressive infection of the left lung, whereas the right lung of the same animal did not develop an infection. Untreated rats died 4-7 days after bacterial inoculation. Administration of free gentamicin at a dose of 5 mg/kg prolonged the survival of infected rats. However, bacterial numbers in the left lung increased progressively leading to the death of the animals. In contrast, when gentamicin is encapsulated in PEG-liposomes, a strong bactericidal effect (>> 99% killing) was observed without occurrence of death. Apparently, the encapsulated drug becomes 'therapeutically available' within the infected lung. This superior efficacy is the result of liposome encapsulation: free drug in combination with empty liposomes had the same effect as free drug alone. Administration of empty liposomes did not show a therapeutic effect.

Improved therapy results were also obtained with the cefalosporin ceftazidime encapsulated in PEG-liposomes (Bakker-Woudenberg et al., 1995). Pharmacokinetic studies showed that liposomal encapsulation results in relatively high and sustained blood concentrations of liposome-bound gentamicin or ceftazidime. In addition, it was found that the encapsulated antibiotics did not leak from the long-circulating liposomes to a substantial extent. Remarkably, bactericidal activity of both liposomal antibiotics was minimal under *in vitro* conditions. From these observations we deduce that it is likely that degradation of the liposomes within the target site is needed for achieving sufficient release and thereby adequate therapeutic availability of the encapsulated antibiotic.

In conclusion, the data obtained thus far indicate that the superior antibacterial activities of gentamicin and ceftazidime when given encapsulated in PEG-liposomes are the consequence of preferential localization and local release of encapsulated antibiotic within the site of infection. The present results indicate that PEG-liposomes can have a major impact on the clinical application of antibiotics, in particular in the case of the immunocompromised patient where the site of localization as well as the etiology of the infection are often unknown.

REFERENCES

Baba, A.A., McKillop, J.H., Cuthbert, G.F., Neilson, W., Gray, H.W., and Anderson, J.R., 1990, ^{111}Indium leucocyte scintigraphy in abdominal sepsis: do the results affect management?, *Eur. J. Nucl. Med.* 16: 307.

Bakker-Woudenberg, I.A.J.M., Lokerse, A.F., ten Kate, M.T., and Storm, G., 1992, Enhanced localization of liposomes with prolonged blood circulation time in infected lung tissue, *Biochim. Biophys. Acta* 1138: 318..

Bakker-Woudenberg, I.A.J.M., Lokerse, A.F., ten Kate, M.T., Mouton, J.W., Woodle, M.C., and Storm, G., 1993, Liposomes with prolonged blood circulation and selective localization in *Klebsiella Pneumoniae*-infected lung tissue, *J. Infect. Dis.* 168: 164.

Bakker-Woudenberg, I.A.J.M., ten Kate, M.T., Stearne-Cullen, L.E.T., and Woodle, M.C., 1995, Efficacy of gentamicin or ceftazidime entrapped in liposomes with prolonged blood circulation and enhanced localization in *Klebsiella pneumoniae*-infected lung tissue, *J. Infect. Dis.* 171:938.

Blok, D., van Ogtrop, M., Arndt, J.W., Camps, J.A.J., Feitsma, R.I.J., Goedemans, W., and Pauwels, E.K.J., 1990, Detection of inflammatory lesions with radiolabelled immunoglobulins, *Eur. J. Nucl. Med.* 16: 303.

Buscombe, J.R., Lui, D., Ensing, G., de Jong, R., and Ell, P.J., 1990, ^{99m}Tc-human immunoglobulin (HIG) - first results of a new agent for the localization of infection and inflammation, *Eur. J. Nucl. Med.* 16: 649.

Boerman, O.C., Storm, G., Oyen, W.J.G., van Bloois, L., van der Meer, J.W.M., Claessens, R.A.M.J., Crommelin, D.J.A., and Corstens, F.H.M., 1995, Sterically stabilized liposomes labeled with indium-111 to image focal infection in rats, *J. Nucl. Med.* 36: 1639.

Boerman, O.C., Oyen, W.J.G., van Bloois, L., van der Meer, J.W.M., Koenders, E.B., Corstens, F.H.M., and Storm, G., 1997, Optimization of ^{99m}Tc-labeled PEG-liposomes to image focal infection: effects of particle size and circulation time, J. Nucl. Med. 38: 489.

Boerman, O.C., Oyen, W.J.G., Corstens, F.H.M., and Storm, G., 1997, Sterically stabilized liposomes to image infection and inflammation, in: *Long-Circulating Liposomes: Old Drugs, New Therapeutics*, M.C. Woodle and G. Storm, eds., Landes Bioscience Pub., Springer, Austin, Texas

Buscombe, J.R., Lui, D., Ensing, G., de Jong, R., and Ell, P.J., 1990, ^{99m}Tc-human immunoglobulin (HIG) - first results of a new agent for the localization of infection and inflammation, *Eur. J. Nucl. Med.* 16: 649.

Bone, R.C., 1993, Gram-negative sepsis: a dilemma of modern medicine, *Clin. Microbiol. Rev.* 6:57.

Claessens, R.A.M.J., Boerman, O.C., Koenders, E.B., Oyen, W.J.G., van der Meer, J.W.M., and Corstens, F.H.M., 1996, Technetium-^{99m} labelled hydrazinonicotinamido human non-specific polyclonal immunoglobulin G for detection of infectious foci: a comparison with two other technetium-labelled immunoglobulin preparations, *Eur. J. Nucl. Med.* 23, 414.

Coleman, R.E., 1982, Radiolabeled leukocytes, *Nuclear Medicine Annual,* L.M. Freeman and H.S. Weissmann, eds., New York: Raven Press, p. 119.

Ellner, J.J., Hinman, A.R., Dooley, S.W., Fischl, M.A., Sepkowitz, K.A., Goldberger, M.J., Shinnick, T.M., Iseman, M.D., and Jacobs, W.R., 1993, Tuberculosis symposium: emerging problems and promise, *J. Infect. Dis.* 168:537.

Froelich, J.W., and Swanson, D., 1984, Imaging of inflammatory processes with labeled cells, *Semin. Nucl. Med.* 14:128.

Ito, Y., Okuyama, S., Awano, T., Takahashi, K., Sato, T., and Kanno, 1971, I. Diagnostic evaluation of Ga-67 scanning of lung cancer and other diseases, *Radiology* 101: 355.

Lavender, J.P., Lowe, J., Barker, J.R., Burn, J.I., and Chaudri, M.A., 1971, Gallium-67 citrate scanning in neoplastic and inflammatory lesions, *Brit. J. Radiol.* 44: 361.

McAfee, J.G., and Thakur, M.L., 1976, Survey of radioactive agents for the *in vitro* labeling of phagocytic leucocytes. I. Soluble agents. II. Particles, *J. Nucl. Med.* 17: 480.

Oyen, W.J.G., Claessens, R.A.M.J, van der Meer, J.W.M., and Corstens, F.H.M., 1991, Detection of subacute infectious foci with indium-111-labeled autologous leukocytes and indium-111- labeled human nonspecific immunoglobulin G: a prospective comparative study, *J. Nucl. Med.* 32: 1854.

Oyen, W.J.G., Claessens, R.A.M.J., van der Meer, J.W.M., Rubin, R.H., Strauss, H.W., and Corstens, F.H.M., 1992a, ^{111}Indium-labeled human nonspecificimmunoglobulin G: A new radiopharmaceutical for imaging infectious and inflammatory foci, *Clin. Inf. Dis.* 14: 1110.

Oyen, W.J.G., van Horn, J.R., Claessens, R.A.M.J., Slooff, T.J.J.H., van der Meer, J.W.M., and Corstens, F.H.M., 1992b, Diagnosis of bone, joint and joint prothesis infections with indium-111-labeled nonspecific human immunoglobulin G scintigraphy, *Radiology* 182: 195.

Oyen, W.J.G., Claessens, R.A.M.J., Raemaekers, J.M.M., de Pauw, B.E., van der Meer, J.W.M., and Corstens, F.H.M., 1992c, Diagnosing infection in febrile granulocytopenic patients with indium-111- labeled human IgG, *J. Clin. Oncol.* 10: 61.

Oyen, W.J.G., Netten, P.M., Lemmen, A.M., Claessens, R.A.M.J., Lutterman, J.A., van der Vliet, J.A., Goris, J.A., van der Meer, J.W.M., and Corstens, F.H.M., 1992d, Evaluation of infectious diabetic foot complications with Indium-111-labeled human nonspecific immunoglobulin G, *J. Nucl .Med.* 33: 1330.

Oyen, W.J.G., Boerman, O.C., Storm, G., van Bloois, L., Koenders, E.B., Claessens, R.A.M.J., Perenboom, R.M., Crommelin, D.J.A., van der Meer, J.W.M., and Corstens, F.H.M., 1996, Detecting infection and inflammation with Tc99m-labeled Stealth liposomes, *J. Nucl. Med.* 37: 1392.

Oyen, W.J.G., Boerman, O.C., Storm, G., van Bloois, L., Koenders, E.B., Crommelin, D.J.A., van der Meer, J.W.M., and Corstens, F.H.M., 1996, Labeled Stealth^{R} liposomes in experimental infection: an alternative to leukocyte scintigraphy?, *Nuclear Med. Commun.* 17; 742.

Peters, A.M., Danpure, H.J., Osman, S., Hawker, R.J., Henderson, B.L., Hodgson, H.J., Kelly, J.D., Neirinckx, R.D., and Lavender, J.P., 1986, Clinical experience with Tc-^{99m}-HMPAO for labelling leucocytes and imaging infection, *Lancet* 2: 946.

Phillips, W.T., Rudolf, A.S., Goins, B., Timmons, J.H., Klipper, R., and Blumhardt, R., 1992, A simple method for producing a technetium-99m-labeled liposome which is stable *in vivo*, *Nucl. Med. Biol.* 19: 539.

Rubin, R.HH, Fischman, A.J., Callahan, R.J., Khaw, B.A., Keech, F., Ahmad, M., Wilkinson, R., and Strauss, H.W., 1989, In-111-labeled nonspecific immunoglobulin scanning in the detection of focal infection, *N. Eng. J. Med.* 321: 935.

Storm, G., Belliot, S.O., Daemen, T., and Lasic, D.D., 1995, Surface modification of nanoparticles to oppose uptake by the mononuclear phagocyte system, *Adv. Drug Delivery Rev.* 17: 31.

BIOLOGICALLY ACTIVE LIGAND-BEARING POLYMER-GRAFTED LIPOSOMES

Samuel Zalipsky, Joshua Gittelman, Nasreen Mullah, Masoud M. Qazen, and Jennifer A. Harding

SEQUUS Pharmaceuticals, Inc., 960 Hamilton Court, Menlo Park, California 94025

INTRODUCTION

Introduction of polymer-grafted lipid vesicles allowed to overcome some of the shortcomings of classical liposomes pertaining to their use in drug delivery, e.g. short circulating lifetimes *in vivo* with concomitant uptake by liver and spleen (Woodle, 1997). Optimal formulations of polymer-grafted liposomes contain 3-7 mole % of methoxypolyethylene glycol (MW 2000)-distearoylphosphatiyl-ethanolamine (mPEG-DSPE) in addition to various amounts of lecithin and cholesterol (Lasic and Martin, 1995).

The water-solvated, conformationally flexible, highly mobile polyethylene glycol (PEG) chains mask the surface of liposomes and thus minimize the non specific interaction with various plasma components. As a result, liposomes covalently coated with PEG chains have extended half-lives in the bloodstream (terminal $t_{1/2}$ >48 h in humans) and drastically reduced liver uptake (Gabizon et al., 1994; Woodle, 1997; Papahadjopoulos, et al., 1991). Such liposomes were named Stealth® (registered trademark of SEQUUS Pharmaceuticals, Inc.) for their ability to avoid uptake by the reticuloendothelial system (RES). In scientific literature they have often been referred to as sterically stabilized liposomes on the basis of a proposed mechanism where protein adsorption is inhibited by a steric barrier created by the surface-grafted polymer chains (Woodle and Lasic, 1992). It is pertinent to emphasize that the high chain mobility and conformational flexibility of the grafted PEG chains play the crucial role in exerting the steric stabilization effect, while the hydrophilicity of the polymer *per se* is of secondary importance (Woodle et al., 1994; Blume and Cevc, 1993; Torchilin and Papisov, 1994). Taking advantage of these properties, PEG grafting has been widely used in other systems as a method for reduction of various undesirable consequences of biological recognition manifested by immunogenicity and antigenicity in the case of proteins, and thrombogenicity, cell adherence, and protein adsorption in the case of artificial biomaterials (Harris and Zalipsky, 1997).

Targeting of Drugs 6: Strategies for Stealth Therapeutic Systems
Edited by Gregoriadis and McCormack, Plenum Press, New York, 1998

Extended blood circulation of PEG-grafted liposomes opened up numerous opportunities in ligand-mediated targeting of such vehicles. Furthermore, PEG-liposomes could be utilized as platforms on which biologically active ligands are presented in a multivalent array. Low molecular weight ligands, which in their free form are cleared from the systemic circulation rapidly, might have an opportunity to exhibit their biological activity due to their presentation on a long-circulating liposomal platform. There are a variety of biologically-relevant ligands that could be used for these applications. These include: immunoglobulins (Zalipsky et al., 1996) and their fragments (Kirpotin et al., 1996), other proteins (Blume et al., 1993), oligo- (Shimada et al., 1997; DeFrees et al., 1996; Zalipisky et al., 1997) and polysaccharides (Sato and Sunamoto, 1992), peptides (Zalipsky et al., 1997; Zalipsky et al., 1995), and vitamins (Lee and Low, 1994, Wong et al., 1996; Goren et al., 1997).

We will briefly overview herein some of our recently developed approaches to the systems containing ligands bound to the termini of liposomal surface-grafted PEG chains and comment on some of the properties of these constructs.

METHODS OF ASSEMBLY OF LIGAND-PEG-LIPOSOMES

To assemble ligand-bearing PEG-grafted liposomes a few approaches have been used. Several attempts to link ligands directly to polar head groups of lipids incorporated into mPEG-liposomes have been described in the literature. It was found that PEG chains interfere with binding of ligands and their target receptors (Zalipsky et al., 1996; Klibanov et al., 1991; Hansen et al., 1995). The highly mobile PEG chains also inhibit the conjugation reactions between reactive groups on the surface of a liposome and a ligand molecule. Due to these considerations we found it conceptually more appealing to link potential ligands to the far end of PEG chains. For this purpose we developed a number of derivatives of general formula X-PEG-DSPE (Zalipsky, 1995a,b), where X represents a reactive functional group-containing moiety, while PEG-DSPE represents a urethane-linked conjugate of distearoylphosphatidyl-ethanolamine and PEG (Table 1). Most of the end-group functionalized PEG-lipids were synthesized from a few heterobifunctional PEG derivatives containing hydroxyl and carboxyl or amino groups (Zalipsky and Barany, 1986; Zalipsky and Barany, 1990; Zalipsky et al., 1994a; Zalipsky, 1993; Zalipsky, 1995a,b). Typically the hydroxyl end-group of PEG was utilized to form a urethane attachment with the hydrophobic lipid anchor, DSPE, while the amino or carboxyl groups were utilized for conjugation or further functionalization reactions (Zalipsky, 1995a,b; Zalipsky, 1993; Allen et al., 1995; Zalipsky et al., 1994b). The nature of the hydrophobic anchor is very important for the retention of the macromolecular lipid-conjugates in the liposomal bilayer (Silvius and Zuckermann, 1993). DSPE seems to offer both, optimal bilayer retention properties as well as a primary amino group for the attachment of functionalized PEG derivatives.

We prepared a variety of ligand-bearing liposomes. Examples of various ligands attached to extremities of PEG chains on liposomes are summarized in Table 2. These formulations are currently being studied in a number of biological systems. Two major areas of applications are currently being pursued: (A) ligand-mediated targeting of liposomes and (B) presentation of low molecular weight ligands on a long circulating liposomal platform. In most cases longevity in systemic circulation could be well balanced with good preservation of biological activity.

Three general strategies have been employed to assemble ligand-bearing Stealth® liposomes. In some instances end group-functionalized PEG-lipids (structures **1 - 8**; Table 1) can be incorporated into liposomes and then conjugated to a specific ligand (Fig. 1). This approach seems more suitable for macromolecular ligands, e.g. immunoglobulins. However, it leads to a ligand-bearing liposome containing some unreacted end groups. In other cases we synthesized ligand-PEG-lipids first, then formulated these three-component conjugates into

Table 1. Summary of end-group functionalized polymer-lipids of general structure X-PEG-DSPE prepared in our laboratory. The PEG-DSPE residue in all the conjugates listed is urethane-linked and derived from PEG of average molecular weight 2000

No	X	Comments
1	H_2N-	Used for preparation of other end-group functionalized- and ligand-PEG-DSPE via amino-group modification. Forms long-circulating liposomes, behaving as positively-charged particles (Zalipsky et al.,1994).
2	HO_2C-	Useful starting material for further modification / conjugation reactions.
3	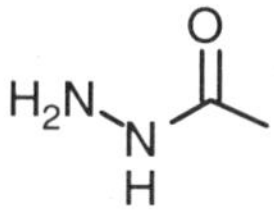 O H_2N N H	Used for conjugation of antibodies, oxidized on their carbohydrate residues, to PEG chains of liposomes. Conjugation of N-terminal Ser or Thr peptides (Zalipsky et al., 1995; Zalipsky 1993).
4	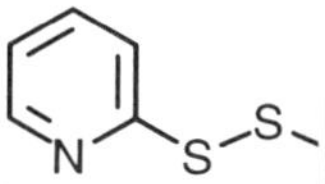 S N S	Used for binding thiol-containing ligands through disulfide linkage. Precursor for HS-PEG-DSPE, for attachment of maleimide-containing ligands to liposomes (Allen at al., 1995).
5	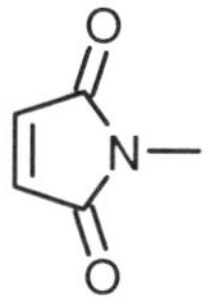 O N— O	Used for attachment of Fab' fragments (Kirpotin et al., 1996) and other free thiol, Cys-containing peptides.
6	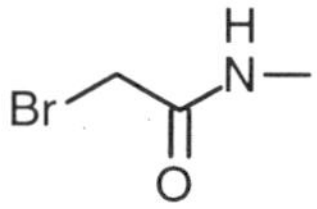 H N— Br O	Used for attachment of Fab' and other thiol-containing ligands (Zalipsky, 1995).
7	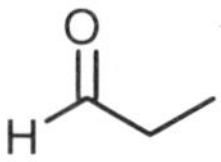 O H	Useful reactive PEG-lipid for reductive alkylation of primary amine-containing ligands.
8	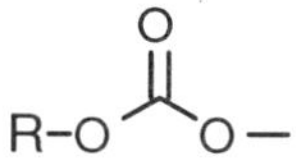 O R-O O—	Active carbonates (R = nitrophenyl or succinimidyl) are useful derivative for binding amino-containing ligands and other functional residues forming urethane attachments.

Table 2. Summary of various ligands conjugated to the end groups of liposomal surface grafted PEG chains

Ligand	Comments
Immunoglobulins and Fab' fragments	IgG specific towards various tumor epitopes were successfully linked to PEG-grafted liposomes utilizing derivatives **3** (Goren et al., 1996; Harding et al., 1997) and **4** (Allen et al., 1995) Derivatives **5** (Kirpotin et al., 1997) and **6** were utilized for attachment of Fab' fragments (see Table 1 for identification of derivatives)
Peptides	Various peptide sequences were linked to derivatives **3** - **7** (Zalipsky et al., 1997; Zalipsky et al., 1995)
Oligosaccharide	Sialyl Lewis X pentasaccharide was conjugated to derivative **4**-derived thiol. The liposomal formulation of the conjugate exhibited three orders of magnitude enhanced binding to E-selectin (DeFrees et al., 1996; Zalipsky et al., 1997)
Vitamins	Active conjugates of folic acid (Goren et al., 1997) and biotin (Wong et al., 1997) were obtained from derivative **1**. Pyridoxal and pyridoxal phosphate conjugates were prepared from derivative 3

ligand-bearing vesicles (Fig. 2). The second approach is more attractive in the case of low molecular weight ligands, peptides, oligosaccharides, and vitamins. Both methods have their advantages and drawbacks, which are important to recognize.

The first approach involves conjugation of preformed liposomes and thus does not require preparation of pure ligand-PEG-lipids. Synthesis of such conjugates is still a formidable challenge, and purification of the products often necessitates employment of tedious and labor intensive protocols. Since the first approach involves conjugation after the formation of

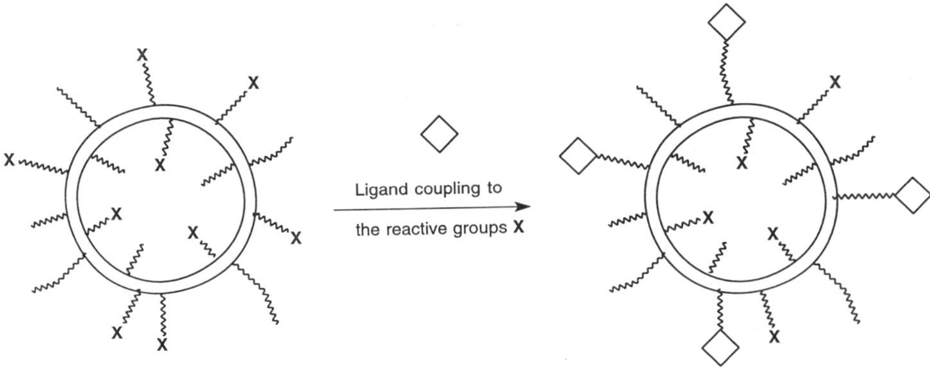

Figure 1. Ligands covalently linked to extremities of PEG-grafted liposomes: First approach. Ligands coupled to preformed liposomes containing an appropriate end group functionalized conjugate, X-PEG-lipid.

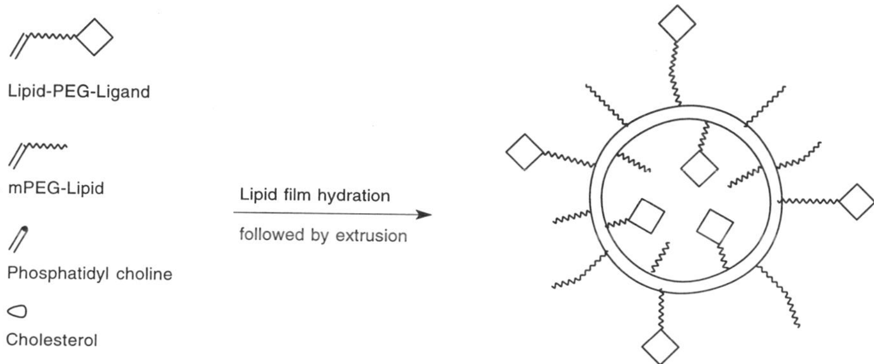

Figure 2. Ligands covalently linked to extremities of PEG-grafted liposomes: Second approach. Preformed lipid-PEG-ligand conjugates mixed with other lipid matrix components and then liposomes are made.

liposomes, some of the reactive end groups might remain on the outer surface. In the case of a protein ligand, this creates a possibility of crosslinking through the multiple attachment to a single ligand molecule. Additionally, the reactive groups at the termini of the PEG chains on the inner surface of liposomes remain unutilized (Fig. 1). The potential of the unutilized reactive functionalities on either the inner or outer monolayers to undergo side reactions with water, drug molecules, if present, or other lipid components is a source of concern. In some instances, quenching of the unreacted end groups is advisable. For example, after conjugation of Fab-thiols to maleimido-PEG-liposomes, excess of β-mercaptoethanol was used to consume the remaining maleimide residues (Kirpotin et al., 1996).

All of the above issues are circumvented by the second approach in which pure ligand-PEG-lipid is mixed with other liposomal matrix-forming components, e.g. lecithin and cholesterol, and then made into unilamellar vesicles (Fig. 2). Three-component conjugates containing various ligands: eg. vitamins (Wong et al., 1996; Goren et al., 1997), peptides and oligosaccharides (DeFrees et al., 1996; Zalipsky et al., 1997) were recently synthesized by reacting the appropriate X-PEG-DSPE with suitably functionalized ligands. The conjugates were chromatographically purified and characterized by NMR, HPLC, and matrix-assisted laser desorption / ionization time-of-flight mass spectrometry (MALDI-TOFMS). MALDI, in particular, is a powerful technique for the characterization of this type of conjugate (Zalipsky et al., 1997). A straightforward formulation process is among the main attractive features of the total lipid mixing / extrusion approach to ligand-bearing PEG-liposomes. On the other hand, incorporation of the preformed three-component conjugates into liposomes by this method results in slightly less than half of all the ligands facing the inner aqueous compartment and the remaining 55-60% being positioned on the external surface (for 100 nm liposome) (Zalipsky et al., 1997), where the ligands should exert their biological activity. Although the partition of the ligand-containing conjugate between the inner and outer monolayers of liposomal membrane depends, among other variables, on the vesicle size, charge and bulkiness of the ligand residue, in any case, a substantial portion of the total ligand-PEG-lipid that is inner leaflet bound, is wasted in this approach.

It was demonstrated that mPEG-DSPE can be inserted into preformed liposomes achieving similar external surface densities and *in vivo* performance as PEGylated liposomes prepared by the lipid mixing / extrusion approach (Uster et al., 1996). We recently applied a similar insertion strategy to several ligand-PEG-DSPE conjugates. We found that incubation of ligand-PEG-DSPE (1.2-1.5 mole %) with preformed liposomes containing mPEG-DSPE (2-

3 mole %) at 37° C resulted in complete insertion of the three-component conjugate (Zalipsky et al., 1997). This, aggregation-free process positioned the PEG-tethered ligands (oligosaccharide and peptide conjugates were tested) exclusively on the outer leaflet of the liposomal bilayer. This was demonstrated by a combination of a specific enzymatic cleavage of outer-monolayer bound ligand and HPLC-mediated quantitation of both starting and modified ligand-PEG-lipids. The insertion methodology allows for achievement of the same external surface densities of both PEG-tethered ligands and mPEG chains as by the lipid mixing / extrusion process (Zalipsky et al., 1997). Therefore this approach, schematically depicted in Fig. 3, overcomes the drawbacks of both of the methods discussed above.

A great deal of effort has been directed toward the development of one important class of ligands, immunoglobulins (Zalipsky et al., 1996; Kirpotin et al., 1996; Hansen et al., 1995; Allen et al., 1995; Goren et al., 1996). Ideally IgG bound to the surface of PEG-liposomes would direct the liposomal drugload to the target tissues. We recently examined immunogenicity of immunoliposomes prepared by site-specific conjugation of IgG through its oxidized oligosaccharide residue to liposomes utilizing hydrazide-PEG-DSPE linker (structure **3**; Table 1) (Harding et al., 1997). The immunoliposomes studied were composed of hydrogenated soy phosphatidyl choline, cholesterol, mPEG-DSPE, and **3** or compound 3 (120 nm, mole ratio 57 : 38 : 3.3 : 1.7), and contained on average 18 IgG residues per vesicle. We observed that while the immunoliposomes were relatively long circulating following the first i.v. injection, they exhibited drastically faster blood clearance after a second and subsequent administrations in rats. Presence of specific antiserum recognizing the ligand-IgG in the plasma of these animals was detected by ELISA. In contrast, repeated injections of the free IgG following otherwise the same protocol showed no evidence of an immune response. A single i.v. injection of the immunoliposomes was adequate for efficient recognition and clearance of subsequently injected either free IgG-ligand or its liposomal conjugate. These results indicate that our immunoliposomes, although containing relatively high density of grafted mPEG chains, and constructed by a well defined aggregation-free methodology, were effective potentiators of the immune response directed toward the bound IgG-ligand. The result of this study (Harding et al., 1997) is in agreement with previously made observations of Phillips et al, (1994) and strongly suggest that immunogenicity might be an important factor limiting the development of PEG-grafted immunoliposomes as targeted drug delivery constructs. It is important to determine how general this phenomenon is, and whether it applies to other classes of ligands.

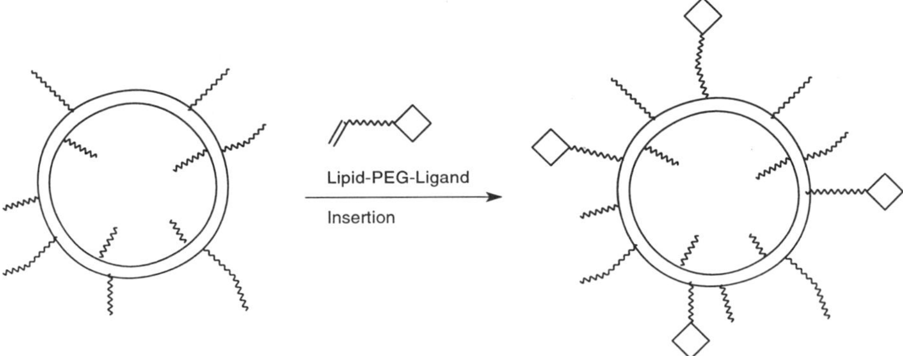

Figure 3. Ligands covalently linked to extremities of PEG-grafted liposomes: Third approach. Ligand-PEG-lipid conjugate inserted into the outer leaflet of the membrane of a preformed liposome.

SUMMARY

Various synthetic methods for the preparation of end group-functionalized PEG-DSPE derivatives were established using heterobifunctional PEG derivatives. These new materials are used for the preparation of PEG-grafted liposomes, which contain biologically relevant ligands covalently fixed at the periphery of the polymeric "brush" covering the vesicles. Among the various methods of assembly of ligand-bearing liposomes, the insertion of pure ligand-PEG-lipids into preformed liposomes appears to be the method of choice. A balance between preservation of binding activity of the ligands and reasonably long circulating plasma lifetime of the constructs could be found in almost every case. Formation of IgG-ligand-specific neutralizing antibodies, in the case of immunoliposomes, is likely to impede the development of such constructs if intended for repeated administration. The technology of PEG-grafted liposomes bearing ligands linked to the extremities of the polymer chains is currently being explored in both targeted liposomal drug delivery as well as in platform presentation of biologically active ligands.

REFERENCES

Allen, T.M., Brandeis, E., Hansen, C.B., Kao, G.Y., and Zalipsky, S., 1995, A new strategy for attachment of antibodies to sterically stabilized liposomes resulting in efficient targeting to cancer cells, *Biochim. Biophys. Acta* 1237:99.

Blume, G., and Cevc, G., 1993, Molecular mechanism of the lipid vesicle longevity *in vivo*, *Biochim. Biophys. Acta* 1146:157.

Blume, G., Cevc, G., Crommelin, M.D.J.A., Bakker-Woudenberg, I.A.J.M., Kluft, C., and Storm, G., 1993, Specific targeting with poly(ethylene glycol)-modified liposomes: coupling of homing devices to the ends of the polymeric chains combines effective target binding with long circulation times, *Biochim. Biophys. Acta* 1149:180.

DeFrees, S.A., Phillips, L., Guo, L., and Zalipsky, S., 1996, Dramatically increased binding to E-selectin of liposomal preparation of Sialyl-Lewis^x-polyethylene glycol-distearoylphosphatidylethanolamine conjugate, *J. Am. Chem. Soc.* 118:6101.

Gabizon, A., Catane, R., Uziely, B., Kaufman, B., Safra, T., Cohen, R., Martin, F., Huang, A., and Barenholz, Y., 1994, Prolonged circulation time and enhanced accumulation in malignant exudates of doxorubicin encapsulated in polyethylene glycol coated liposomes, *Cancer. Res.* 54:987.

Goren, D., Horowitz, A.T., Zalipsky, S., Woodle, M.C., Yarden, Y., and Gabizon, A., 1996,Targeting of Stealth liposomes to erbB2 (Her/2) receptor: *In vitro* and *in vivo* studies, *Br. J. Cancer* 74:1749.

Goren, D., Horowitz, A.T., Mandelbaum-Shavit, F., Tzemach, D., Zalipsky,S., and Gabizon, A., 1997, *In vitro* and in *vivo* studies of folate-targeted liposomes, *Proc.Int.Symp. Control. Rel. Bioact. Mater.* 24:865.

Hansen, C.B., Kao, G.Y., Moase, E.H., Zalipsky, S., and Allen, T.M.,1995, Attachment of antibodies to sterically stabilized liposomes: Evaluation, comparison and optimization of coupling procedures, *Biochim. Biophys. Acta* 1239:133.

Harding, J.A., Engbers, C.M., Newman, M.S., Goldstein, N.I., and Zalipsky, S., 1997, Immunogenicity and pharmacokinetic attributes of poly(ethylene glycol)-grafted immunoliposomes, *Biochim. Biophys. Acta* 1327:181.

Harris, J.M., and Zalipsky, 1997, *S., Poly(ethylene glycol): Chemistry and Biological Applications,* ACS Symposium series 680, Washington.

Kirpotin, D., Park, J.W., Hong, K., Zalipsky, S., Li, W.-L., Carter, P., Benz, C.C., and Papahadjopoulos, D., 1996, Sterically stabilized anti-Her2 immunoliposomes: Design and targeting to human breast cancer cells *in vitro*, *Biochemistry* 36:66.

Klibanov, A.L., Maruyama, K., Beckerleg, A.M., Torchilin, V.P., and Huang, L., 1991, Activity of amphiphathic poly(ethylene glycol) 5000 to prolong the circulation time of liposomes depends on the liposome size and is unfavorable for immunoliposome binding to target, *Biochim. Biophys. Acta* 1062:142.

Lasic, D.D., and Martin, F., 1995, eds., *Stealth Liposomes*, CRC Press, Boca Raton.

Lee, R.J., and Low, P.S., 1994, Delivery of liposomes into cultured KB cells via folate receptor-mediated endocytosis, *J. Biol. Chem.* 269:3198.

Papahadjopoulos, D., Allen, T.M., Gabizon, A., Mayhew, E., Matthay, K., Huang, S.K., Lee, K.-D., Woodle, M.C., Lasic, D.D., Redemann, C., and Martin, F.J., 1991, Sterically stabilized liposomes: Improvements in pharmacokinetics and antitumor therapeutic efficacy, *Proc.Natl. Acad.Sci. USA*, 88:11460.

Phillips, N.C., Gagne, L., Tsoukas, C., and Dahman, J., 1994, Immunoliposome targeting to murine $CD4^+$ leucocytes is dependent on immune status, *J. Immunol.* 152:3168.

Sato, T., and Sunamoto, J., 1992, Recent aspects in the use of liposomes in biotechnology and medicine, *Prog. Lipid Res.* 31:345.

Shimada, K., Kamps, J.A.A.M., Regts, J., Ikeda, K., Shiozawa, T., Hirota, S., and Scherphof, G.L., 1997, Bio-distribution of liposomes containing synthetic galactose-terminated diacylglycerol-poly(ethylene glycol)s, *Biochim. Biophys. Acta* 1326:329.

Silvius, J.R., and Zuckermann, M.J., 1993, Interbilayer transfer of phospholipid-anchored macromolecules via monomer diffusion, *Biochemistry* 32:3153.

Torchilin, V.P., and Papisov, M.I., 1994, Why do polyethylene glycol-coated liposomes circulate so long?, *J. Liposome Res.* 4:725.

Uster, P.S., Allen, T.M., Daniels, B.E., Mendez, C.J., Newman, M.S., and Zhu, G.Z., 1996, Insertion of poly-(ethylene glycol) derivatized phospholipid into pre-formed liposomes results in prolonged *in vivo* circulation time, *FEBS Lett.* 386:243.

Wong, J., Kuhl, T.L., Israelashvili, J.N., Mullah, N., and Zalipsky, S., 1996, Direct measurement of tethered ligand-receptor interaction potential, *Science* 275:820.

Woodle, M.C., 1997, PEG-grafted liposome therapeutics, in: *Poly(ethylene glycol): Chemistry and Biological Applications,* J.M. Harris and S. Zalipsky, eds., ACS Symposium series 680, Washington.

Woodle, M.C., and Lasic, D.D., 1992, Sterically stabilized liposomes, *Biochim. Biophys. Acta*, 1113:171.

Woodle, M.C., Engbers, C.M., and Zalipsky, S., 1994, New amphipathic polymer-lipid conjugates forming long-circulating reticuloendothelial system-evading liposomes, *Bioconjugate Chem.* 5:493.

Zalipsky, S., 1993, Synthesis of end-group functionalized polyethylene glycol-lipid conjugate for preparation of polymer-grafted liposomes, *Bioconjugate Chem.* 4:296.

Zalipsky, S., 1995a, PEG-lipid conjugates, in: *Stealth Liposomes,* Lasic, D. and Martin, F., eds., CRC Press, Boca Raton.

Zalipsky, S., 1995b, Functionalized poly(ethylene glycol) for preparation of biologically relevant conjugates, *Bioconjugate Chem.* 6:150.

Zalipsky, S., and Barany, G., 1986, Preparation of polyethylene glycol derivatives with two different groups at the termini, *Polym. Prepr. Am. Chem. Soc. Div. Polym. Chem.* 27:1.

Zalipsky, S., and Barany, G., 1990, Facile synthesis of α-hydroxy-ω-carboxymethyl-polyethylene oxide, *J. Bioact. Compatible Polym.* 5:227.

Zalipsky, S., Chang, J.L., Albericio, F., and Barany, G., 1994a, Preparation and applications of polyethylene glycol-polystyrene graft resin supports for solid-phase peptide synthesis, *Reactive Polym.* 22:243.

Zalipsky, S., Brandeis, E., Newman, M.S., and Woodle, M.C., 1994b, Long-circulating, cationic liposomes containing amino-PEG-phosphatidylethanolamine, *FEBS Lett.* 353:71

Zalipsky, S., Puntambekar, B., Bolikas, P., Engbers, C.M., and Woodle, M.C., 1995, Peptide attachment to extremities of liposomal surface grafted PEG chains: Preparation of the long-circulating form of laminin pentapeptide, YIGSR, *Bioconjugate Chem.*, 6:705.

Zalipsky, S., Hansen, C.B., Lopes de Menezes, D.E., and Allen, T.M., 1996, Long-circulating, polyethylene glycol-grafted immunoliposomes, *J. Controlled Release* 39:153

Zalipsky, S., Mullah, N., Harding, J.A., Gittelman, J., Guo, L., and DeFrees, S.A., 1997, Poly(ethylene glycol)-grafted liposomes with oligosaccharide or oligopeptide ligands appended to the tethered termini of the polymer chains, *Bioconjugate Chem.* 8:111.

ENGINEERING STEALTH™ LIPOSOME SURFACES:

EXERCISES IN COLLOID CHEMISTRY PRINCIPLES

Kostas Kostarelos

Department of Chemical Engineering & Chemical Technology,
Imperial College of Science, Technology and Medicine,
University of London, London, UK
Present address: Farmeco Co., 32 Ag. Glykerias Str., Galatsi 11146,
Athens, Greece

INTRODUCTION

Liposome bilayers have been attractive tools mainly as models of biological membranes and as delivery devices. Almost immediately after being widely accepted as membrane mimicking models, numerous studies appeared which used liposomes as carrier vehicles of other molecules (Gregoriadis, 1988). Due to their ability to act as "solvents" for both lipophilic (lipid bilayer phase) and hydrophilic (inner aqueous phase) molecules, liposomes have been used to deliver enzymes, genetic material, anticancer drugs (eg. doxorubicin), agents for diagnostic imaging and antibacterials, to name only a few. Liposomes have also proved particularly useful as general vaccine adjuvants (for example, there is already a liposome-based vaccine against hepatitis A), surfactant providers for newborn babies suffering from lung surfactant deficiency, and in the formulation of cosmetics as evidenced by the plethora of commercial cosmetic products claiming to contain liposomes.

Another area of research concerned with liposomes is the examination of their interaction with various biological surfaces and solutes. Considering liposomes as models of cell surfaces and studying such processes as the liposome-liposome, cell-liposome interactions, fusion and engulfment, can provide valuable information about complex biological functions. From this point of view, some examples of the numerous studies carried out, dealt with the interaction between liposomes and macrophage cells, viruses, bacterial surfaces and skin towards dermatological applications.

However, all lipid vesicles once formed appear to be stable only for a limited time. Aggregation and reversion into a phospholipid lamellar phase in water occurs. This destructive process, for any liposome dispersion, towards larger and more polydisperse entities takes place relatively slowly because of the slow exchange rates between the lipids in the vesicle and the monomers in the surrounding medium. Critical parameters on which the stability of liposome dispersions is judged upon and need to be assessed (Szoka and Papahadjopoulos, 1980),

Targeting of Drugs 6: Strategies for Stealth Therapeutic Systems
Edited by Gregoriadis and McCormack, Plenum Press, New York, 1998

include: chemical stability of the participating lipids, maintenance of homogenous vesicle size, including examination of the extent of aggregation, polydispersity index, maintenance of vesicle structure, retention of entrapped contents and influence of biological fluids on vesicle integrity and permeability. Currently, the most common method to store and contain liposome dispersions, particularly for commercial purposes, is by freeze drying them (Crommelin and van Bommel, 1984). During lyophilisation liposome size distribution and content retention is maintained for long periods of time, and also following liposome reconstitution.

The manipulation of phospholipid vesicles has been directed primarily towards improving the delivery of active molecules and their controlled release at target sites for therapeutic or other purposes. Manipulation of liposomes aims at both increasing their physical and chemical stability and also rendering their surfaces recognizable by specific receptors of the target sites. This optimization of phospholipid vesicle systems towards the needs of the particular application has been attempted by following certain "manipulation routes". These include:

1. Strategies to increase the rigidity and physicochemical stability of the *liposome bilayer* by (a) hydrogenation of double bonds within liposomes. The hydrogenation of double bonds can be readily accomplished, providing a number of potential applications, for e.g. saturated lipid liposomes can be used above their Tc (fluid state), and with inclusion of other molecules specific reactions can occur by means of a homogenous catalyst; (b) polymerization of the bilayer using synthesized polymerizable amphiphiles (e.g. lipids) to construct vesicles followed by polymerization; (c) mimicking the cytoskeleton in natural biomembranes, liposome dispersions can be prepared from negatively charged lipids containing polymerizable counter-ions. The resulting polymerized structures are called "liposomes in a net"; (d) inclusion of cholesterol to rigidify the bilayer and also imitate biological membranes.

2. Strategies to modify the *liposome surface* by (a) physical adsorption of macromolecules onto the liposome surface (eg. proteins, copolymers); (b) covalent bonding of macromolecules to lipids and subsequent formation of vesicles (usually using mixtures of the lipids of choice bound with glycoproteins, antibodies, polyethylene glycol); (c) incorporation of hydrophobic segments of macromolecules within the lipid bilayer, allowing coating of the liposome surface to take place by the hydrophilic part of the guest macromolecule (eg. glycoproteins, glycolipids, polymers).

The latter systems, (i.e. where the liposomal surface is modified) have had a leading role in the development of liposome research in the recent decade or so (Lasic, 1993, Jones, 1995). The present chapter will focus on systems which are prepared by placing an effective barrier (most commonly a polyethylene glycol or oxide chain) protruding from the liposome surface. Irrespective of the way in which the polymer chains are attached onto the liposome surface, which is discussed in the last section of the chapter, the mechanism by which these systems operate *in vivo* is usually referred as Stealth. The term has been coined to describe these systems, stemming from their proposed *in vivo* behaviour leading to disguise from the blood serum macrophages and consequently, delayed recognition and sequestration by the RES (mainly the mononuclear phagocytic cells of the liver and spleen). From a physicochemical point of view, these systems can be classified as sterically stabilized colloid particles.

STABILIZATION OF LIPOSOMES AS COLLOID PARTICLES

Since liposomes are colloidal particles (size range between 1nm to 10 μm) the stability of their dispersions is determined by a variety of forces, namely the dispersion (Van der Waals force), electrostatic (e.g. double layer repulsion force), undulation, hydration, steric, and hydrophobic forces. Thus, the stability of a liposome dispersion depends on a fine interplay between these forces. It has to be noted at this point that liposome stability is a relative term; depending on the application, a liposome system can be stable enough or not. For example, positively charged (cationic) liposomes can be prepared to form a 'stable', electrostatically stabilised dispersion. Following intravenous administration of thi it will be removed quickly from the blood circulation due its surface characteristics, hence being totally 'unstable' for the particular application. However, the same cationic liposome system mentioned has been found to form the most 'stable' liposome-DNA complexes todate, and has been used extensively to test gene transfer via liposomal vectors. Therefore, as the above example clearly illustrates, one

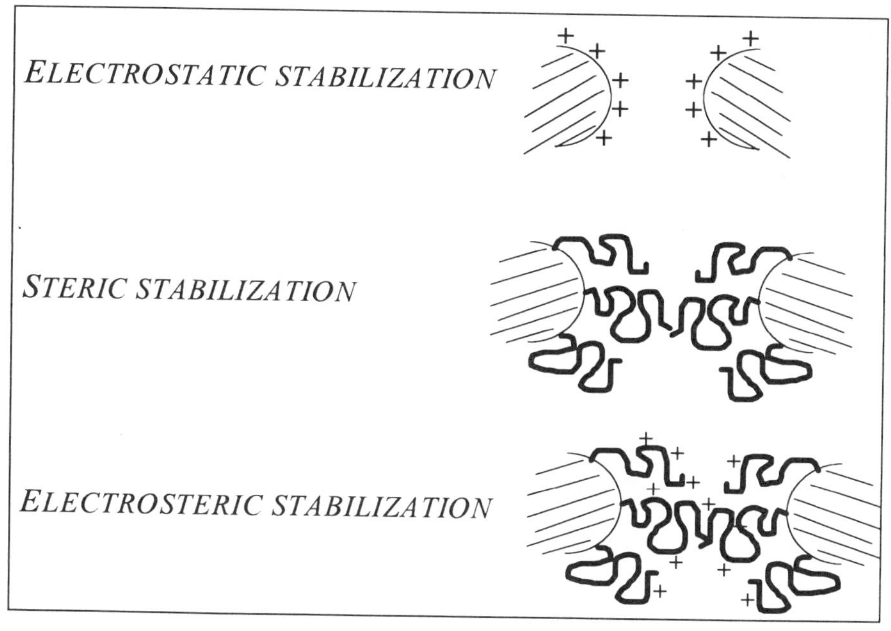

Figure 1. Possible routes to achieve stabilization of colloidal and liposome dispersions

has to define the term 'liposome stability', depending on the nature of the study and/or application. In the present chapter the term stability refers solely to the physicochemical stability of the liposome dispersions.

There are three routes that can be followed when attempting to stabilize a colloidal dispersion (Figure 1), all based on the central idea of inducing an additional repulsive force that will be able to maintain the dispersed particles well apart. Thus, aggregation, coalescence, bridging flocculation, Ostwald ripening and other processes that are catastrophic for the dispersion can be surely avoided. *Electrostatic stabilization* can be achieved by using charged particles to form dispersions (and, of course, appropriate adjustments of the electrolyte in the surrounding medium). *Steric stabilization* is attained by attaching polymer chains onto the surface of dispersed particles, acting as effective barriers. *Electrosteric stabilization* is achieved when polyelectrolyte chains are bound onto the particle surfaces, offering both electrostatic and steric repulsion, resulting however in much more complicated behaviour and structures.

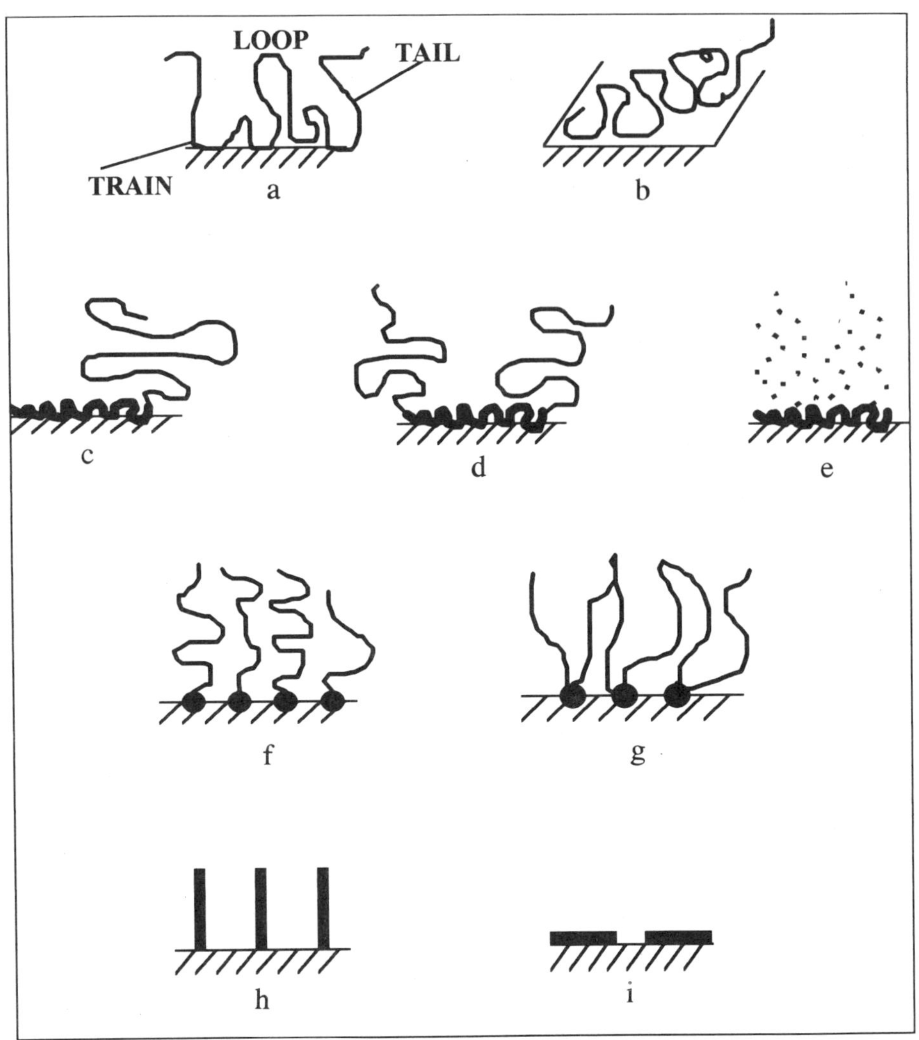

Figure 2. Theoretical types of macromolecule adsorption and conformation onto a plane rigid surface. (a) homopolymer (loop-train-tail); (b) homopolymer (flat conformation); (c) AB and (d) ABA block copolymer (preferential adsorption of certain groups); graft copolymer ("comb" adsorption); (f) polymer tails covalently grafted at terminal positions; (g) polymer covalently grafted at random points; rigid polymers adsorbed (h) vertically and (i) horizontally.

TYPES OF ADSORPTION AND CONFORMATION OF MACROMOLECULES AT INTERFACES

As mentioned previously in this chapter, Stealth liposomes refer to sterically stabilized vesicles. The way in which the polymer chain is attached onto the liposome surface and the conformations it will attain are of vital importance for the effectiveness of the chain as a protective barrier, hence the "Stealth" character of the vesicle. In addition to the usual adsorption considerations such as polymer-surface, polymer-polymer, solvent-surface interactions (which are the main forces involved in surfactant adsorption), one of the main problems to be resolved is the conformation of the polymer molecule on the surface. Theoretically, various conformations of the macromolecule can be envisaged depending on its interaction with the surface and the solvent (Tadros, 1982). A schematic representation of the theoretical conformations for a variety of macromolecules onto a plane rigid surface is presented in Figure 2. Also in theory, these schematic structures may apply in the case of the lipid bilayer / water interface. However, in the latter case a portion of the molecule may penetrate the fluid interface, depending on its solubility in the lipid bilayer and the geometrical restrictions of the vesicle structure (Israelachvili et al., 1980, Israelachvili, 1985).

Figure 2(a) represents the case for adsorption of a homopolymer, showing a random sequence of tails, trains and loops. Not only are homopolymers ineffective in stabilizing liposomes, but such hydrophilic homopolymers as polyethylene glycol, have proven to lead to liposome aggregation and eventual fusion. Presumably, the PEG chains adopt the flat configuration shown in Fig.2(b). In Figures 2(c-e), the structures have the same principle, i.e. the molecule is made from two blocks A and B, one with high affinity to the surface (the B block which may be termed the 'anchoring' block) and one with high affinity to the medium (the A block which may be termed the 'stabilizing' chain). These molecules are usually designated AB, ABA block copolymers and BA_n graft copolymers (sometimes referred to as 'comb' stabilizers). With the AB block copolymer, the B chain usually forms small loops at the surface, leaving a long dangling tail in the medium. The same with the ABA block copolymers, which consist of a B chain with small loops and two long chains strongly solvated by the medium. With the BA_n graft, several tails ('teeth') that are strongly solvated by the medium are attached to the B block. The structures shown in Fig.2(f and g), represent the case where a specific group, that may become chemically bound to the surface, is introduced to ensure the strong anchoring of the chain. These types of molecules are specifically used for the production of sterically stabilized lattices and liposomes, whereby a macromonomer such as methoxy polyethylene glycol methacrylate is introduced by a polymerization process. This produces lattices and liposomes with chemically bound PEG chains which are effective steric stabilizers in aqueous media. The vast majority of studies concerning Stealth (sterically stabilized) liposomes have used the latter method of polymer coating. The structures shown in Fig.2(h and i), are those of rigid polymers such as proteins that are widely used in the food industry, while their interaction with liposomes has been studied (de Kruijff et al., 1990).

It is clear from the above description of polymer conformation that for a full description of a polymer-coated surface, it is necessary to obtain information on the following parameters: the amount of polymer adsorbed per unit area of the surface (Γ), the fraction of segments in direct contact with the surface (p), the distribution of the polymer segments from the surface $\rho(z)$, and the adsorbed layer thickness (δ). It is important to know how these parameters change with polymer concentration / coverage, the structure of the polymer and its molecular weight as well as any change in its environment such as temperature and solvency. Several theories are available describing steric stabilization and polymer conformation at interfaces (de Gennes, 1987) which have been adequately presented and reviewed previously. Summarizing the above, the following conditions must be achieved: in order to optimize steric stabilization, and thus the "Stealth" character of a liposome dispersion, a) full surface coverage by the coating polymer; b) thick polymer layers to act as effective barriers; c) good solvent conditions for the stabilizing moieties; and d) firm anchoring of the macromolecule at the interface.

STERIC STABILIZATION OF LIPOSOMES BY PHYSICALLY GRAFTING ABA BLOCK COPOLYMERS

The construction of novel sterically stabilized vesicle systems with the addition of ABA tri-block copolymers was pursued in our laboratory a few years ago. The essence of the work carried out is accurately indicated in a review by Lasic (1994): 'Liposomes aggregate and fuse in the presence of hydrophilic polymers but their properties were difficult to explain when

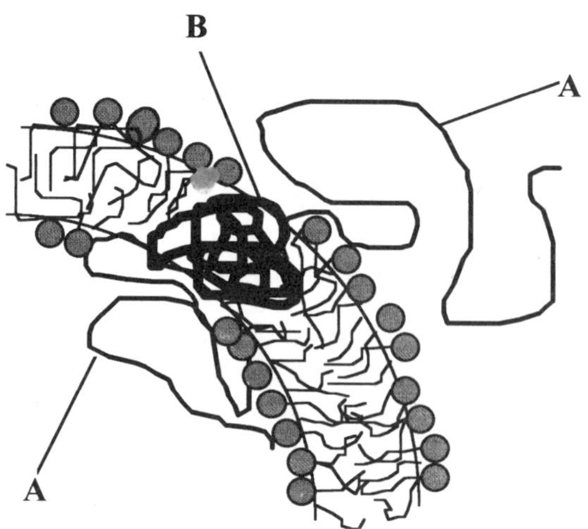

Figure 3. An ABA block copolymer molecule as an integral component of a lipid bilayer. The concept of liposome 'physical grafting' is hereby exemplified.

block copolymers or surfactants with larger polymeric polar heads were inserted into the liposome membrane, because such systems are inherently ill defined'. Exactly this ill definition we aimed at studying. The work is based on the recognition that inadequate anchoring would be achieved by using block copolymer molecules to adsorb onto liposome surfaces. Addition of block copolymers onto preformed liposomes would result in weak steric stabilization, and also almost spontaneous and complete desorption of the copolymer molecules from the liposome surface if administered *in vivo*, whereby a considerable dilution shock occurs. Reported blood circulation data (Woodle et.al., 1992) of the tri-block copolymers PF127 and PF108 added to preformed liposomes showed in fact better than expected data, of prolonged circulation and sustained release of the entrapped marker molecules.

The addition of ABA block copolymers during their formation is being proposed as an alternative novel method to sterically stabilize phospholipid bilayer vesicles. By allowing the anchoring B block of the copolymer to participate in the vesicle formation process a novel structure was obtained. The B block is physically grafted inside the bilayer, thus firmly anchoring the two A stabilizing chains dangling in solution. From the various experimental methods used to elucidate the effect that the block copolymer molecules had on the vesicle systems, it was demonstrated that the copolymer : lipid molar ratio was an extremely important factor determining aggregate structure (Kostarelos et al., 1998a). Interaction between block copolymer molecules and lipids in a vesicular form takes place at low to moderate molar ratios. For higher copolymer : lipid ratios destruction of the vesicle structures occurs, with a concomitant formation of mixed micelles between the copolymers and the lipids, similar to the phase behaviour reported for the poly(ethyleneglycol)-phosphatidylethanolamine (PEG-PE) conjugates (Edwards et al., 1997). Both the vesicle surface characterization (Kostarelos et al., 1998a) and the bilayer structural and morphological studies (Kostarelos et al., 1997) revealed the physical grafting of the hydrophobic block of the copolymer.

Figure 3 illustrates the novel structural features of the vesicles by allowing the B block of the copolymer to participate in the vesicle formation and thus form an integral component of the bilayer. The resulting sterically stabilized vesicles offer improved anchoring of the polymer chains onto the vesicle surface, as opposed to the weak binding of the adsorbed copolymer molecules. Combining this with the effective protection against coagulant and fusogenic agents (divalent cations) (Kostarelos et al., 1998b) and the evidence of a more rigid bilayer, the vesicle systems containing physically grafted copolymer molecules inside their bilayers can be proposed as a novel tool resulting from the application of colloid chemistry principles to liposome research.

The concept of 'physical grafting' can thus be introduced. The pronounced advantages of association of the B block as an integral part of the bilayer are evident. By anchoring the polymer molecule inside the lipid bilayer, surface coating is achieved in a much more efficient way than just allowing for copolymer adsorption. The steric barriers are much more robust and less susceptible to desorption (caused by interparticle collisions, dilution effects etc.), therefore offering permanent and resistant polymer coating to the vesicle. This is attainable purely physically, thus, advantage is taken of the unique ability of liposomes to be manipulated by physical means, which is another valuable aspect of their nature which is seldom mentioned, even less exploited. Most of the widely used ways to alter their surface characteristics in order to form Stealth systems involve chemical conjugation of a polymer chain to the headgroup of a particular lipid (usually PE). This process apart from being restricted by the ability of certain groups to be covalently linked to lipid headgroups, requires knowledge of organic synthesis, and involves the usually cumbersome step of covalent conjugation, thus limiting the types of modifications that can be readily applied to liposomal surfaces, and also form vesicles of specific, rigid, inflexible characteristics. By 'physical grafting'; apart from avoiding chemical manipulation of the liposome-participating lipids, any macromolecule, synthetic or biological, can act as a physical anchor, provided its structure allows for bilayer incorporation.

REFERENCES

Crommelin, D.J.A. and van Bommel, E.M.G. 1984. Stability of liposomes on storage: freeze dried, frozen or as an aqueous dispersion. *Pharm. Res.* 4:159.

Edwards, K., Johnsson, M., Karlsson, G., and Silvander, M., 1997, Effect of polyethyleneglycol-phospholipids on aggregate structure in preparations of small unilamellar liposomes, *Biophys. J.* 73:258.

de Gennes, P. G., 1987, Polymers at an interface; a simplified view, *Adv. Coll. Int. Sci.* 27:189.

Gregoriadis, G., e.d., 1988, "Liposomes as Drug Carriers - Recent Trends and Progress", John Wiley & Sons Ltd., London.

Israelachvili, J. N., Marcelja, S. and Horn, R. G., 1980, Physical principles of membrane organization, *Quart. Rev. Biophys.* 13:121.

Israelachvili, J. N., "Intermolecular and Surface Forces", 1985, Academic Press, New York.

Jones, M. N., 1995, The surface properties of phospholipid liposome systems and their characterization, *Adv. Coll. Int. Sci.* 54:93.

Kostarelos K., Tadros Th.F., and Luckham P.F., 1997, Molecular morphology and structure of phospholipid vesicles sterically stabilized by (tri-) block copolymers using hydrophobic dye molecules as bilayer probes, *J.Coll.Int.Sci.* 191:291.

Kostarelos, K., Luckham, P.F., and Tadros, Th.F., 1998a, Interaction of block copolymers of the polyethylene oxide - polypropylene oxide type with soybean lecithin vesicles, *Langmuir,* in press.

Kostarelos, K., Luckham, P.F., and Tadros, Th.F., 1998b, Investigating the steric stabilization of phospholipid vesicles by block copolymers: studies of vesicle flocculation and osmotic swelling caused by monovalent and divalent cations, *J.Chem.Soc. Faraday Trans.* in press.

de Kruijff, B., de Gier, J., van Hoogevest, P., van der Steen, N., Taraschi, T.F., and de Kroon, T., 1991, Effects of an integral membrane glycoprotein on phospholipid vesicle fusion, *in* "Membrane Fusion", J. Wilschut and D. Hoekstra, eds., Marcel Dekker, New York.

Lasic, D.D, "Liposomes: from Physics to Applications", 1993, Elsevier, Amsterdam.

Lasic, D.D., 1994, Sterically stabilized vesicles, *Angew. Chem. Int. Ed. Engl.*, 33:1685.

Szoka, F. and Papahadjopoulos, D., 1980, Comparative properties and methods of preparation of lipid vesicles (liposomes), *Ann. Rev. Biophys. Bioeng.* 9:467.

Tadros, Th.F., ed., 1982, "The Effect of Polymers on Dispersion Properties", Academic Press, London.

Woodle, M.C., Newman, M.S., and Martin, F.J., 1992, Liposome leakage and blood circulation: comparison of adsorbed block copolymers with covalent attachment of PEG, *Int. J. Pharm.* 88:327.

THE ROLE OF HYDRATION IN STABILIZATION OF LIPOSOMES: RESISTANCE TO OXIDATIVE DAMAGE OF PEG-GRAFTED LIPOSOMES

Aba Priev, Ayelet M. Samuni, Oren Tirosh, and
Yechezkel Barenholz

Department of Biochemistry, The Hebrew University-Hadassah
Medical School, P.O.B. 12272, Jerusalem 91120, Israel

INTRODUCTION

Phospholipids are amphipathic compounds which form structures such as micelles and liposomes in aqueous media as a result of minimizing hydrophobic interaction. Thermodynamic stability of the lipid assemblies is due to a balance between the weak attractive van der Waals and hydrophobic forces, and repulsive electrostatic, steric, and hydration ones (Lasic and Martin, 1995). Lipid peroxidation (LPO) is a major damaging process in membranes and in liposomal dispersions. Oxidation processes in membranes are initiated by reactive oxygen species (ROS) formed in the aqueous phase and in the lipids, leading to chemical degradation, especially of polyunsaturated acyl chains of phospholipids.

An earlier study using egg phosphatidylcholine (EPC) liposomes exposed to γ-irradiation, and two antioxidants having different lipophilicities, dealt with the role of the aqueous phase in oxidative damage. The results of the study suggest that the protection of lipid vesicles against radiation-induced LPO is mediated by scavenging of reactive oxygen species formed in the aqueous phase, near the lipid bilayer (Samuni and Barenholz, 1997). γ-Irradiation of frozen liposomal samples was also used to study the role of water in the liquid versus solid state in the damaging process. Prevention of γ-irradiation-induced chemical degradation by freezing served as evidence for the aqueous phase being the source of ROS involved in the damaging process (Samuni et al., 1997). Another work demonstrated that highly hydrated polyethylene glycol (PEG) covalently attached to a phospholipid reduces, even at low concentrations, oxidative damage from γ-irradiation (Tirosh et al., 1997a).

Since water has been shown to play a major role in oxidative damage, more effort is being invested in understanding the contribution of bilayer hydration to liposome resistance to oxidative damage; this is one of the reasons why the stability and hydration of liposomes having PEG grafted to their surface are currently receiving considerable attention. PEG is a nontoxic, water-soluble, and chemically inert synthetic polymer and is preferentially excluded from the surface of proteins (Arakawa and Timasheff, 1985), DNA (Jordan et al., 1972), and

Targeting of Drugs 6: Strategies for Stealth Therapeutic Systems
Edited by Gregoriadis and McCormack, Plenum Press, New York, 1998

liposomes (Arnold et al., 1990). PEG grafted on the liposome surface does not cause remarkable changes in cohesive and mechanical properties of liposomes (Martin and Lasic, 1991, Needham et al., 1992) but greatly increases their biological stability, reduces leakage of an encapsulated drug, and allows major advances in therapeutic application of liposomes to be achieved (Barenholz and Lasic, 1996; Lasic, 1996; Lasic and Martin, 1995; Torchilin et al., 1994; Woodle, 1993; Woodle, 1997). Despite intensive experimental efforts, the mechanism of stabilization of the liposomes by PEG is still a controversial subject. It was shown that attachment of PEG molecules to the surface of liposomes both strongly reduces the attractive forces (van der Waals) while increases the repulsive ones (steric and hydration) between liposomes to each other and to cells (Kenworthy et al., 1995a, Kenworthy et al., 1995b, Kuhl et al., 1994, Lasic and Martin, 1995). Blume and Cevc (1993) and Torchilin et al. (1994) suggested that the high mobility of attached PEG molecules is primarily responsible for the prolongation of the circulation time and stability of liposomes *in vivo*. It appears that the stabilizing effect of grafted PEG may be described as a general physical phenomenon rather than the result of a unique biochemical property of PEG (Woodle, 1997). Thermodynamic properties of solutes, such as heat capacity, partial volume, and compressibility, are known to be sensitive to the degree of their hydration and have been widely used to estimate the hydration of proteins (Almagor et al., 1997, Kharakoz, 1989, Priev et al., 1990, Sarvazyan, 1991), DNA (Buckin et al., 1989, Buckin et al., 1994, Chalikian et al., 1994), and liposomes (Barenholz et al., 1983, Buckin et al., 1979, Mitaku and Aruga, 1982). It was shown that hydration of biopolymers could be satisfactorily described using experimental data on the hydration of individual atomic groups.

The goals of this work were: (1) to investigate the influence of freezing and PEGylation of liposomes on damage induced by γ-irradiation; (2) to investigate the thermodynamic properties of large unilamellar liposomes (LUV) composed of a mixture of two phospholipids, one of them containing grafted PEG^{2000}, and compare the degree of hydration of these liposomes, and (3) to identify the configuration and hydration of PEG^{2000} when it is in the bulk solution and when it is covalently attached to the amphiphilic surface.

MATERIALS AND METHODS

Lipids

Racemic dihexadecylphosphatidyl (DHP)-PEG^{2000} was synthesized in our laboratory and was characterized by ^{31}P NMR, ^{1}H NMR, thin layer chromatography, and element analysis, as described elsewhere (Tirosh et al., 1997a). Distearoylphosphatidylethanolamine (DSPE)-PEG^{2000} was a gift from SEQUUS Pharmaceuticals, Menlo Park, CA, USA. Dihexadecylphosphatidylcholine (DHPC) was a gift from Genzyme, Cambridge, UK. EPC, egg phosphatidylglycerol (EPG), and fully hydrogenated soy PC (HPC) of purity higher than 98% were obtained from Lipoid KG, Ludwigshafen, Germany. Dipalmitoylphosphatidylcholine (DPPC, >99% pure) and dipalmitoylphosphatidylglycerol (DPPG, >98% pure) were gifts from Nattermann Phospholipid GmbH (Cologne, Germany). All chemicals were used without further purification.

Determination of Phospholipid Concentration

Phospholipid concentration of the liposome preparation was determined using a modification of the Bartlett procedure (Barenholz and Amselem, 1993).

Particle Size Determination

The distribution of the particle size in the liposomal dispersion was measured by photon correlation spectroscopy using a Coulter Model N4 SD apparatus (Barenholz and Amselem, 1993).

Liposome Preparation

Effect of freezing. The liposomes (200 nm) were prepared by the "thin film hydration" method. Mixtures of the lipids were dissolved in chloroform/methanol (1:1, v/v) in a round-bottom flask and the organic solvent was removed under vacuum by rotary evaporation. The thin film obtained was dried for at least 3 h under reduced pressure and then hydrated with 10 mM phosphate buffer, pH 7.4, in 0.13 M NaCl, at 10°C above the temperature of the maximum main phase transition (T_m) for DPPC:DPPG mixtures, and at room temperature for EPC:EPG mixtures. The liposome dispersions were extruded with a Sartorius extrusion system (Göttingen, Germany) once through a 0.6-μm and three times through a 200-nm pore-size polycarbonate filter, respectively (Nuclepore, Costar Corporation, Cambridge, MA, USA). The liposomes produced by this process are oligolamellar, having on average 2-3 bilayers per vesicle (Goren et al., 1990). The pH of the dispersion was measured before and after extrusion and was adjusted, if necessary, to 7.4 and the final phospholipid concentration was 22 mM. Aliquots of 2 ml of the liposome dispersions were dispensed into 10-ml vials, some of which were frozen at dry-ice temperature (-79 °C).

PEG-grafted liposomes. The following large unilamellar liposomes, having 100 nm average diameter were prepared with various mixtures/concentrations of EPC and DHP-PEG by extrusion through a polycarbonate filter, according to an already published procedure (MacDonald et al., 1991). Several preparations were made, one composed only of EPC (EPC LUV), the others a mixture of EPC containing 2, 4, 7, 10, or 12 mole% of DHP-PEG (DHP-PEG LUV).

γ-Irradiation

Effect of freezing. The liposome samples were irradiated using a ^{60}Co source, Gammaster B.V. (Ede, Netherlands), at ambient temperature or at -79 °C, with 32 kGy at a dose rate of 3.4 kGy/h. The absorbed dose reported here is the minimum dose. Dosimetry was done using dosimeters consisting of red Perspex having a precision of 5% (as stated by Gammaster; Marino and Benjamin, 1984).

PEG-grafted liposomes. LUV (200 μl; 8 mM lipids) of either EPC or 3 mole% DHP-PEG/EPC were γ-irradiated with a dose of 9.2 kGy by ^{137}Cs in a Gamma-Cell 220 (Atomic Energy of Canada Inc., Ottawa) at a dose rate of 8.5 Gy min^{-1} under atmospheric conditions.

Micellar Suspension Preparation

DHP-PEG or DSPE-PEG was lyophilized overnight before preparation of micellar suspensions. The lipids were dissolved at 10% w/w in water. Optically clear solution was obtained after sonication for 1 min at 8 kHz using a bath sonicator.

Phospholipid Peroxidation

Phospholipids were analyzed and quantified by HPLC as described earlier (Grit et al.,

1991); samples for the HPLC analysis were prepared by Bligh and Dyer extraction (Bligh and Dyer, 1959) and the phospholipids were collected in the chloroform phase. After dilution of the chloroform phase in methanol, aliquots of 100 µl were directly injected into the column. The HPLC system consisted of a type 400 solvent delivery system (Kratos, Ramsey, NJ, USA), a Kontron sampler MSI 660 (Kontron AG, Zurich, Switzerland), and a Waters 410 RI detector (Waters Associates, Milford, MA, USA). Chromatograms were collected and analyzed with a computerized data system (WOW, Thermo Separation Products, Riviera Beach, FL, USA). The separation of the phospholipids was carried out on a Zorbax aminophase column (25 cm x 4.6 mm, I.D., 5µm particle size, Du Pont, Wilmington, DE, USA) at 35°C. An Adsorbosphere NH_2 5µ-guard column (Alltech Associates, Deerfield, IL, USA) was connected before the Zorbax aminophase column. The mobile phase consisted of acetonitrile/methanol/5 mM ammonium dihydrogen phosphate solution, pH 4.8 (64/26/5, v/v). The flow rate was 1.5 ml/min.

Determination of Phospholipid Acyl Chain Composition in Liposomes

LUV (100µl; 8 mM lipids) before and after exposure to oxidative stress were diluted with 900 µl of water. The lipids were extracted from the aqueous phase by adding 1 ml of ethanol and 1 ml of chloroform and vortexing for 1 min. Two phases were formed, and the chloroformic phase was collected and dried by a stream of nitrogen followed by 2 h of lyophilization to remove all traces of water. The residue was dissolved in 50 µl of toluene, 10 µl of methanol, and 20 µl of the methanol esterification reagent Meth-Prep II (Alltech). The methyl esters formed were analyzed by GC (using a Perkin-Elmer 1020 plus GC, with a Silar 10C Alltech chromatographic column using a temperature gradient of 5°C min^{-1} from 140 to 240°C (Barenholz and Amselem, 1993).

Oxidative Stress by 2,2'-Azo(2-Amidinopropane) Dihydrochloride (AAPH)

LUV (200 µl; 8 mM lipids) of either EPC or EPC/DHP-PEG were incubated in two different experiments with 20 mM AAPH at 37°C for 24 h.

Quantification of Bound Water by Differential Scanning Calorimetry

Evaluation of bound water to PEG or to PEG-lipids was done by the differential scanning calorimetry (DSC) technique using a Mettler thermal analyzer model 4000. The ice-to-water fusion heat enthalpy was evaluated by the peak integral at around 0° C. Scanning was conducted from -30°C to 10°C at a rate of 2°C/min and was repeated 3 times for each sample until a plateau was reached. The amounts of water bound to free PEG^{2000} and to PEG covalently attached to dihexadecylphosphatidic acid were calculated from the fusion enthalpy ($\Delta H_{(fu)}$) of the water solutions containing PEG (or PEG-lipids) and that of water without the PEG (Barenholz et al., 1983):

$$n_h = [100(\Delta H_{(fu)}\ water - \Delta H_{(fu)}\ PEG/water)]\ /(\Delta H_{(fu)} water\ N_{PEG}) \qquad (1)$$

where n_h is the hydration number or moles of water bound to one mole PEG, and N_{PEG} is the moles of PEG^{2000} in the solution.

Volumetric Measurements

The density ρ_c of solutions at the selected solute concentration c was determined using the vibrating tube densitometer DMA-60/DMA-601 (Anton Paar, Gratz, Austria), with a precision of $\pm 3\times10^{-6}$ g/ml. The partial molar volume of the solute V is defined by the change in the solution volume resulting from dissolving 1 mole of solute at infinite dilution. It has been shown in the review by Zamyatnin (1984) that for diluted solutions:

$$V = 1/\rho_0 - \lim_{c\to 0}[(\rho_c - \rho_0)/(\rho_0 c)] \tag{2}$$

where ρ_0 is the density at 0 solute concentration (solvent density).

All solutions were degassed in vacuum for at least 12 h at 4°C before performing volumetric as well as ultrasonic measurements. The value for the density of water at different temperatures was taken from Kell (1975). The experimental error in the determination of the solute concentration was about 0.5%. For each evaluation of V, three independent measurements were carried out within a concentration range of 0.4-4 %.

Ultrasonic Measurements

Measurements of ultrasound velocity and absorption in liposome suspensions were performed using the resonator method (Eggers and Funk, 1973) in its differential version (Sarvazyan, 1982). Due to the formation of standing waves, resonance is produced in the measuring cell at frequency f at which the distance l between the acoustic transducers in the measuring cell equals $n\ l/2$, where n is a positive integer and $l/2$ is a half-wavelength. By parallel experiments with a sample-filled and reference cell, excess ultrasonic velocity and absorption can be determined. For dilute solutions in which ultrasound velocity U does not differ substantially from that measured in the solvent, U_0 (Sarvazyan and Chalikian, 1991),

$$\Delta U = U_0 \Delta f_n / f_n \tag{3}$$

where f_n is the frequency of the *n-th* resonance peak of the acoustic resonator measured in the solvent.

For those solutions in which ultrasound absorption per wavelength, $\alpha\lambda$, is sufficiently high (Sarvazyan and Chalikian, 1991),

$$\Delta(\alpha\lambda) = \Delta\ (\pi\ \delta f_n / f_n) \tag{4}$$

where δf_n is the half-power width of the resonance peak.

The acoustic cells contained 0.7 ml volume and were thermostated within ± 0.01°C. The experimental error in determining relative changes in U was about 1×10^{-3}% and in determining relative changes in $\alpha\lambda$ was about 2%. Temperature dependencies of U and $\alpha\lambda$ were measured by changing the temperature in steps: after setting a temperature, the samples were equilibrated and ultrasonic data were taken. The procedure was repeated for each temperature at 3, 10, 20, 30, and 45°C. All the measurments were performed at a frequency of about 7.7 MHz.

Compressibility Calculation

The adiabatic compressibility, K, which is defined by $-(dV/dP)$, where P is pressure applied at constant entropy, can be calculated from the relation $K = \beta\ V$, where β is the coefficient of adiabatic compressibility. Note, that in contrast to K (measured in ml/g-atm), β (measured in 1/atm) is independent of the apparent volume of the solute. β can be determined from the measured density ρ and sound velocity U using the relation $\beta = 1/\rho U^2$. It has been shown in the review by Sarvazyan (1991) that

$$K = \beta_0\ (2V - 2\,[(U_c - U_0)/(U_0\, c)] - 1/\rho_0) \tag{5}$$

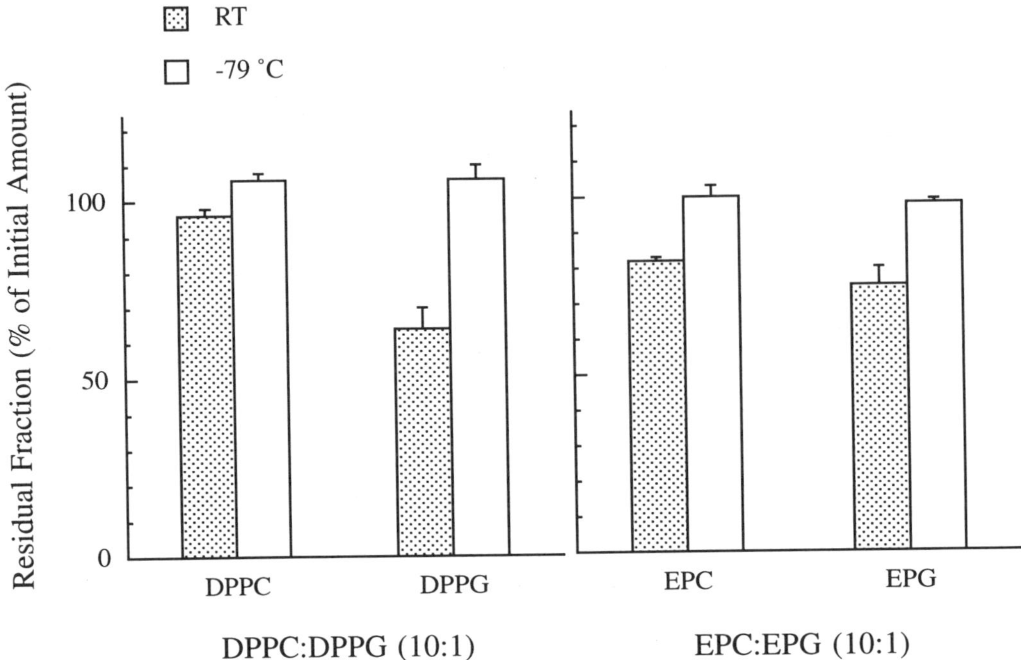

Figure 1. Effect of freezing on liposomal phospholipid degradation upon γ-irradiation. Liposomal dispersions in 10mM phosphate buffer, 0.13 M NaCl, were γ-irradiated by 32 kGy either at room temperature or in a frozen state (-79° C). Phospholipid levels of irradiated samples were quantified by HPLC and are presented as a residual fraction of their original levels.
Values represent the mean ± SD of three different liposome batches.

For each determination of K, three independent measurements of U and V were carried out within a concentration range of 0.4-4 %.

RESULTS

Radiation Effect on Phospholipids (Backbone Damage)

Liposomes (200 nm) of the following lipid compositions (mole ratios) were made: DPPC:DPPG 10:1 and EPC:EPG 10:1. Part of the liposomal dispersion batch was γ-irradiated in a frozen state at -79°C. The quantification of each of the liposomal phospholipids was done using HPLC, before and after exposure to a dose of 32 kGy. The results of HPLC analysis of liposomal phospholipids are shown in Figure 1. The extent of degradation after γ-irradiation was in the following order: DPPC < EPC≈ EPG < DPPG. When liposomes were irradiated in the frozen state no degradation of phospholipid backbone was found (Figure 1).

Effect of Liposome Grafted PEG^{2000} on Acyl Chain Degradation in the Liposomes

Liposomes containing and lacking DHP-PEG were exposed to γ-irradiation. The oxidative damage was analyzed from radiation-induced changes in their acyl chain composition as described in the experimental section. Four species of PUFA were followed (16:0; 18:2,

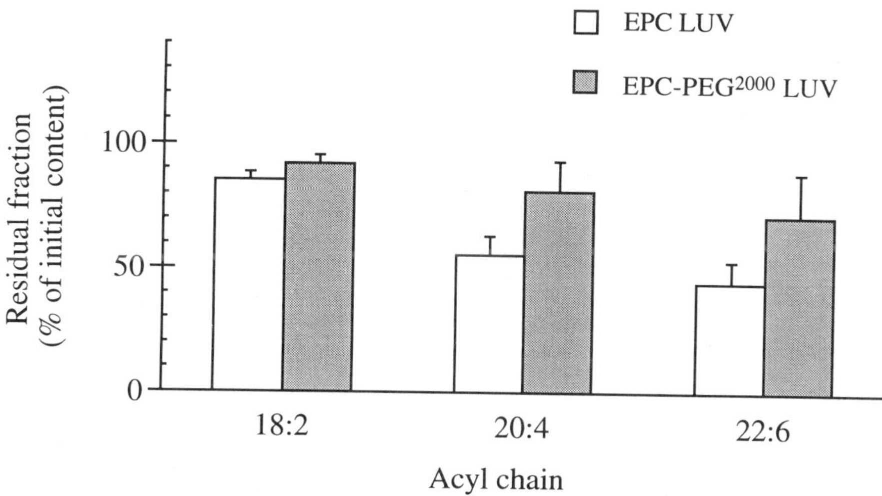

Figure 2. Effect of γ-irradiation on fatty acid composition in EPC and EPC/DHP-PEG^{2000} LUV. Liposomal dispersions of EPC and EPC/DHP-PEG^{2000} LUV were γ-irradiated at room temperature with 9.2 kGy. Fatty acid composition was determined after methyl esterification followed by GC separation. The new levels of fatty acids, related to palmitic acid which served as internal standard, are presented as a fraction of the original levels

20:4, 22:6). As no damage was found in C16:0, it was used as an internal standard which enables calibration of loss of PUFA in the liposomes. The results are shown in Figure 2, and are presented as the residual fraction of the initial content before irradiation. γ-Irradiation at a dose of 9.2 kGy caused a significant loss of PUFA, which increased with increasing degree of unsaturation. However, degradation of PUFA was higher for the liposomes lacking PEG, especially for 20:4 and 22:6 acyl chains (Figure 2).

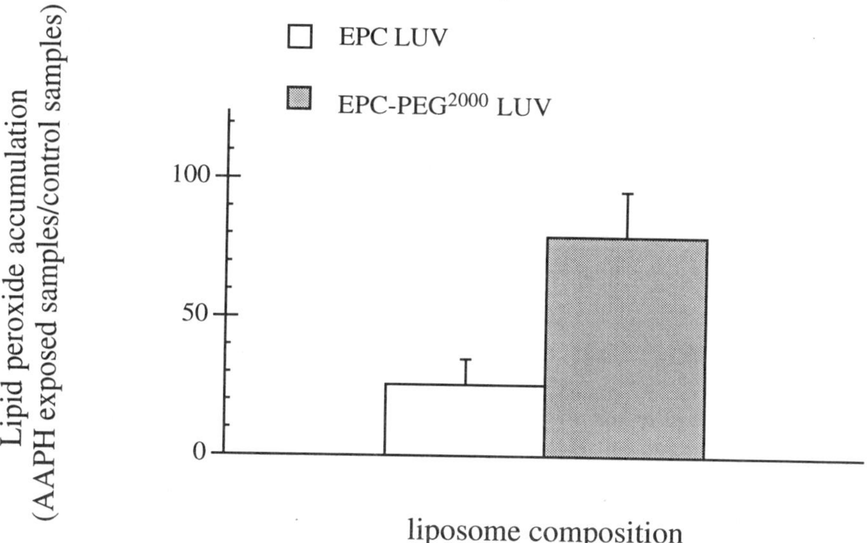

Figure 3. Effect of AAPH on lipid peroxide accumulation in EPC and EPC/DHP-PEG^{2000} LUV. Accumulation of lipid peroxides in liposomes composed of EPC and in liposomes composed of EPC/DHP-PEG^{2000} LUV following 24 h of incubation in AAPH (20 mM) at 37°C vs. control of 24 h at 37°C. Data presented as the ratio of AAPH samples/control samples

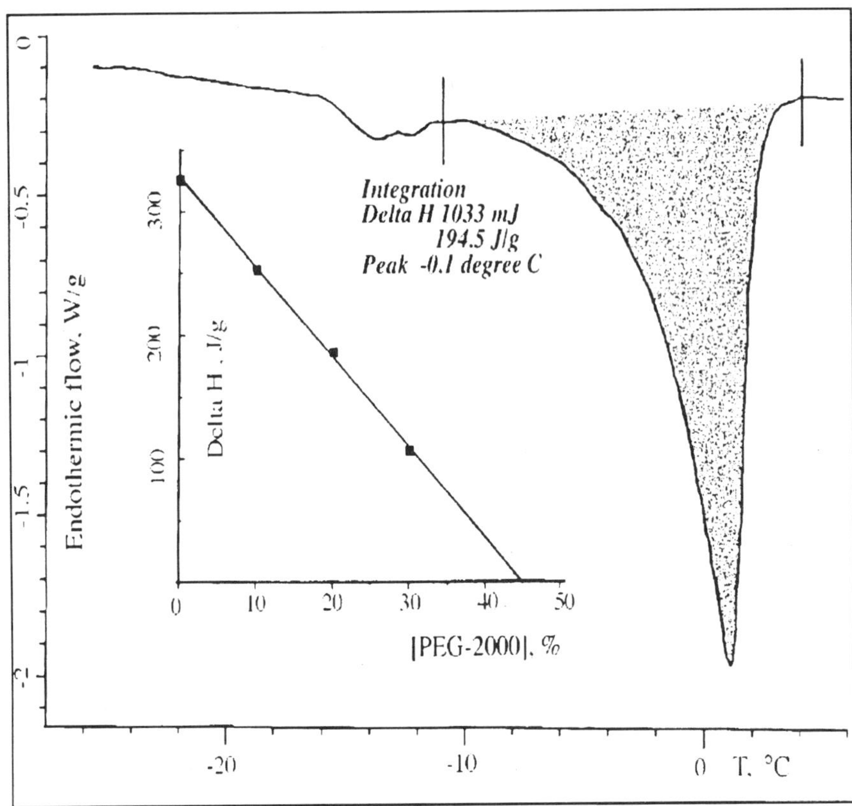

Figure 4. Effect of 20% PEG^{2000} on $\Delta H_{(fu)}$ of water and calibration curve of the fusion enthalpy of free water $\Delta H_{(fu)}$ (insert). The amount of bound water was calculated from the decrease in the $\Delta H_{(fu)}$ of water at about 0°C.

AAPH Oxidation of Liposome Lipids

Liposomes (8 mM lipids) with and without DHP-PEG were incubated with and without (control) AAPH (20 mM) at 37 °C for 24 h and peroxide accumulation was determined. The results are presented in Figure 3, showing lipid peroxide accumulation as a fraction of AAPH-exposed samples/control samples. In the presence of AAPH both EPC liposomes and EPC/DHP-PEG liposomes significantly accumulated peroxides. The liposomes containing DHP-PEG showed greater accumulation of peroxides than liposomes composed only of EPC (Figure 3).

DSC Results

Increasing concentrations of PEG in water induced large changes in the DSC plot of water. Scans were performed from -30°C to 10°C for water and for solutions containing increasing concentrations of PEG^{2000}. The fusion enthalpy of water, $\Delta H_{(fu)}$ water, was 330 J/g. Twenty percent (w/w) of PEG induced an enormous decrease in the free water peak (Figure 4): this is an indication that water became tightly bound to the PEG. From a calibration curve (Figure 4 insert) of the $\Delta H_{(fu)}$ of free water versus PEG concentrations at 45% PEG, there is no free water in the solution. Based on this, the hydration number, n_h, or the amount of water bound to PEG was calculated. The amount of water bound to DHP-PEG or DSPE-PEG was estimated from 3 separate measurements of the $\Delta H_{(fu)}$ of 10% PEG-lipid micellar solutions. DSC measurements of the hydration number of free and grafted DHP-PEG or DSPE-PEG, as well as of DHPC (a control to the DHP-PEG), are summarized in Table 1.

Table 1. Hydration number (molecules of water per molecule of solute) of DHPC liposome and PEG^{2000}, free and grafted onto the surface of micelles.

	PEG configuration	Hydration number
DHPC LUV		11 ± 2
PEG^{2000}	random coil	136 ± 4
DHP-PEG^{2000} micelles	brush	200 ± 6
DSPE-PEG^{2000} micelles	brush	210 ± 6

Volumetric and Acoustic Results

Ultrasound absorption, sound velocity, and density of LUV composed of a mixture of EPC and DHP-PEG^{2000} were measured as a function of the DHP-PEG concentration (0-12 mole%), lipid concentration (0-4%), and temperature (3-45°C).

Figure 5 represents the behavior of density (top), sound velocity (middle), and ultrasound absorption (bottom) at 7.7 MHz of EPC LUV and DHP-PEG LUV containing 7 mole% of PEG-lipids at 30°C as a function of lipid concentration. In this system, and in all the other

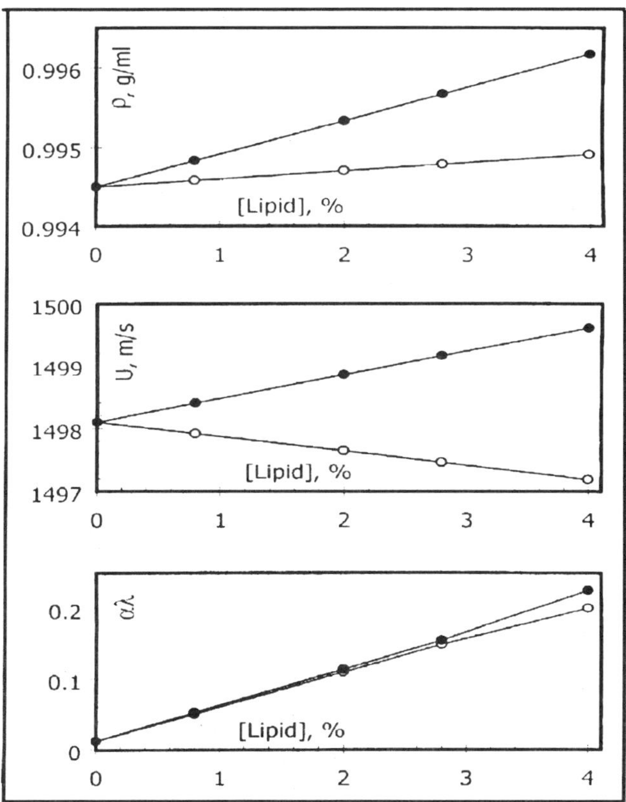

Figure 5. Density (top), sound velocity (middle), and ultrasound absorption (bottom), at 7.7 Mhz of EPC LUV (empty circles) and DHP-PEG^{2000} LUV containing 7 mole% of DHP-PEG (filled circles) at 30°C as a function of liposome concentration

systems studied in this work, there was a linear relationship between density, ultrasound absorption, sound velocity, and the lipid concentrations; thus such parameters as density number, $[\rho] = lim_{c \to 0} [(\rho_c - \rho_0) / (\rho_0 c)]$, sound velocity number, $[U] = lim_{c-} [(U_c - U_0) / (U_0 c)]$, and molecular absorption number, $[A] = lim_{c \to} [(\alpha\lambda_c - \alpha\lambda_0) / c]$, can be used to characterize the thermodynamic properties of the liposomal solutions. A good linearity of the dependence of ρ, U, and $\alpha\lambda$ on c indicates that the contribution of liposome-liposome interaction to V, K, and $[A]$ is negligible.

The apparent molar volume of the EPC liposomes at infinite dilution (that is, the partial molar volume) was 740.5 ml/mole in water solution at 30 °C. The partial molar compressibility was 32.84 10^{-6} ml/mole-atm. These data are in good agreement with the literature (Barenholz et al., 1977, Buckin et al., 1979, Mitaku and Aruga, 1982). Addition of PEG-lipids to the EPC liposomes leads to a decrease in partial molar volume and compressibility due mostly to the contribution of PEG, which has a relatively small value of partial molar volume and compressibility (36.77 ml/mole and 1.5 10^{-6} ml/mole-atm, respectively, per PEG repeated unit).

The values of sound velocity number, specific volume, specific adiabatic compressibility, and molecular absorption number for the EPC liposomes containing PEG-lipids in water solution at different temperatures are given in Figure 6. Errors were estimated by taking into account uncertainties due to apparatus limitations and concentration determination. The temperature dependencies of the partial specific volumes, V, and adiabatic compressibilities, K, of PEG-modified liposomes can be approximated by second-order polynomial functions (correlation coefficient higher than 0.99).

DISCUSSION

Hydration of Polyethylene Glycol

Calorimetric and volumetric studies on a large variety of low molecular weight compounds indicate that solute-induced changes in the properties of water in the vicinity of a solute atomic group extend to a distance of about 0.4 nm, which corresponds to an average of 1-1.5 layers of water molecules (Chalikian et al., 1993, Sarvazyan, 1991). We can assume that in the case of high molecular weight compounds, hydration also consists primarily of those water molecules that contact the solute directly. Thus, the number of water molecules in the first layer around the solute can be considered as a lower limit for the hydration number, n_h. n_h can be calculated as the ratio of the solvent-accessible surface area of a solute S_m to the effective cross section of a water molecule S_w. Thus, n_h is equal to S_m/S_w, where the effective cross section of a water molecule is 0.09 nm^2 (Chalikian et al., 1994). Our calculation of the accessible surface area of PEG from the specific volume data for its components (ethylene glycol monomer, dimer, etc. (Harada et al., 1978, Kiyosawa, 1991)) showed that about three water molecules are bound per PEG repeated unit, which has a solvent-accessible surface area of about 0.3 nm^2. Thus, the whole PEG^{2000} molecule, having a degree of polymerization 46, binds about 142 water molecules. This calculation, based on volumetric studies, is in good agreement with our DSC results.

The hydration number of a PEG molecule depends on the configuration of the PEG. This was demonstrated by the DSC measurements of free PEG (a random coil configuration) compared to PEG covalently attached to the headgroup of a phospholipid (a brush configuration). Our results show that in the brush configuration, PEG is more accessible to water, binding about 30 % more water than PEG in the random coil configuration.

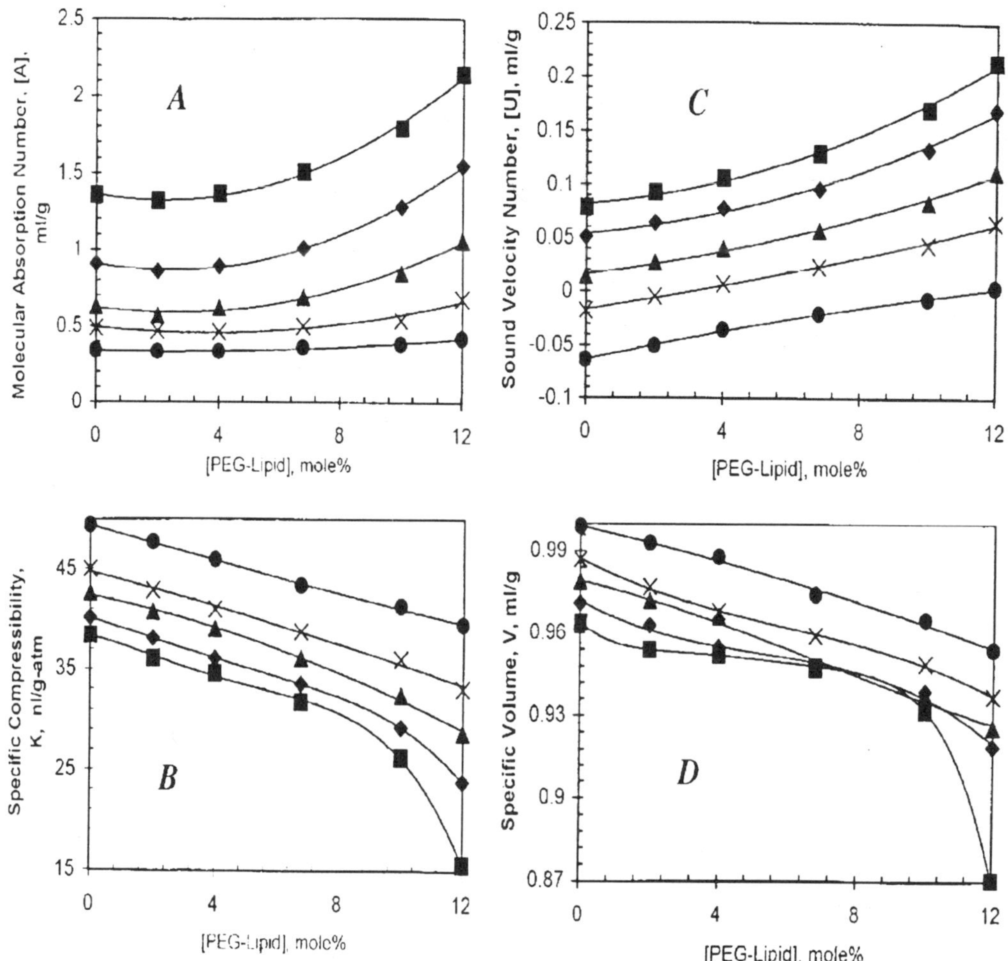

Figure 6. PEG-grafted liposomes properties: (A) molecular absorption number, (B) specific adiabatic compressibility, (C) sound velocity number, and (D) specific volume, as a function of PEG-lipid concentration at temperatures ■ 3°C, ♦ 10°C, ▲ 20°C, ✚ 30°C, ● 45°C. Experimental errors for *[U], V, K* and *[A]*, are 0.001, 0.002, 0.3, and 0.002 in the corresponding units

Adiabatic Compressibility and Specific Volume of PEG-Grafted Liposomes

The values of specific volume, specific adiabatic compressibility, sound velocity number, and molecular absorption number given in figure 6 include the contributions of three main components: lipid bilayer acyl chains, headgroup region, and PEG chains. In the case of acoustic measurements we have an additional contribution due to relaxational processes. Let us analyze all these contributions and their relation with the hydration and conformation of the main components of the liposome.

Relaxational processes of liposomes. Acoustical properties such as adiabatic compressibility and molecular absorption number are the sum of an instantaneous component (K_∞ or $[A]_\infty{}^\circ$), which is mainly caused by the intermolecular interactions, and a relaxational component (K_{rel} or $[A]_{rel}$), caused by pressure-induced changes in the structure of the system:

$$K = K_{\infty} + K_{rel} \quad (6)$$

$$[A]=[A]_{\infty}+[A]_{rel} \quad (7)$$

The main relaxational processes in liposomes that are displayed at the frequencies of our ultrasonic measurements, ≈7.7 MHz, are trans-gauche transformation with relaxation times τ of the order of 1 ns, heterophase fluctuations in liquid crystalline phase, $20< \tau < 50$ ns, and headgroup rotation, $2 < \tau < 4$ ns (Colotto et al., 1993, Kharakoz et al., 1993, Mitaku et al., 1983).

Our measurements of molecular absorption number obtained in EPC LUV containing DHP-PEG^{2000} and in EPC LUV containing HPC are in close agreement (Figure 7). This is due to the fact that the main relaxation process in the PEG-modified liposomes, as well as in EPC LUV containing HPC, is the heterophase fluctuation of bilayer density. HPC was used as a control to distinguish between the contribution of the phosphatidyl part of the DHP-PEG and the contribution of the attached PEG^{2000}. The mechanism of this relaxation was described by Kharakoz et al. (1993) and shown to be detectable acoustically even at temperatures very far from the transition temperature, T_m. For EPC, $T_m \approx$ -5 °C; for DHP-PEG^{2000} and for HPC, $T_m \approx$ 50°C (Lichtenberg and Barenholz, 1988, Marsh, 1990, Tirosh et al., 1997b).

The PEG-induced excess ultrasound absorption $[A]_{exc}$ (difference in the ultrasound absorption number between the DHP-PEG^{2000}/EPC LUV and HPC/EPC LUV) is shown in Figure 7 (insert). $[A]_{exc}$ has a distinctly higher value for PEG-lipid concentrations over 5 mole%, which points to the existence of an additional fast process in the PEG-modified liposomes. The most likely process responsible for $[A]_{exc}$ is an increase in the relaxation time of the headgroup rotation with increases in PEG-lipid concentration. This is in good agreement

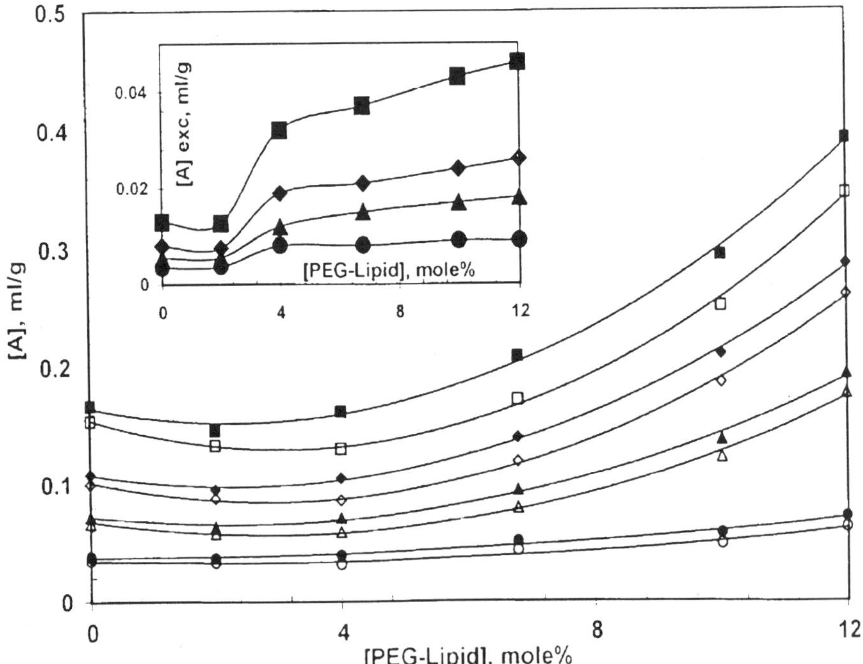

Figure 7. Molecular absorption number of DHP-PEG^{2000} LUV (filled symbols) and EPC/HPC LUV (empty symbols) at 3 (■), 10 (♦), 20 (▲), and 45 (●)°C. PEG-induced excess ultrasound absorption number $[A]_{exc}$ of PEG-grafted liposomes (insert)

with the current scaling models (deGennes, 1987, Hristova and Needham, 1994) for polymers at interfaces, which predict mushroom-brush transition in PEG conformation at 5 mole% of PEG-lipid, when PEG coils start to repel each other and extend out from the surface on which they are grafted (Figure 8). This figure shows a schematic diagram of a PEG-grafted bilayer at low grafting concentration (mushrooms) and a PEG-grafted bilayer at high grafting concentration (brush). The theory allows one to predict the extension L of the polymer from the surface. For the mushroom regime it is given by :

$$L_{mush} \approx a N^{3/5} \tag{8}$$

where a is monomer size ($a = 0.35$ nm for PEG^{2000}) and N is the degree of polymerization ($N = 46$ for PEG^{2000}). According to this equation, L_{mush} for PEG^{2000} is about 3.5 nm. For the brush regime it is given by:

$$L_{brush} \approx N a^{5/3} / D^{2/3} \tag{9}$$

where D is the distance between points of grafting of PEG.

According to this equation, L_{brush}, for example, for 12 mole% of PEG^{2000} -lipids is about 4.5 nm; (when the distance between two points of grafting of PEG, D, is approx. 2.5 nm), and it can be increased to up to 10 nm at higher concentrations of PEG^{2000}. Thus, when PEG exists on the liposome surface in the brush regime it is more efficient as a repulsive barrier against the close approach of other particle surfaces, which could destabilize the liposome.

As shown by experimental studies (Baekmark et al., 1997, Kenworthy et al., 1995a,b; Needham et al., 1992), increasing the concentration of grafted PEG, as well as its molecular weight, improves the repulsive properties of the lipid bilayer surfaces, creating a denser, larger brush. Further increasing concentrations of grafted PEG will cause a transition from liposomes to micelles.

The predictions, based on free energy considerations involving lipid bilayer cohesion and lateral pressure induced by grafted polymer globules, show (Hristova and Needham, 1994) that in the case of PEG^{2000} the threshold concentration n_t, (concentration above which the vesicles will start to break down and form micellar lipid phases) is about 8 mole%. Increasing the PEG-lipid concentration above this value decreases the elastic constant and the tensile strength of the bilayer. These changes occurred in the packing of the lipid bilayer with high concentrations of grafted PEG, as verified by X-ray and calorimetric studies (Baekmark et al., 1997, Kenworthy et al., 1995a,b). Based on these results and ours, it seems that for the sake of stability it is better if the grafted PEG globules barely touch and interact, i.e., they are at the borderline between mushroom and brush, which occurs at PEG-lipid concentrations between 5 and 8 mole%.

Hydration of the lipid bilayer of the PEG-grafted liposomes. The partial volume of a solute, V, can be considered to be the sum of the three elements:

$$V = V_i + V_v + V_h \tag{10}$$

where V_i is the intrinsic volume of the solute atoms, as determined from their van der Waals dimensions; V_v is the total free volume, including voids formed by the folding of the lipid chain ($V_i + V_v$ form the liposome "core", which is free of solvent molecules); and V_h is the volume change of the solution caused by the liposome hydration.

Similarly, an instantaneous part of the adiabatic compressibility of the solute can be expressed as:

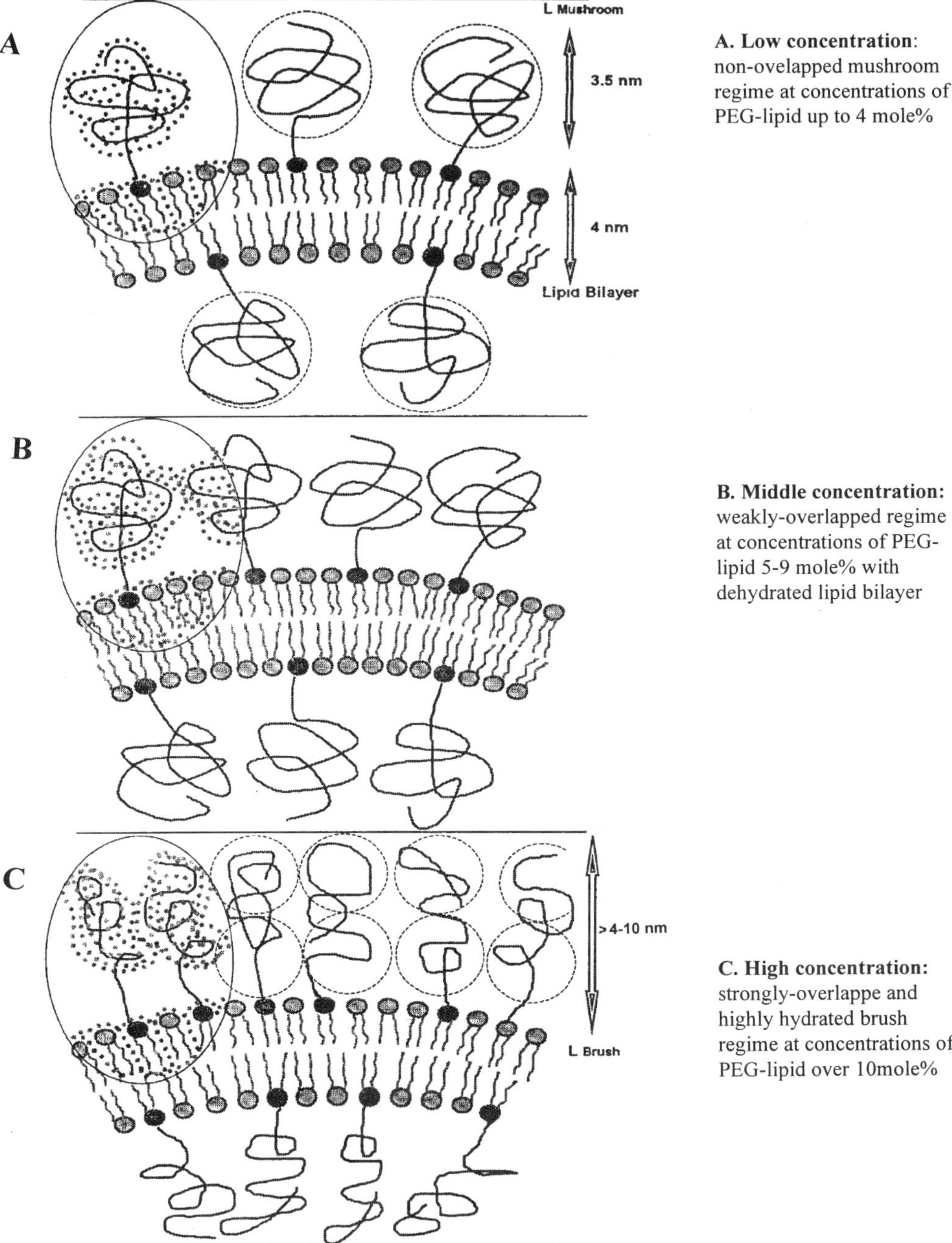

Figure 8. Schematic diagrams of the changes in the outer layer and bilayer of PEG-grafted liposomes at different concentrations of PEG-lipid

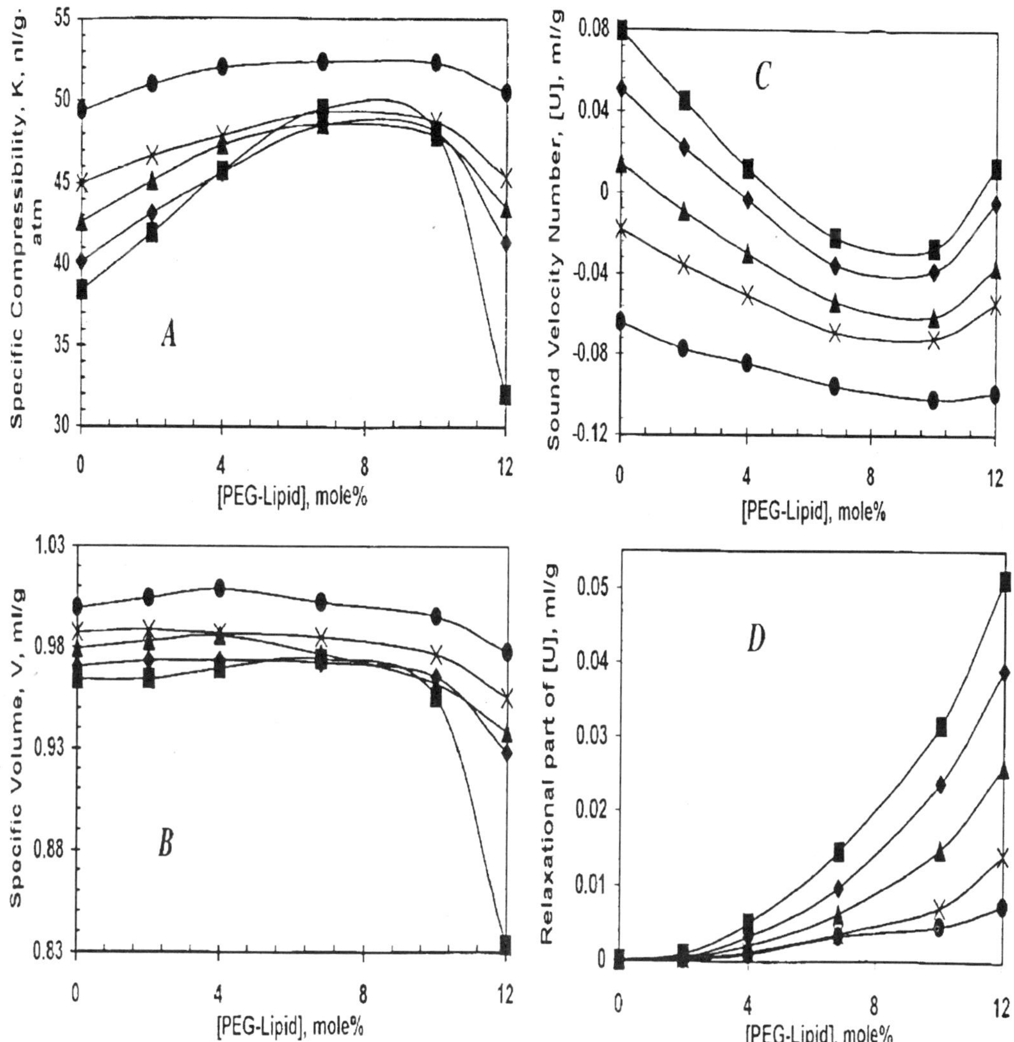

Figure 9. Lipid bilayer properties: (A) specific adiabatic compressibility, (B) specific volume, (C) sound velocity number, and (D) contribution of relaxational processes to sound velocity number of liposomes, each as a function of PEG-lipid concentration at temperatures ■ 3°C, ♦ 10°C, ▲ 20°C, ✚ 30°C, ● 45°C.

$$K = K_i + K_v + K_h \tag{11}$$

where K_i is the intrinsic compressibility of lipid bilayer, K_v is the compressibility of voids in the liposomes, and K_h is the compressibility change of the solution caused by liposome hydration. These components of liposome compressibility correspond, respectively, to the change of the volumes V_i, V_v, and V_h with pressure.

In order to evaluate the intrinsic compressibility of the lipid bilayer of EPC LUV the relation between K_i and K_h was taken from Buckin et al. (1979). K_i for the lipid bilayer is about (70±4) 10^{-6} ml/mole-atm at 30°C, which is 6, 45, or 60 times the K_i for protein (Kharakoz and Sarvazyan, 1993), DNA (Chalikian et al., 1994), or carbohydrate (Shilnikov et al., 1991), respectively, and corresponds to amplitude of thermal fluctuations of atoms of about 3-3.5 nm.

The partial volume and the adiabatic compressibility of the lipid bilayer of the liposome increase monotonically with PEG-lipid concentration up to concentrations of about 7 mole% of PEG-lipid (Figure 9) . Partial volume of the lipid bilayer was calculated by subtracting the contribution of PEG^{2000} (Harada et al., 1978) from results in Figure 6. Partial compressibility of the lipid bilayer was calculated by subtracting from the results in Figure 6 both the PEG^{2000} contribution (Harada et al., 1978) and the contribution of the relaxational processes, according to their relationship with ultrasound absorption data (Kharakoz et al., 1993). Analysis of the results shows that grafted PEG makes the lipid phase less dense (by a maximum of 1.4%) and more compressible (by a maximum of 40%). At PEG-lipid concentrations higher then 7 mole%, K and V of the lipid phase of the samples decrease, reflecting considerable increases in hydration of the lipid headgroup. This effect may be related to an increase in distance between lipid headgroups due to the repulsion of PEG chains.

To determine the hydration number for these headgroups we have applied an approach in which it is assumed that strong electrostatic headgroup-water interactions abolish the unique anomalous properties inherent in bulk water (Chalikian et al., 1993, Kharakoz, 1989). As a result, water in the hydration layer of charged atomic groups exhibits a linear temperature dependence of compressibility, like any "normal" liquid. In other words, the second temperature derivative of compressibility of water in the hydration shell of charged groups, d^2K_h/dT^2, is equal to 0. Water near the uncharged groups has the same anomalous properties as bulk water with a slightly shifted temperature minimum (Figure 10). Electrostatic interactions near b charged groups considerably distorted the water and decrease the water

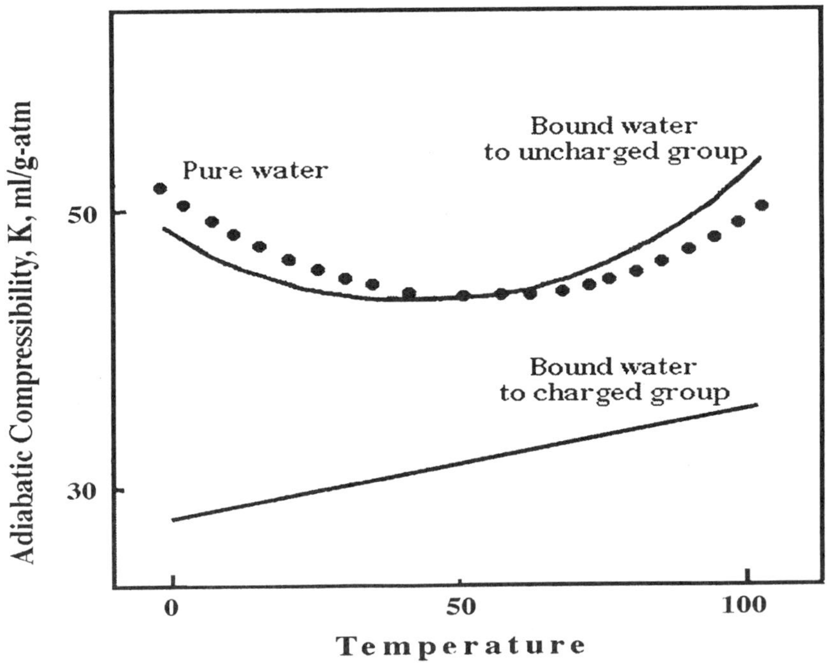

Figure 10. Qualitative picture of temperature dependence of compressibility of pure water and the water in the hydration layer of charged and uncharged atomic groups

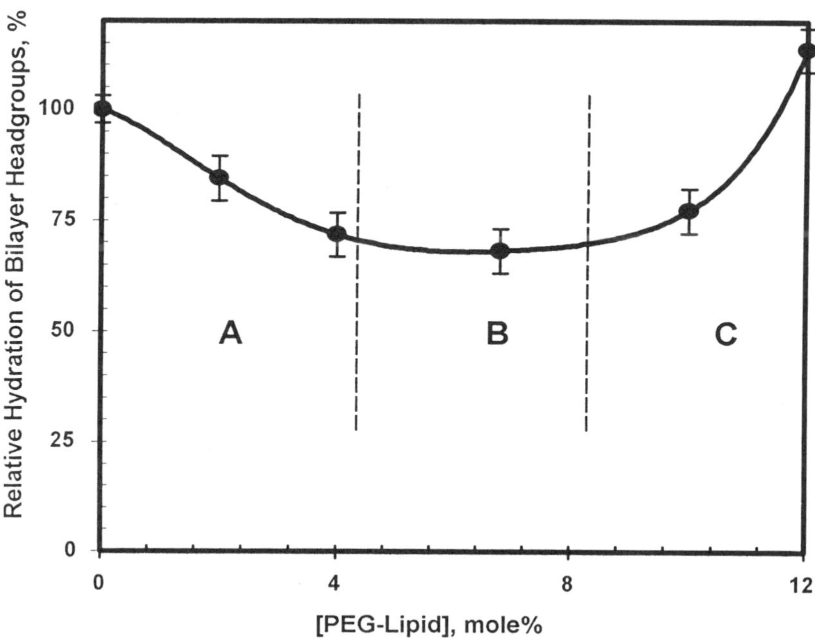

Figure 11. Relative hydration of bilayer headgroups at different concentrations of PEG-lipid. Regions A, B, and C correspond to those in Figure 8: non-overlapped, weakly overlapped, and strongly overlapped regimes

compressibility. In this context, equation 11 can be differentiated as follows, assuming that the temperature dependence of the intrinsic compressibility of the lipid bilayer and the voids, (K_i, + K_v) is negligible:

$$d^2K/dT^2 = -n_h d^2K_w/dT^2 \tag{12}$$

Thus, the hydration number, n_h, can then be evaluated if the second temperature derivative of the partial compressibility is determined. Analysis of the temperature dependence of the compressibility made it possible to determine quantitatively the amount of water bound to liposomes after adding different concentrations of attached PEG (Figure 11). These results suggest that the compressibility of the lipid phase is mainly contributed by the bound water. This water interacts with the lipid in a way that its elimination leads to a decrease in the total volume of the cavities, and to enhanced intermolecular bonding.

The hydration number of the lipid bilayer has a minimum at concentrations of PEG-lipid of about 5-9 mole%. At the same concentrations of PEG-lipid, liposomes have maximum stability (Baekmark et al., 1997). This allows one to assume that hydration plays an important role in stabilization of the PEG-modified liposomes. Of course, the main reason for stabilization is related to the ability of the grafted PEG to act as a repulsive barrier against the close approach of another surface (Lasic, 1996, Martin and Lasic, 1991, Needham et al., 1992, Torchilin et al., 1994). But the removal of water from the hydration shell of the lipid diminishes the effective size of the polar headgroup, which subsequently reduces bilayer defects, enhancing the lateral packing of the acyl chains. It is concluded that bound water plays a crucial role in destabilizing the lipid bilayer. Water plays a similarly crucial role in the stability of proteins, as shown by Priev et al. (1996) and Almagor et al. (1997). In general, lipid assemblies, like other biomolecules, are not highly stable structures and conformational changes induced by small changes of bound water can considerably stabilize them and improve their biological functions. We assume that dehydration of the lipid headgroup region, in conjunction with the increase of the hydration of the outer layer by grafting PEG, is responsible for increasing thermodynamic stability of the liposomes at 5-9 mole% of PEG-lipid.

Mechanism of Dehydration of Lipid Bilayer

As mentioned before, free PEG is preferentially excluded from the surface of biopolymers and liposomes (Arakawa and Timasheff, 1985; Arnold et al., 1990). By dialysis equilibrium coupled with density measurements, it was shown that the thickness of the effective shell relatively impenetrable to free PEG around the biomolecule increases with increasing radius of gyration, R_g, of PEG, which is a function of PEG molecular weight (Bhat and Timasheff, 1992). It seems that the preferential exclusion of free PEG from liposomes is principally due to the steric exclusion of PEG from the hydrated surface. PEG chemically attached to the lipid headgroup undergoes steric exclusion from the liposome surface as well, resulting in greater density of the grafted PEG further from the surface. Thus, the local concentration gradient of PEG chains from the liposome surface leads to osmotic imbalance and changes in the thermodynamic properties and hydration of the lipids.

Role of Water in Lipid Peroxidation

The use of freezing is an important step in a number of liposome preparation methods, as well as a means of protection and preservation of liposomes. In the present work we used freezing to achieve a better understanding of the mechanisms underlying lipid peroxidation and the role of bilayer hydration in the damaging process. HPLC results indicate that irradiation-induced damage to the liposomal phospholipid backbone was prevented completely by freezing (Figure 1). Freezing prevented lipid peroxidation by reducing the formation and slowing down the movement of radiation-derived radicals. These results imply the importance of bilayer hydration to lipid peroxidation. Further support for the role of bilayer hydration in lipid peroxidation was found by using PEG^{2000}-grafted liposomes. Damage to EPC liposomes and EPC/DHP-PEG^{2000} liposomes was induced in two different ways: (1) γ-irradiation, which ionizes water molecules; and (2) addition of AAPH, a positively charged hydrophilic azo compound which generates a constant flux of peroxyl radicals at a temperature-dependent rate. As can be seen in Figure 2, PEG^{2000} covalently attached to a phospholipid, strongly reduces oxidative damage from γ-irradiation. The results in Figure 11, indicating a reduction in the lipid-bilayer hydration of PEGylated liposomes, also point to the role of water molecules in oxidative damage, suggesting that PEG provides protection by causing bilayer dehydration. Measurements of rate constants for $^{\bullet}OH$ reactions show that this radical is one of the most reactive chemical species, having extremely high rate constants, with many molecules found in living cells. Most of the free radicals, including $^{\bullet}OH$, have very short half-lives (generally less than 10^{-10} s) (Halliwell and Gutteridge, 1989). Based on these facts and $^{\bullet}OH$ diffusion coefficient (2.3 x 10^{-5} cm^2/s), it can be shown that $^{\bullet}OH$ radicals formed in our irradiated test system would diffuse to a distance not more than 1 nm. These facts indicate that the oxidative damage is caused by radicals derived from the water in the vicinity of the bilayer acyl chains and not from more distant outer hydration layers of grafted PEG. Initiating oxidative damage using AAPH gave results opposite to those obtained by using γ-irradiation; EPC-PEG liposomes showed greater peroxide accumulation than EPC liposomes. Since in the case of AAPH the damaging species does not originate from water molecules, but from the AAPH itself, the presence of PEG and its influence on the degree of bilayer hydration did not prevent or reduce oxidative damage. On the contrary, the reduction in bilayer hydration caused by PEGylation can allow vacant sites in the membrane to be occupied by AAPH molecules.

AAPH, a positively charged small molecule, is attracted to the liposome surface due to the negatively charged DHP-PEG (Tirosh et al., 1998), a fact which can further explain the enhancement of oxidative damage in the EPC-PEG liposomes.

Acknowledgments

This work was supported in part by a grant of BSF (95/318) to Y. Barenholz. We thank Mr. Sigmund Geller for his help in editing the manuscript.

REFERENCES

Almagor, A., Priev, A., Barshtein, G., Gavish, B., and Yedgar, S., 1997, Reduction of protein volume and compressibility by macromolecular cosolvents:dependence on the cosolvent molecular weight, *Biochim. Biophys. Acta* :in press.

Arakawa, T., and Timasheff, S.N., 1985, Mechanism of polyethylene glycol interaction with proteins, 24:6756.

Arnold, K., Zschoemig, O., Barthel, D., and Herold, W., 1990, Exclusion of polyethylene glycol from liposome surfaces, *Biochim. Biophys. Acta* 1022:303.

Baekmark, T.R., Pedersen, S., Jorgensen, K., and Mouritsen, O.G., 1997, The effect of ethylene oxide containing lipopolymers and tri-block copolymers on lipid bilayers of dipalmitoylphosphatidylcholine, *Biophys. J.* 73:1479.

Barenholz, Y., and Amselem, S., 1993, Quality control assays in the development and clinical use of liposome-based formulations, in: *Liposome Technology*, G. Gregoriadis, ed., CRC Press, Boca Raton.

Barenholz, Y., Freire, E., Thompson, T.E., Correa-Freire, M.C., Bach, D., and Millar, I.R., 1983, Thermotropic behavior of aqueous dispersions of glucosylceramide-dipalmitoylphosphatidylcholine mixtures, *Biochemistry* 22:3497.

Barenholz, Y., Gibbes, D., Litman, B J., Goll, J., Thompson, T.E., and Carlson, F.D., 1977, A simple method for the preparation of homogeneous phospholipid vesicles, *Biochemistry* 16:2806.

Barenholz, Y., and Lasic, D.D., 1996, *Handbook of Nonmedical Applications of Liposomes*, CRC Press, Boca Raton.

Bhat, R., and Timasheff, S.N., 1992, Steric exclusion is the principal source of the preferential hydration of proteins in the presence of polyethylene glycols, *Protein Sci.* 1:1133.

Bligh, E.G., and Dyer, W. J., 1959, A rapid method of total lipid extraction and purification, *Can. J. Biochem. Physiol.* 37:911.

Blume, G., and Cevc, G., 1993, Molecular mechanism of the lipid vesicle longevity *in vivo*, *Biochim. Biophys. Acta.* 1146:157.

Buckin, V.A., Kankiya, B.I., Bulichov, N.V., Lebedev, A.V., Gukovsky, I.Y., Chuprina, V.P., Sarvazyan, A.P., and Williams, A.R., 1989, Measurement of anomalously high hydration of (dA)n(dT)n double helices in dilute solution, *Nature* 340:321.

Buckin, V.A., Kankiya, B.I., Rentzeperis, D., and Marky, L.A., 1994, Mg^{2+} recognizes the sequence of DNA through its hydration shell, *J. Am. Chem. Soc.* 116:9423.

Buckin, V.A., Sarvazyan, A.P., and Pasechnik, V. I., 1979, Study of vesicular lipid membranes by the ultrasound method, *Biofizika* 24:61.

Chalikian, T.V., Sarvazyan, A.P., and Breslauer, K. J., 1993, Partial molar volumes, expansibilities, and compressibilities of a,w-amonocarboxylic acids in aqueous solutions between 18 and 55° C., *J. Phys. Chem.* 97:13017.

Chalikian, T.V., Sarvazyan, A.P., Plum, G.E., and Breslauer, K.J., 1994, Influence of base composition, base sequence, and duplex structure on DNA hydration: apparent molar volumes and apparent molar adiabatic compressibilities of synthetic and natural DNA duplexes at 25° C, *Biochemistry* 33:2394.

Colotto, A., Kharakoz, D.P., Lohner, K., and Laggner, P., 1993, Ultrasonic study of melittin effects on phospholipid model membranes, *Biophys. J.* 65:2360.

deGennes, P.G., 1987, Polymers at an interface: a simplified view, *Adv. Colloid Interface Sci.* 27:189-209.

Eggers, F., and Funk, T., 1973, Ultrasonic measurements with milliliter liquid samples in the 0.5-100 MHz range, *Rev. Sci. Instrum.* 44:969.

Goren, D., Gabizon, A., and Barenholz, Y., 1990, The influence of physical characteristics of liposomes containing doxorubicin on their pharmacological behavior, *Biochim. Biophys. Acta* 1029:285.

Grit, M., Crommelin, D.J.A., and Lang, J.K., 1991, Quantitative determination of phosphatidylcholine, phosphatidylglycerol and their lyso forms from liposome dispersions by high performance liquid chromatography (HPLC) using high sensitivity refractive index detection, *J. Chromatogr.* 585:239.

Halliwell, B., and Gutteridge, J.M.C., 1989, *Free Radicals in Biology and Medicine*, Clarendon Press, Oxford.

Harada, S., Nakajima, T., Komatsu, T., and Nakagawa, T., 1978, Apparent molar volumes and adiabatic compressibilities of ethylene glycol derivatives in water at 5, 25, and 45 °C, *J. Solut. Chem.* 7:463.

Hristova, K., and Needham, D., 1994, The influence of polymer-grafted lipids on the physical properties of lipid bilayers: a theoretical study, *J. Colloid Interface Sci.* 168:302.

Jordan, C.F., Lerman, L.S., and Venable, J.H., 1972, Structure and circular dichroism of DNA in concentrated polymer solutions, *Nature* 236:67.

Kell, G.S.J., 1975, Volume properties of ordinary water, *Chem.Eng. Data*, 20:97.

Kenworthy, A.K., Hristova, K., Needham, D., and McIntosh, T.J., 1995a, Range and magnitude of the steric pressure between bilayers containing phospholipids with covalently attached polyethylene glycol., *Biophys. J.* 68:1921.

Kenworthy, A.K., Simon, S.A., and McIntosh, T.J., 1995b, Structure and phase behavior of lipid suspensions containing phospholipids with covalently attached polyethylene glycol, *Biophys. J.* 68:1903.

Kharakoz, D.P., 1989, Volumetric properties of proteins and their analogs in diluted water solutions. 1. Partial volumes of amino acids at 15-55 °C., *Biophys. Chem.* 34:115.

Kharakoz, D.P., Colotto, A., Lohner, K., and Laggner, P., 1993, Fluid-gel interphase line and density fluctuations in dipalmitoylphosphatidylcholine multilamellar vesicles. An ultrasound study, *J. Phys. Chem.* 97:9844.

Kharakoz, D.P., and Sarvazyan, A., 1993, Hydrational and intrinsic compressibilities proteins, *Biopolymers* 33:11.

Kiyosawa, K., 1991, Volumetric properties of polyols (ethylene gly, glycerol, meso-erythritol, xilitol and mannitol) in relation to their membrane permeability: group additivity and estimation of the maximum radius of the molecules, *Biochim. Biophys. Acta,* 1064:251.

Kuhl, T.L., Leckband, D.E., Lasic, D.D., and Israelachvili, J.N., 1994, Modulation of interaction forces between bilayers exposing short-chained ethylene oxide headgroups, *Biophys. J.* 66:1479.

Lasic, D.D., 1996, Doxorubicin in sterically stabilized liposomes, *Nature,* 380:561.

Lasic, D.D., and Martin, F., 1995, *Stealth Liposomes*, CRC Press, Boca Raton.

Lichtenberg, D., and Barenholz, Y., 1988, Liposomes: preparation, characterization, and preservation, in: *Methods of Biochemical Analysis*, D. Glick, ed., John Wiley & Sons, New York.

MacDonald, R.C., MacDonald, R.I., Menco, B.P., Takeshita, K., Subbarao, N.K., and Hu, L., 1991, Small-volume extrusion apparatus for preparation of large, unilamellar vesicles, *Biochim. Biophys. Acta,* 1061:297.

Marino, F.J., and Benjamin, F., 1984, A review of current principles and practices, in: *Pharmaceutical Dosage Forms: Parenteral Medications*, K. E. Avis, L. Lachman and H. A. Lieverman, ed., Marcel Dekker, New York.

Marsh, D., 1990, Phase behavior and hydration, in: *Handbook of Lipid Bilayers*, D. Marsh, ed., CRC Press, Boca Raton.

Martin, F.J., and Lasic, D.D., 1991, The molecular origin of the long blood circulation time of Stealth liposomes., *Biophys. J.* 59:497.

Mitaku, S., and Aruga, S., 1982, Effect of calcium ion on the mechanical properties of lipid bilayer membrane, *Biorheology* 19:185.

Mitaku, S., Jippo, T., and Kataoka, R., 1983, Thermodynamic properties of the lipid bilayer transition. Pseudocritical phenomena, *Biophys. J.* 42:137.

Needham, D., McIntosh, T.J., and Lasic, D., 1992, Repulsive interactions and mechanical stability of polymer-grafted lipid membranes, *Biochim. Biophys. Acta.* 1108:40.

Priev, A., Almagor, A., Yedgar, S., and Gavish, B., 1996, Glycerol decreases the volume and compressibility of protein interior. *Biochemistry.* 35:2061.

Priev, A., Sarvazyan, A.P., Dzantiev, B.B., Zherdev, A.V., and Cherednikova, T.V., 1990, Changes in the adiabatic compressibility of mono- and polyclonal antibodies during interaction with antigens, *Mol. Biol.* 24:629.

Samuni, A.M., and Barenholz, Y., 1997, Stable nitroxide radicals protect lipid acyl chains from radiation damage, *Free Radic. Biol. Med.* 22:1165.

Samuni, A.M., Barenholz, Y., Crommelin, D.J.A., and Zuidam, N. J., 1997, γ-irradiation damage to

liposomes differing in composition and their protection by nitroxides, *Free Radic. Biol. Med.* 23:972.
Sarvazyan, A.P., 1982, Development of methods of precise ultrasonic measurements in small volumes of liquids, *Ultrasonics* 20:151.
Sarvazyan, A.P., 1991, Ultrasonic velocimetry of biological compounds, *Annu. Rev. Biophys. Biophys. Chem.* 20:321.
Sarvazyan, A.P., and Chalikian, T.V., 1991, Theoretical analysis of an ultrasonic interferometer for precise measurements at high pressure, *Ultrasonics* 29:119.
Shilnikov, G.V., Priev, A., and Achmedov, A., 1991, Bulk-elastic properties of aqueous solutions of certain carbohydrates, *Biofizika* 36:276.
Tirosh, O., Barenholz, Y., Katzhendler, J., and Priev, A., 1998, Hydration of polyethylene glycol-grafted liposomes, *Biophys. J.* 74:in press.
Tirosh, O., Kohen, R., Alon, A., Katzhendler, J., and Barenholz, Y., 1997a, Novel synthetic phospholipid protects lipid bilayers against oxidative damage: role of hydration layer and bound water, *J. Chem. Soc. Perkin Trans. 2.* 383.
Tirosh, O., Kohen, R., Katzhendler, J., Alon, A., and Barenholz, Y., 1997b, Oxidative stress effect on the integrity of lipid bilayers is modulated by cholesterol level of bilayers, *Chem. Phys. Lipid.* 87:17.
Torchilin, V.P., Omelyanenko, V.G., Papisov, M.I., Bogdanov, A.A., Trubetskoy, V.S., Herron, J.N., and Gentry, C.A., 1994, Polyethylene glycol on the liposome surface: on the mechanism of polymer-coated liposome longevity, *Biochim. Biophys. Acta.* 1195:11.
Woodle, M.C., 1993, Surface-modified liposomes: assessment and characterization for increased stability and prolonged blood circulation., *Chem. Phys. Lipids* 64:249 .
Woodle, M.C., 1997, Polyethylene glycol-grafted liposome therapeutics, in: *Polyethyleneglycol Chemistry and Biological Applications*, S. Z. J.M. Harris, ed., American Chemical Society, Washington.

PHYSICOCHEMICAL CHARACTERIZATION OF DOTAP-CONTAINING LIPOPLEXES BY FLUORESCENT PROBES: RELEVANCE TO LIPOFECTION

Danielle Hirsch-Lerner[1], Nicolaas J. Zuidam[2] and Yechezkel Barenholz[1]

[1]Department of Biochemistry, The Hebrew University–Hadassah Medical School, PO Box 12272, Jerusalem, Israel
[2]Department of Pharmaceutics, Utrecht University, Utrecht, The Netherlands

INTRODUCTION

Delivering therapeutic genes to cells by their complexation with cationic liposomes to form lipoplexes has been shown to be efficient *in vitro* and *in vivo* (Behr, 1994; Ledley, 1995). The lipoplexes (plasmid DNA complexed with cationic liposomes) are safer than viral vectors (Mulligan, 1993; Crystal, 1995) for the following reasons: the absence of viral DNA, no constraint on DNA size, protection of DNA from degradation, and ability to target recombinant genes to specific cells. Therefore, lipoplexes might become the mainstream of research for gene therapy. The successful use of these lipoplexes will depend on their efficient delivery to cells and their ability to produce therapeutic levels of gene expression.

Cationic liposomes are composed of an amphipathic (mono or poly) cationic lipid and, in many cases, a helper lipid. Since DOTMA was first used by Felgner et al. (1987), other cationic lipids have been developed to improve gene transfer efficacy and decrease toxicity. Among the monocationic lipids are DOTAP (Leventis and Silvius, 1990); DC-CHOL (Gao and Huang, 1991); and DMRIE (Felgner et al., 1994). The polycationic lipids tested include lipopoly(L-lysine) (Zhou and Huang, 1994), DOSPA (Eastman et al., 1997) and DOGS (Behr et al., 1989).

High levels of transfection into eukaryotic cells for transient or stable gene expression using a liposome formulation of DOTAP were obtained (Buci et al., 1992; Levine et al., 1992; Strauss et al., 1993). In addition, DOTAP was successfully used in the transfer of other negatively charged molecules such as RNA and oligonucleotides (Walker et al., 1992; Chen et al., 1993) into eukaryotic cells.

The helper lipid is neutral and the best results were obtained with DOPE at a mole ratio

Targeting of Drugs 6: Strategies for Stealth Therapeutic Systems
Edited by Gregoriadis and McCormack, Plenum Press, New York, 1998

of cationic lipid/DOPE of ~1 (Felgner et al., 1994; Zhou and Huang, 1994).

Relationships between the physicochemical properties of lipoplexes and the transfection efficiency of the lipoplexes are still not clear. The formation of the lipoplexes is a multifactorial process as it depends on thermodynamic as well as kinetic parameters. Most lipoplexes used so far result from a spontaneous interaction between the cationic liposomes and the negatively charged DNA. The gain in electrostatic energy due to the neutralization is the main driving force: when the gain is larger than the bending energy needed for the lipids to adapt their organization so they can interact with the DNA, lipoplexes will be formed (May and Ben-Shaul, 1997).

The electrostatic energy and bending energy described above are determined by such physicochemical factors as charge ratio, concentration, DNA structure, ionic strength, sequence of mixing and incubation time before transfection. The lipoplexes demonstrate polymorphism (Zabner et al., 1995), which can be determined by static light-scattering (Zuidam and Barenholz, 1998). In cases where lipoplexes are highly unstable, kinetic phenomena are also taking place.

To better understand the relevance of lipoplexes to gene delivery we focus on one of the most studied cationic lipids DOTAP in the presence or absence of the helper lipid DOPE. In this study we characterize, by using various fluorescent probes, DOTAP and DOTAP/DOPE (mole:mole) large unilamellar vesicles (LUV) before and after complexation with DNA. In this way we follow the electrostatics at the plane of interaction (lipid/water interface), the lateral organization and defect formation in the lipid bilayer, and the size and size instability of lipoplexes. The relevance of these properties to lipofection efficiency will be discussed.

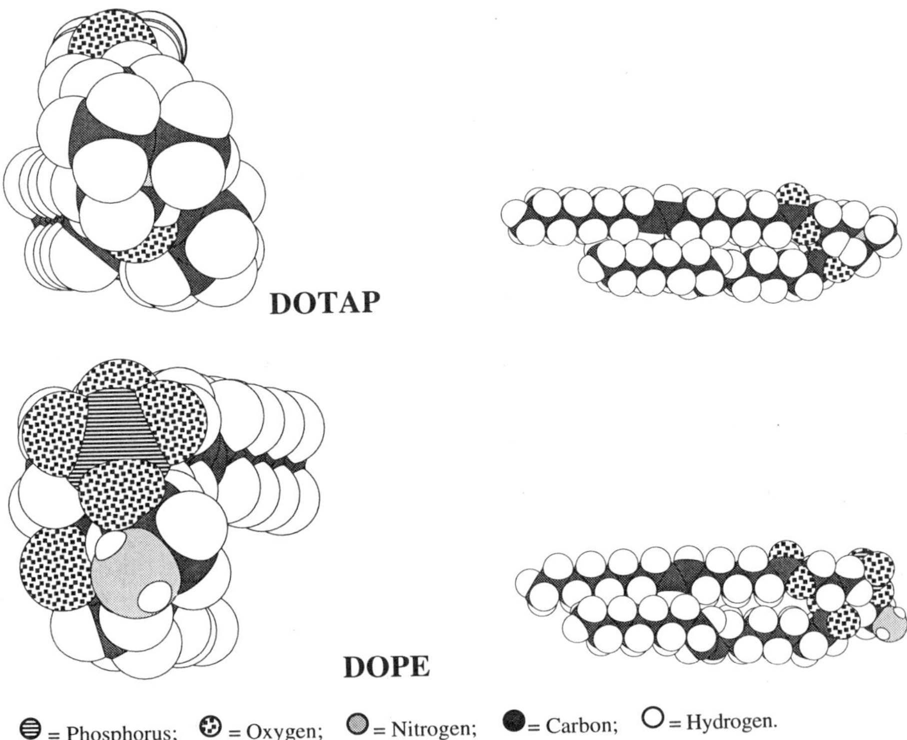

Figure 1. Space-filling models of the molecules DOTAP and DOPE viewed from the headgroup side and from the lateral plane (for more details see text)

CHARACTERIZATION OF CATIONIC LIPOSOMES

All the liposomes used in our studies (Zuidam and Barenholz, 1997, 1998; Hirsch-Lerner and Barenholz, 1998) were downsized by extrusion through 0.1-μm size polycarbonate filters to form ~100 nm LUV. The T_m (the temperature of maximum change in the heat capacity during the gel-to-liquid phase transition) of DOTAP (Hirsch-Lerner and Barenholz, 1998) and the T_c (the onset temperature during the gel-to-liquid phase transition) of DOPE (Cullis et al., 1978) are -11 °C and -16 °C respectively. Therefore, it is expected that at room temperature DOTAP-containing cationic liposomes were in the fluid phase. This was confirmed by the 1-(4-(trimethylammoniumphenyl)-6-phenylhexa-1,3,5-triene (TMADPH) steady-state fluorescence anisotropy of DOTAP/DOPE liposomes of ~0.2 at room temperature.

The properties of the lipid headgroup region are important for the neutralization by DNA, for the level of hydration at the lipid/water interface, and for the bending energy. In comparing DOTAP and DOPE, each has two identical acyl chains (oleoyl chains); the difference between the two molecules is in their headgroups. Figure 1 shows a space-filling model of the three lipids after free energy minimization (using CSC Chem3D/Plus software, Cambridge, MA); this figure demonstrates that the headgroup area of DOPE is larger than that of DOTAP. However, other parameters which relate to the intermolecular interactions should also be considered, such as hydrogen bonding at the headgroup region and its effect on headgroup hydration are very important. Weaker hydration was observed for phosphatidylethanolamines (PE), which was explained by the network of highly stable hydrogen bonds between the primary amino group of PE and the phosphate group of either phosphatidylcholine or neighbor PE molecules (Gruen et al., 1984; Wilschut and Hoekstra, 1986).

Determination of Electrostatic Parameters of LUV

The electrostatic properties of cationic liposomes and lipoplexes were studied by including 0.5 mole% of the fluorophore 4-heptadecyl-7-hydroxycoumarin (HC) which is sensitive to surface pH and surface potential. HC is a weak acid. Proton binding to a molecule which is present at the water/lipid interface, such as HC, can be described by :

$$pK_a = pK_H + \Delta pK_{pol} + \Delta pK_{el}$$

where pK_H is the intrinsic proton binding constant, ΔpK_{pol} is the shift in pK_a due to a change in surface polarity (dielectric constant), and ΔpK_{el} is the shift in pK_a due to a change in surface potential (for more details see Zuidam and Barenholz, 1997). At $pH<pK_a$ the maximal fluorescence intensity is found at an excitation wavelength of about 320 nm (emission 450 nm), and at $pH>pK_a$ the excitation maximum is shifted to about 380 nm. The fluorescence intensity at excitation wavelength of 330 nm is independent of surface pH and surface potential, thus 330 nm is the isosbestic point. This fluorescence intensity reflects only the concentration of HC in the lipid bilayer. Therefore, the dissociation degree of HC incorporated into cationic liposomes can be monitored by the ratio of the excitation fluorescence intensities at 380 and 330 nm or 320 and 330 nm (emission at 450 nm). The 380 nm/330 nm ratio is preferable due to much higher fluorescence intensity after excitation at 380 nm than at 320 nm; another reason is that we were dealing with cationic liposomes in which the surface pH was higher than the pK_a of HC.

Changes in electrostatic properties of LUV as a measure of LUV stability. The effect of liposome storage time on the dissociation degree of HC (380 nm/330 nm ratio) in cationic liposomes is described in Figure 2. Without dilution (at 40 mM phospholipids) for DOTAP/DOPE 1:1 the change was very slow (<10% in 48 h). However, when these LUV were diluted to 80 μM, large changes in the dissociation degree of HC with time were observed. A

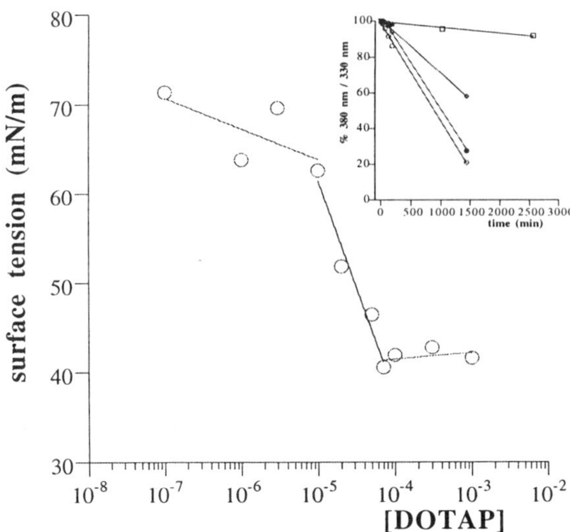

Figure 2. The relationship between the crtitical aggregation concentration (CAC) or DOTAP (calculated from the main figure) and the stability of DOTAP-containing liposomes as determined from the change in 380nm/330nm fluorescence ratio (insert). The concentration and composition of the cationic liposome dispersions during storage were:○, 80μM DOTAP/DOPE (1/1); •, 80μM DOTAP/DOPC (1/1); ◇, 80μM DOTAP; □, 40 mM DOTAP/DOPE (1/1) (measured immediately after dilution to 80μM) (n=2)

large linear increase represented by the dissociation degree of HC occurred with time (80% reduction in 380 nm/330 nm fluorescence ratio after 24 h). The intensity of fluorescence at 330 nm (the isosbestic point, which reflects the amount of HC) barely changed with time, indicating that the major changes in the dissociation degree of HC cannot be explained by a leak of the probe from the lipid or its degradation. The changes at 40 mM are due to a chemical hydrolysis of the liposomal ester lipids DOTAP and DOPE (Zuidam and Barenholz, 1997). The experimental hydrolysis rate constants indicate a degradation rate of 2% of DOTAP and 3% of DOPE after 24 h storage at 25°C, which contributes up to 5% nonesterified negatively charged oleate. The decrease in the percent dissociation degree of 40 mM DOTAP/DOPE (1/1) liposomes of about 8% after 16 h storage (see Figure 2) supports the findings made above that the hydrolysis producing free fatty acids which is due to the high surface pH reduces the surface potential of cationic bilayers.

However, the fast changes in surface potential cannot be explained by lipid hydrolysis. This instability was found to be related to the cationic lipid critical aggregation concentration (CAC). DOTAP has a relatively high CAC (7×10^{-5} M; see Figure 2) and dilution to concentrations below its CAC (to 4×10^{-5} M) induced instability, probably due to desorption of DOTAP, as was also supported by changes in static light-scattering (Zuidam and Barenholz, 1997).

The stability properties of the complexes should not change after dilution. However, at a certain stage in the transfection process, lipids have to be released from the DNA–lipid complexes (Zabner et al., 1995; Szoka et al., 1996). It was previously suggested that after internalization by endocytosis, anionic lipids of the cell membrane induce the release of DNA from the lipoplexes (Szoka et al., 1996). Here, we suggest an additional mechanism based on the dissociation of the lipoplex due to the release of DOTAP from the complex at concentrations below its CAC.

Ionization Parameters of DOTAP and DOTAP/DOPE LUV. The dissociation degree of HC incorporated into the cationic liposomes was followed by the ratio of fluorescence

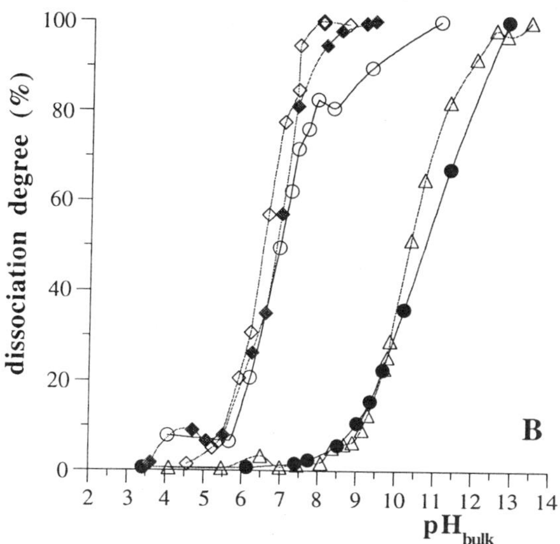

Figure 3. The dissociation degree of HC in liposomes as translated into percentages of the maximum value against the pH_{bulk}. The liposomes were composed of DOTAP/DOPE (1/1) (○), DOTAP/DOPC (1/1) (♦), DOTAP (◇), DOPC (△), and DOPC/DOPE (1/1) (●). The curves shown are not fitted

intensities at 380 nm/330 nm as a function of pH_{bulk}. Titration curves of cationic LUV, presented in Figure 3, show single sigmoids differing from each other in pH dependency. DOPC liposomes are neutral in the pH range used in this study (Tatulian, 1993). Therefore, ΔpK_{el} of HC in charged liposomal membranes can be estimated by taking the pK_a of HC in DOPC liposomes as a neutral reference, assuming that there is no change in surface polarity (see eq. above).

Figure 3 shows that the pK_a's of HC in liposomes of 100% DOTAP and of DOTAP/DOPE (1:1) were 4.2 and 3.8 pH units lower (respectively) than the pK_a of HC in neutral DOPC liposomes (Tatulian, 1993). This shift means that the ΔpK_{el} of these cationic liposomes is negative, therefore their surface potential is highly positive (240 mV and 217 mV for DOTAP and DOTAP/DOPE LUV, respectively) and their $pH_{surface}$ is high (see Table 1).

The dissociation curve of HC in DOPC/DOPE (1/1) liposomes (Fig. 3) was slightly shifted to higher pH values (relative to HC in DOPC liposomes), probably due to a partial dissociation of the primary amine group of DOPE at high pH_{bulk} compared with the complete dissociation of the choline DOPC quaternary amine. This indicates that the helper lipid (DOPE or DOPC) affects some electrostatic characteristics of the cationic LUV (as shown in Table 1). Between pH_{bulk} ~4 to 5.5 the dissociation degree of HC in DOTAP/DOPE and DOTAP/DOPC (all 1/1) showed a plateau; i.e., at pH_{bulk} 5 an ionizable group is present, probably the phosphate group of DOPE or DOPC. Generally, the pK_a of the phosphate group of PE or PC is recorded between pH 1 and pH 3.5, depending on the method of measurement and the medium (Cevc, 1990; Tocanne and Teissié, 1990; Tatulian, 1993). We suggest a salt bridge between the quaternary amine of DOTAP and the phosphate group of DOPE or DOPC as responsible for this pK_a shift of the phosphate group. The presence of such salt bridges does not affect the overall positive charge which is compensated by the contribution of the positively charged groups of DOPE and DOPC, although the exact location of the positive charge might be changed.

Table 1. The pK_a of a MU or HC in micelles or liposomes and the electric properties of the lipid surfaces in 20 mM Hepes buffer ($pH_{bulk} = 7.4$)[a]

Probe	Micellar or liposome dispersion	pK_a	Ψ_0^{HC} (mV)	Ψ_0^{GC} (mV)	$pH_{surface}^{HC}$ measured	$pH_{surface}^{HC}$ calculated	$pH_{surface}^{GC}$ calculated
MU		7.9					
HC	hydrogenated Triton X-100	9.0	0	0	7.4 [b]	7.4 [b]	7.4 [b]
HC	DOPC [b]	10.5	0	0	7.4 [b]	7.4 [b]	7.4 [b]
HC	DOPC/DOPE 1/1 [b]	10.7	0	0	7.4 [b]	7.4 [b]	7.4 [b]
HC	DOTAP/DOPE 1/1	6.8	217	200	10.9	11.1	10.8
HC	DOTAP/DOPC 1/1	6.7	222	193	11.2	11.2	10.7
HC	DOTAP	6.3	240	235	11.6	11.5	11.4

Abbreviations: MU, methylumbelliferone; Ψ_0^{HC}, surface potential determined by HC fluorescence; Ψ_0^{GC}, surface potential calculated by Gouy-Chapman approximation; $pH_{surface}^{HC}$, $pH_{surface}$ determined by HC fluorescence; $pH_{surface}^{GC}$, $pH_{surface}$ calculated by Gouy-Chapman approximation.

a See text for symbols and determination of values. The pK_a, Ψ_0^{HC}, and Ψ_0^{GC} are determined using Eqs. 1, 2, and 3, respectively (see appendix). The $pH_{surface}^{HC}$ was measured by comparing the value of the dissociation degree of HC in the cationic liposomes with the value of the dissociation degree of HC in DOPC-liposomes (using Eq. 1). The $pH_{surface}^{HC}$ was calculated using Eqs. 2 and 3. The $pH_{surface}^{GC}$ was calculated using Eqs. 4 and 6. Note that both $pH_{surface}$ and Ψ_0 also depend on the distance from the boundary of the membrane (which is λ–in the present study, 0.74 nm). Gouy–Chapman calculations with ε_r of 78 (as done by many others) will increase λ and decrease the values for both $pH_{surface}^{GC}$ and Ψ_0^{GC}. Also note that $pH_{surface}$ is actually an apparent quantity, because it is also affected by both a degeneration of the proton activity due to the low ε_r of the medium close to the liposomal bilayer (Harned and Owen, 1958; Fernández and Fromherz, 1977) and by a decrease in the proton concentration due to non-electrostatic effects (Cevc, 1990). Theoretical correction for the latter effect is poorly understood (Cevc, personal communication). We assume that these two effects are equal for both neutral and charged liposomes (no change in surface polarity), so it will not influence the electrostatic calculations.

b The $pH_{surface}^{HC}$ cannot be measured here because the measurement of pH values by use of a pH-sensitive probe is limited to the range where the fluorescence is sensitive to changes in pH. It has been assumed that Triton X-100-micelles and DOPC and DOPC/DOPE (1/1) liposomes are neutral at pH 7.4.

CHARACTERIZATION OF THE DNA–CATIONIC LIPOSOME COMPLEXES

Neutralization of DOTAP by DNA

Neutralization of cationic lipids (Zuidam and Barenholz, 1998) in the assembly by DNA was determined using the HC approach described above combined with Gouy–Chapman calculations. The measurements were performed 15 min post mixing of DNA and LUV. It is worth noting that the limitation of this method is that it can be only used in the pH range in which the fluorophore is responsive. As shown in Figure 2, this range fits well in the current study.

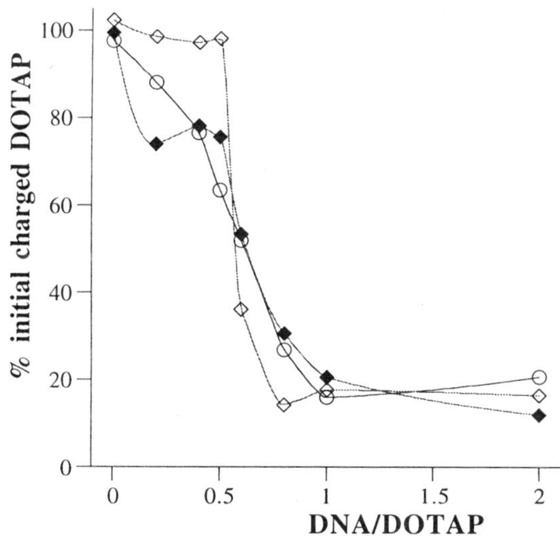

Figure 4. Percentage of charged cationic lipids in the lipid assemblies upon addition of plasmid DNA to cationic liposomes as a function of the mole ratio of DNA/DOTAP at 15 min (n = 2). The data were calculated from the measurements shown in Fig. 3. See text for further details. The composition of the cationic liposomes were DOTAP/DOPE 1/1 (○), DOTAP/DOPC 1/1 (♦), 100% DOTAP (◇)

Figure 4 shows that all three LUV compositions used in this study differ in their neutralization curves in which the DOTAP was titrated with DNA. For 100% DOTAP, DNA has only minimal effect until the DNA/DOTAP ratio of 0.5 is reached. At higher DNA/DOTAP ratio DOTAP neutralization increased linearly, reaching a plateau of ~80% neutralization at DNA/DOTAP ratio of ~1.0. For DOTAP in DOTAP/DOPC LUV the neutralization by DNA was already 25% at DNA/DOTAP ratio of 0.2. Above the ratio of 0.5 the neutralization responded linearly to the increase in DNA/DOTAP, reaching a plateau of ~80% DOTAP neutralization at DNA/DOTAP ~1.0. For DOTAP/DOPE LUV, titration of DOTAP by DNA was continuous until reaching a plateau of ~ 80% DOTAP neutralization at DNA/DOTAP ~1.0. That is for all DOTAP–containing LUV, DOTAP neutralization reached a maximal level of 80% at DNA/DOTAP 1:1.

The cationic LUV neutralization did not reach its minimum value of zero after the addition of an excess of DNA. This result points out that not all the positive charges of the cationic LUV are reached and/or fully neutralized by the DNA phosphates, although the explanation that the probe HC is located closer to the cationic lipid than to DNA phosphates cannot be ruled out.

As shown by others, the DNA/cationic lipid charge ratio of 1.0 is a critical ratio (Gershon et al., 1993; Eastman et al., 1997). A plateau of maximal percent neutralization as measured by zeta potential was reached at $DNA^{-}/L^{+}=1$. The study of Minsky and coworkers (Gershon et al., 1993) suggests that at the ratio of ~1, DNA induces fusion of liposomes so that it is encapsulated by a lipid bilayer, while at the same time the DNA collapses.

Exposure of Lipid Acyl Chains to Water

The fluorescent probes TMADPH and 1,6-diphenylhexa-1,3,5-triene (DPH) were used in cationic LUV to follow lipoplex formation after addition of DNA. TMADPH was chosen because its positively-charged quaternary amine resides close to the lipid/water interface (Ben-

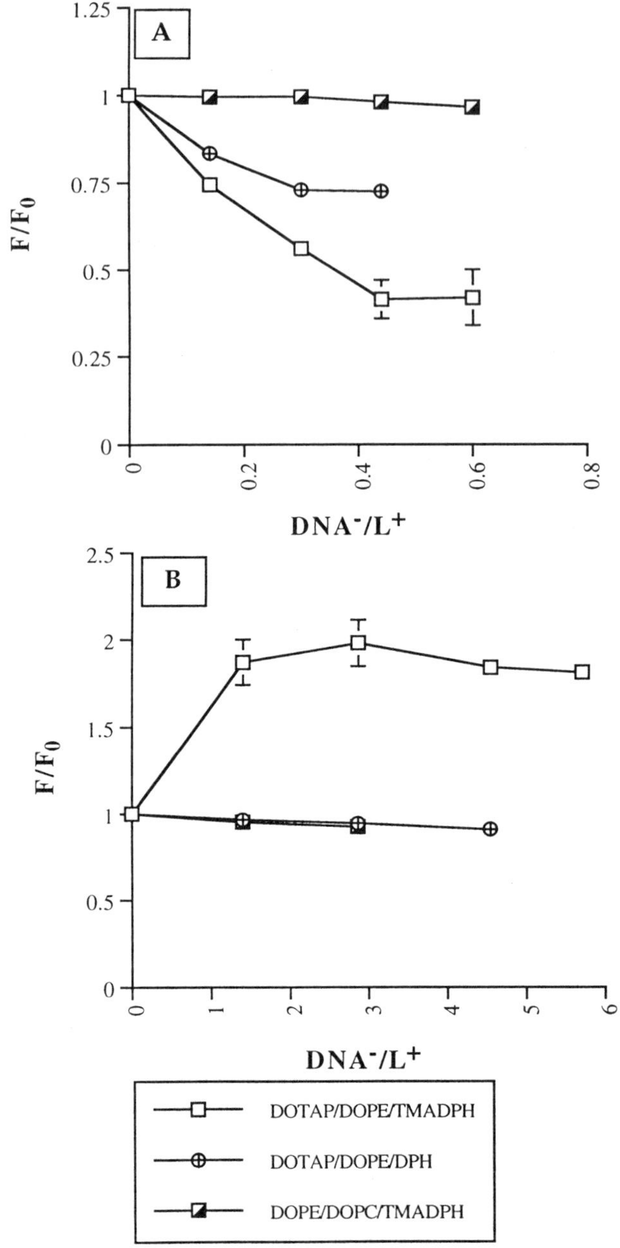

Figure 5. Effect of mole charge ratio of DNA^-/L^+ on relative fluorescence intensity of various liposomes labeled with either TMADPH or DPH. Lipid concentrations were 0.78 mM for the range of DNA^-/L^+ from 0 to 1 (A) and 0.08 nM for the range of DNA^-/L^+ from 1 to 5.7 (B). F_0 is the fluorescence intensity without DNA

Yashar and Barenholz, 1991) and its fluorescence yield (but not spectrum) is sensitive to the polarity of the environment (Lentz, 1989; Ho and Stubbs, 1992; Borenstain and Barenholz, 1993). On the other hand, DPH has no polar group and therefore is located deeper in the hydrophobic region of the bilayer (Borenstain and Barenholz, 1993). Figure 5 represents the changes in fluorescence intensities upon increasing amounts of DNA; two ranges of DNA^-/L^+

(0 to 0.6 and 1 to 5.7) were studied. DNA induces TMADPH quenching at DNA$^-$/L$^+$ <1 (Figure 5A) and fluorescence enhancement at DNA$^-$/L$^+$ =1 (Figure 5B). Similar results were observed with linear or plasmid DNA (data not shown). No such effects occurred when the LUV were labeled with DPH. Controls of DOPC/DOPE/TMADPH showed that the presence of a cationic lipid is necessary to observe any effect of the DNA.

TMADPH fluorescence measurements indicate major differences between complexes formed at DNA$^-$/L$^+$ <0.6 and DNA$^-$/L$^+$ >1. These results suggest that TMADPH is sensitive to the changes in the organization of the bilayer at the lipid/water interface. At ratios DNA$^-$/L$^+$ <0.6 the fluorescence intensity of TMADPH is quenched, i.e., suggesting an increase in fluorophore exposure to water, while at DNA$^-$/L$^+$ =1, TMADPH fluorescence intensity is increased, suggesting a reduction in the exposure to water molecules (Hirsch-Lerner and Barenholz, 1998).

These results are in good agreement with those of Gershon et al, (1993) who have worked on the interactions of DOTMA/DOPE liposomes with DNA through "DNA eyes" by measuring the reduction in the fluorescence intensity of ethidium bromide intercalated in DNA caused by the increase in the amounts of DOTMA:DOPE (1:1 mole ratio). At a specific mole charge ratio of DNA$^-$/L$^+$=1.1, a significant decrease in fluorescence intensity was monitored, suggesting DNA condensation and/or less DNA exposure to ethidium bromide. At DNA$^-$/L$^+$ >1.1, a constant high fluorescence intensity was observed.

At 150 mM NaCl, TMADPH fluorescence intensity was smaller than at lower (10 mM) NaCl concentration (data not shown). The effect of NaCl emphasizes the role of electrostatics in lipoplex formation.

The degree of fluorescence quenching with increasing DNA was lower with DPH than with TMADPH at DNA$^-$/L$^+$ <1 (Figure 5A). At DNA$^-$/L$^+$ >1 no enhancement was observed with DPH (Figure 5B). The effect of DNA on lipids occurs mainly near the lipid/water interface at the upper region of acyl chain where TMADPH is located. DPH, lacking a polar moiety, is present deeper in the hydrophobic region of the bilayer acyl chains either parallel to the lipid acyl chains or perpendicular to them between the two monolayers and therefore it is much less exposed to the water molecules (Barenholz et al., 1983; Talmon, 1996). On the other hand, TMADPH, being positively charged, is nearer to DNA than is DPH. The result showing no effect of DNA on the fluorescence intensity of DOPC/DOPE/TMADPH in both DNA$^+$/L$^-$ ranges indicates that this interaction requires the presence of cationic lipid. It is also a control showing the validity of this approach, as it indicates that the effect is not related to a direct interaction of the fluorophore with the DNA.

At DNA$^-$/L$^+$ <0.6, an increase in the fluorescence anisotropy was reached which was close to the r_o(fluorescence anisotropy of the probe itself) of TMADPH (data not shown). This results from restriction of TMADPH motion in the lipoplexes as a result of DNA interaction which is more restricted than in the intact LUV. This indicates that the packing density of the lipid acyl chains near the lipid/water interface in the lipid domains increases.

MECHANISMS OF DNA–CATIONIC LIPOSOME INTERACTION

One of the ways to characterize lipoplexes, their formation and their stability is to measure their size distribution and the kinetics of changes in size. The common way to measure size distribution of lipoplexes is based on the practice in liposomology of using photon correlation spectroscopy (PCS), also referred to as dynamic light scattering (DLS) or quasi elastic light scattering (QELS) (Barenholz and Amselem, 1993; Ostrovsky, 1993). Lipoplexes do not fit all the requirements for valid use of this method (Ostrovsky, 1993) under all conditions because of complex polymorphism in shape, size, and possibly composition (Sternberg et al., 1994; Zabner et al., 1995), along with hindered particle motion (Rädler et al.,

1997). Thus the use of this approach may be misleading as regards the actual size distribution of lipoplexes. The problem is well demonstrated by comparing electron micrographs obtained either by negative staining (Zuidam et al., submitted) or by cryotransmission electron microscopy (Poran et al., preliminary results). Therefore, we looked for another method to compare and follow size changes. We found that static light scattering measured by a spectrofluorometer (excitation wavelength=emission wavelength=600 nm) fulfils the requirements. It is highly sensitive (80 nmoles lipids) and linear over a broad range and can be followed with time either continuously or non-continuously. This method does not suffer from all the drawbacks of the PCS though it may give more weight to large particles (for details on static light scattering in liposomology, see Barenholz and Amselem, 1993). We applied this method, along with PCS and electron microscopy (Zuidam et al., submitted) to characterize DOTAP–based lipoplexes.

Following the mixing of DNA with DOTAP–containing LUV, the sizes and the static light-scattering of the lipoplexes based on PCS and electron microscopy were increased at $DNA^-/L^+ <1$ (0.6–0.8) (data not shown). At ratios higher than 1.0, formation of large structures was not observed confirming the results shown by others (Gustafsson et al., 1995; Zabner et al., 1995; Sternberg, 1996; Eastman et al., 1997; Rädler et al., 1997). The large complexes (μm range) seen at $DNA^-/L^+ <0.6$ are in fact aggregates of individual packed particles (Figure 4 in Hirsch-Lerner and Barenholz, 1998), similar to that observed previously with DNA/DC-CHOL/DOPE complexes (Sternberg et al., 1994). With time, at $DNA^-/L^+<0.6$, these aggregates are constantly growing (Zuidam and Barenholz, 1998) and so this system is unstable. This instability can be explained as follows: DNA condenses the lipids by interacting with the cationic lipid at the lipid/water interface; this condensing effect is stronger at $DNA^-/L^+ <1$. At this range 25% neutralization ,or less (for DOTAP/DOPE and DOTAP/DOPC or DOTAP) of the lipid was reached. The amount of DNA is not sufficient to interact with and neutralize all the cationic lipids, even if the DNA is completely expanded. Moreover, it was shown that at a certain neutralization level, DNA molecules condense to form packed shapes (Manning, 1980). These data agree with the model of Dan (1996): DNA electrostatically absorbed on cationic lipid bilayers will form ordered domains and in this way perturbation of the lipid equilibrium will occur. Packing defects at the extremity of the ordered domains will allow water penetration, increasing water content at the upper part of the bilayer acyl chains. Therefore, TMADPH, but not DPH, was quenched. The formation of defects at the outer leaflet of the bilayer could induce aggregation, maybe even fusion, and lipid mixing, as observed by others (Gershon et al., 1993; Felgner et al., 1996; Eastman et al., 1997). This could also explain the lipoplex polymorphism described by Zabner et al. (1995) and others (Rädler et al., 1997).

Three major structures of spontaneously formed lipoplexes are possible: considering that the hydrophobic part of the lipid keeps away from water: (1) DNA is coated by a monolayer of lipid showing a hexagonal array referred to as a "honeycomb" structure (Felgner et al., 1996; May and Ben-Shaul, 1997). The highest level of neutralization is expected to be found in the "honeycomb" structure because all the cationic lipids are in contact with the DNA phosphates. (2) DNA molecules are sandwiched by one side of a lipid bilayer when the other side is interacting with another DNA molecule. This latter structure is referred as multilamellar (Gustafsson et al., 1995; Dan, 1997; Rädler et al., 1997; Eastman et al., 1997). For the multilamellar structure, one would expect a maximal neutralization higher than 50%. (3) Supercoiled plasmid DNA is coated by one cylinder of lipid bilayer, forming a "spaghetti" structure (Sternberg et al., 1994). In this case, the level of maximal neutralization is expected to be 50% according to the distribution of the cationic lipid molecules between the inner and outer monolayer of the "spaghetti" structure. The DC-CHOL/DOPE liposomes forming complexes with DNA have a high probability of forming a "spaghetti" structure; indeed other previous studies (Zuidam and Barenholz, 1998) showed that the maximum neutralization level for these complexes was the lowest among the other cationic LUV. Most of the recent studies

(Gustafsson et al., 1995; Lasic et al., 1997; Rädler et al., 1997) support the finding that, in an excess of lipids, the lipoplexes are multilamellar with respect to their lipids forming aggregated multilamellar structures. A theoretical model of Dan (1998) compared the stability of the complex structures and found that the multilamellar structure has a higher thermodynamic stability than the cylindrical ones. In addition, May and Ben-Shaul (1997) suggest that, based on energy calculations, the honeycomb structures are more stable, even though the spaghetti can exist thermodynamically.

At ratios of $DNA^-/L^+ > 1$, more cationic lipid molecules interact with the DNA. Therefore, higher levels of neutralization were recorded with DOTAP systems, (~80%; Figure 4). The TMADPH anisotropy ($r = 0.245$) in this DNA^-/L^+ range was higher than that of the LUV but was lower than that of the lipoplexes at $DNA^-/L^+ < 0.6$ meaning a smaller condensing effect at the upper part of the lipid bilayer. The DNA interacts more homogeneously with most DOTAP of the lipoplexes. Therefore, no packing defects are present leading to a diminution of exposure of the hydrophobic part of the bilayer to water. This is supported by results showing little or no size increase (data not shown) in this range of DNA^-/L^+, and TMADPH fluorescence intensity enhancement (and no quenching).

RELEVANCE TO LIPOFECTION EFFICIENCY

Although it is clear that the lipofection efficiency is affected by the DNA^-/L^+ ratio, the optimal ratio is still controversial. The complexity of the problem is due to the concentration-dependent toxicity of most cationic lipids and to the lipofection process being multifactorial. The systematic study by Felgner et al. (1994) of many cationic lipids, including the DOTMA/DOPE system (which resembles the system studied here) and pRSVLac Z plasmid in COS7 cells, shows that transfection occurred over a broad range of DNA^-/L^+ ratios. The optimal ratio for transfection was dependent on the DNA^-/L^+ ratio and also on the absolute concentration of DNA. At a DNA concentration range of 0.95–7.5 μM the optimal DNA^-/L^+ ratio was in the range of 0.2–0.6, while at higher DNA concentrations the optimal DNA^-/L^+ ratio was higher, reaching a value of ≥1.5 at DNA concentrations above 15 μM. This shift may be an artefact related to the high cytotoxicity of high concentrations of cationic lipid. This toxicity is reduced when a high concentration of DNA is present. Our results measuring hGH expression after transfection of NIH 3T3 cells with pG16-hGH/(DOTAP/DOPE 1/1) complexes indicate that at a nontoxic cationic lipid concentration, the optimal DNA^-/L^+ is 0.2~0.5, and in such lipoplexes less than 40% of the cationic lipids are neutralized by the DNA (Zuidam and Barenholz, 1998). Our results showed that instability of the lipid–DNA complex leads to aggregates that are heterogeneous in size. Zabner et al. (1995), using transmission electron microscopy, found that DNA–lipid complexes showed a high degree of polymorphism and heterogeneity, a major obstacle in studies attempting to optimize transfection efficiency. It may be that the polymorphism is due in part to different levels of instability of the DNA–lipid complexes.

The relative contributions of level of instability and degree of neutralization to the transfection process remain to be determined.

APPENDIX: Calculation of % charged lipids in the DNA–lipid complexes

The curve of HC in DOPC-liposomes (shown in Zuidam and Barenholz, 1997) was fitted to a modified Henderson–Hasselbach equation:

$$\mathrm{pH_{bulk}} = \mathrm{pK_a} + \mathrm{A}\,\log\left(\frac{\mathrm{D}-\mathrm{D_{min}}}{\mathrm{D_{max}}-\mathrm{D}}\right) + \log\left(\frac{\mathrm{I_a}}{\mathrm{I_b}}\right) \tag{1}$$

where the constant A is ideally 1 since the protonation of HC is a one-to-one event (here 1.26); pK_a is the apparent proton binding constant of HC (here 10.5); D is the dissociation degree; D_{min} and D_{max} are the minimum and maximum values of D, respectively (here 1.0 and 98.7); I_a/I_b is the ratio of the fluorescence intensity of the pH-independent isosbestic point of HC (excitation at 330 nm) and of the unprotonated charged HC (excitation at 380 nm). Log (I_a/I_b) is 0 at an ideal isosbestic point; here we found values between –0.1 and 0.1). By estimating the dissocation degree of HC in lipid bilayers and by using equation 1, the average $pH_{surface}$ of these lipid bilayers can be estimated. This method was used to estimate the change in the average $pH_{surface}$ of the cationic bilayers ($\Delta pH_{surface}$) upon addition of plasmid DNA. The value of $\Delta pH_{surface}$ upon addition of plasmid DNA was used to calculate the change in surface potential, $\Delta\Psi_0$, using the Boltzmann equation for a given potential:

$$\Delta\Psi_0 = \frac{\Delta \mathrm{pH_{surface}}\ kT \ln 10}{e} \tag{2}$$

where k is the Boltzmann constant (1.38×10^{-23} J·K^{-1}), T is the absolute temperature (here 295 K), and e is the electron charge (1.6×10^{-19} C). The initial Ψ_0 of the cationic bilayers in the absence of plasmid DNA was calculated by Gouy–Chapman approximation:

$$\Psi_0 = \left(\frac{2\,kT}{ze}\right) \sinh^{-1}\left(\frac{ze\,\sigma\,\lambda}{2\,\varepsilon_0\,\varepsilon_r\,kT}\right) \tag{3}$$

where z is the valency of the counterions (in the present study, 1), σ is the surface charge density (C·m^{-2}), l is the Debye screening length (m), ε_0 is the relative permittivity of free space (8.85×10^{-12} C^2·J^{-1}·m^{-1}), and ε_r is the dielectric constant at the location of the fluorophore moiety, with a value of 8 (see Zuidam and Barenholz, 1997 for its estimation). The values of σ were calculated taking a molecular surface area of 0.82 nm^2 for DOPC, 0.65 nm^2 for DOPE, 0.30 nm^2 for DC-CHOL, and 0.65 nm^2 for DOTAP. λ is given by:

$$\lambda = \sqrt{\frac{\varepsilon_0\,\varepsilon_r\,kT}{Ne^2 \sum c_i z_i}} \tag{4}$$

where N is the Avogadro constant and c_i is the concentration of all individual ions (here 0.017 mol·m^{-3}). In the present study, a value of 0.74 nm for λ was calculated. The values of Ψ_0 of the cationic bilayers in the absence of plasmid DNA are in Table 1. The Ψ_0 of the cationic bilayers in the presence of plasmid DNA was obtained by adding together its initial value obtained with equation 3 and the $\Delta\Psi_0$ obtained by equation 2. Finally, the value of σ in the presence of plasmid DNA was estimated using a converted form of equation 3:

$$\sigma = \left(\frac{2\,\varepsilon_0\varepsilon_r\,kT}{z\,e\,\lambda}\right) \sinh\left(\frac{\Psi_0\,ze}{2\,kT}\right) \tag{5}$$

Then, the change in the surface charge density, $\Delta\sigma$, and thus the percent of charged lipids, can be calculated. For further information about these calculations and the assumptions made, see Zuidam and Barenholz (1997).

Abbreviations

DC-CHOL, 3β[N-(N',N'-dimethylaminoethane)carbamoyl]cholesterol; DMRIE, 1,2-dimyristyloxypropyl-3-dimethyl-hydroxyethyl ammonium bromide; DOGS, dioctadecylamidoglycylspermine; DOSPA, 2'-(1",2"-dioleoyloxypropyldimethyl-ammonium bromide)-*N*-ethyl-6-amido-spermine tetra trifluoroacetic acid salt; DOPC, 1,2-dioleoyl-*sn*-glycero-3-phosphatidylcholine; DOPE, 1,2-dioleoyl-*sn*-glycero-3-phosphatidylethanolamine; DOTAP, *N*-(1-(2,3-dioleoyloxy)propyl)-*N,N,N*-trimethylammonium chloride; DOTMA, *N*-(1-(2,3-dioleyloxy)propyl)-*N,N,N*-trimethylammonium chloride; DPH, 1,6-diphenylhexa-1,3,5-triene; HC, 4-heptadecyl-7-hydroxycoumarin; Hepes, *N*-(2-hydroxyethyl)piperazine-*N'*-(2-ethanesulfonic acid); L^+, cationic lipid; LUV, large unilamellar vesicle; TMADPH, 1-(4-(trimethylammoniumphenyl)-6-phenylhexa-1,3,5-triene.

Acknowledgements

This work was supported in part by the Israel Science Foundation grant 494/96 to Yechezkel Barenholz. Mr S. Geller is acknowledged for helping with the editing of the manuscript.

REFERENCES

Barenholz, Y., and Amselem, S., 1993, Quality control assays in the development and clinical use of liposome based formulations, in: *Liposome Technology*, Vol. 1, 2nd ed., Gregoriadis, G., ed., CRC Press, Boca Raton

Barenholz, Y., Freire, E., Thompson, T.E., Correa-Freire, M.C., Bach, D., and Miller, I.R., 1983, Thermotropic behavior of aqueous dispersions of monoglucosylceramide (glucocerebroside)-dipalmitoyl phosphatidylcholine mixtures, *Biochemistry* 22:3497.

Behr, J.P., Demeneix, B., Loeffler, J.P., and Mutul, J.P., 1989, Efficient gene transfer into mammalian primary endocrine cells with lipopolyamine coated DNA, *Proc. Natl. Acad. Sci. U.S.A.* 86:6982.

Behr, J.P., 1994, Gene transfer with synthetic cationic amphiphiles: prospects for gene therapy, *Bioconj. Chem.* 5:382.

Ben-Yashar, V., and Barenholz, Y., 1991, Characterization of core and surface of human plasma lipoproteins: A study based on the use of five fluorophores, *Chem. Phys. Lipids* 60:1.

Borenstain, V., and Barenholz, Y., 1993, Characterization of liposomes and other lipid assemblies by multiprobe fluorescence polarization, *Chem. Phys. Lipids* 64:117.

Buci, C., Parton, R.G., Mather, I.H., Stunnenberg, H., Simons, K., Hoflack, B., and Zerial, M., 1992, The small GTPase rab5 functions as a regulatory factor in the early endocytic pathway, *Cell* 70:715.

Cevc, G., 1990, Membrane electrostatics, *Biochim. Biophys. Acta* 1031:311.

Chen, W., Carbone, F.R., and McCluskey, J., 1993, Electroporation and commercial liposomes efficiently deliver soluble protein into the MHC class I presentation pathway. Priming *in vitro* and *in vivo* for class I-restricted recognation of soluble antigen, *J. Immunol. Methods* 160:49.

Crystal, R.G., 1995, Tranfer of genes to humans: early lessons and obstacles to success, *Science* 270:404.

Cullis, P. R., van Dijck, P.W., de Kruijff, B., and de Gier, J., 1978, Effects of cholesterol on the properties of equimolar mixtures of synthetic phosphatidylethanolamine and phosphatidylcholine. A 31P NMR and differential scanning calorimetry study, *Biochim. Biophys. Acta* 513:21.

Dan, N., 1996, Formation of ordered domains in membrane-bound DNA, *Biophys. J.* 71:1267.

Dan, N., 1997, Multilamellar structures of DNA complexes with cationic liposomes, *Biophys. J.* 73:1842.

Dan, N., 1998, The structure of DNA complexes with cationic liposomes — cylindrical or lamellar ? *Biochim. Biophys. Acta*, in press.

Eastman, S.J., Siegel, C., Tousignant, J., Smith, A.E., Cheng, S.H., and Scheule, R.K., 1997, Biophysical characterization of cationic lipid:DNA complexes, *Biochim. Biophys. Acta* 1325:41.

Felgner, P.L., Gadeck, T.R., Holm, M., Roman, R., Chan, H.S., Wenz, M., Northrop, J.P., Ringold, M., and Danielsen, H., 1987, Lipofection: a highly efficient lipid-mediated DNA transfection procedure, *Proc. Natl. Acad. Sci. U.S.A.* 84:7413.

Felgner, J.H., Kumar, R., Sridhar, S.H., Wheeler, C.J., Felgner, J.H., Tsai, Y.J., Border, R., Ramsey, P., Martin, M., and Felgner, P.L., 1994, Enhanced gene delivery and mechanism studies with novel series of cationic lipid formulations, *J. Biol. Chem.* 269:2550.

Felgner, P.L., Tsai, Y.J., and Felgner, J.H., 1996, Advances in the design and application of cytofectin formulations, in: *Handbook of Nonmedical Applications of Liposomes*, Vol. 4, Chapter 4, Lasic, D.D. and Barenholz, Y., eds., CRC Press, Boca Raton

Fernandez, M.S., and Fromherz, P., 1977, Lipoid pH indicators as probes of electrical potential and polarity in micelles, *J. Phys. Chem.* 81:1755.

Gao, X., and Huang, L., 1991, A novel cationic liposome reagent for efficient transfection of mammalian cells, *Biochem. Biophys. Res. Commun.* 179:280.

Gershon, H., Ghirlando, R., Guttman, S.B., and Minsky, A., 1993, Mode of formation and structural features of DNA-cationic liposome complexes used for transfection, *Biochemistry* 32:7143.

Gruen, D.W.R., Marcelja, S., and Parsegian, V.A., 1984, Water structure near the membrane surface, in: *Cell Surface Dynamics: Concepts and Models* ,Perelson, A.S., DeLisi, C. and Wiegel, F.W., eds., Chapter 3, Marcel Dekker, New York

Gustafsson, J., Arvidson, G., Karlsson, G., and Almgren, M., 1995, Complexes between cationic liposomes and DNA visualized by cryo-TEM, *Biochim. Biophys. Acta* 1235:305.

Harned, H.S., and Owen, B.B., 1958, *The Physical Chemistry of Electrolytic Solutions*, Reinhold, New York.

Hirsch-Lerner, D., and Barenholz, Y., 1998, Probing DNA–cationic lipid interactions with the fluorophore trimethylammonium diphenyl-hexatriene (TMADPH), *Biochim. Biophys. Acta*, in press.

Ho, C., and Stubbs, C.D., 1992, Hydration at the membrane protein-lipid interface, *Biophys. J.* 63:897.

Lasic, D.D., Strey, H., Stuart, M.C.A., Podgornick, R., and Fredrick, P.M., 1997, The structure of DNA-liposome complexes, *J. Am. Chem. Soc.* 119:832.

Ledley, F.D., 1995, Nonviral gene therapy: the promise of genes as pharmaceutical products, *Hum. Gene Ther.* 6:1129.

Lentz, B.R., 1989, Membrane "fluidity" as detected by diphenyl hexatriene probes, *Chem. Phys. Lipids* 50:171.

Leventis, R., and Silvius, J.R., 1990, Interactions of mammalian cells with lipid dispersions containing novel metabolizable cationic amphiphiles, *Biochim. Biophys. Acta* 1023:124.

Levine, A., Cantoni, G.L., and Razin, A., 1992, Methylation in the preinitiation domain suppresses gene transcription by an indirect mechanism, *Proc. Natl. Acad. Sci. U.S.A.* 89:10119.

Manning, G.S., 1980, Thermodynamic stability theory for DNA doughnut shapes induced by charge neutralization, *Biopolymers* 19:37.

May, S., and Ben-Shaul, A., 1997, DNA-Lipid complexes: Stability of honeycomb-like and spaghetti-like structures, *Biophys. J.* 73:2427.

Mulligan, R.C., 1993, The basic science of gene therapy, *Science* 260:926.

Ostrovsky, N., 1993, Liposome size measurements by photon correlation spectroscopy, *Chem. Phys. Lipids* 64:45.

Rädler, J.O., Koltover, I., Salditt, T., and Safinya, C.R., 1997, Structure of DNA-cationic liposome complexes: DNA intercalation in multilamellar membranes in distinct interhelical packing regimes, *Science* 275:810.

Sternberg, B., Sorgi, F.L., and Huang, L., 1994, New structures in complex formation between DNA and cationic liposomes visualized by freeze-fracture electron microscopy, *FEBS Lett.* 356:361.

Sternberg, B., 1996, Morphology of cationic liposome/DNA complexes in relation to their chemical composition, *J. Liposome Res.* 6:515.

Strauss, W.M., Dausman, J., Beard, C., Johnson, C., Lawrence, J.B., and Jaenisch, R., 1993, Germ line transmission of a yeast artificial chromosome spanning the murine alpha1 (I) collagen locus, *Science* 259:1904.

Szoka, F.C., Xu, Y., and Zelphati, O., 1996, How are nucleic acids released in cells from cationic lipid-nucleic acid, *J. Liposome Res.* 6:567.

Talmon, Y., 1996, Transmission electron microscopy of complex fluids: the state of the art, *Ber. Bunsenges. Phys. Chem.* 100:364.

Tatulian, S.A., 1993, in: *Phospholipids Handbook*, G. Cevc, ed., Marcel Dekker, New York.

Tocanne, J.F., and Teissié, J., 1990, Ionization of phospholipids and phospholipid-supported interfacial lateral diffusion of protons in membrane model systems, *Biochim. Biophys. Acta* 1031:111.

Walker, C., Selby, M., Erickson, A., Cataldo, D., Valensi, J.P., and Van Nest, G.V., 1992, Cationic lipids direct a viral glycoprotein into the class I major histocompatibility complex antigen–presentation pathway, *Proc. Natl. Acad. Sci. U.S.A.* 89:7915.

Wilschut, J., and Hoekstra, D., 1986, Membrane fusion: lipid vesicles as a model system, *Chem. Phys. Lipids* 40:145.

Zabner, J., Fasbender, A.J., Moninger, T., Poellinger, K.A., and Welsh, M.J., 1995, Cellular and molecular barriers to gene transfer by a cationic lipid, *J. Biol. Chem.* 270:18997.

Zhou, X., and Huang, L., 1994, DNA transfection mediated by cationic liposomes containing lipopolylysine: characterization and mechanism of action, *Biochim. Biophys. Acta* 1189:195.

Zuidam, N.J., and Barenholz, Y., 1997, Electrostatic parameters of cationic liposomes commonly used for

gene delivery as determined by 4-heptadecyl-7-hydroxycoumarin, *Biochim. Biophys. Acta* 1329:211.
Zuidam, N.J., and Barenholz, Y., 1998, Electrostatic and structural properties of complexes involving plasmid DNA and cationic lipids commonly used for gene delivery, *Biochim. Biophys. Acta*, 1368:115.

STERIC STABILIZATION OF CATIONIC LIPOSOME-DNA COMPLEXES:

INFLUENCE ON MORPHOLOGY AND TRANSFECTION ACTIVITY

Brigitte Sternberg[1], Keelung Hong[2], Weiwen Zheng[2], and Demetrios Papahadjopoulos[2]

[1]Research Institute and [2]Liposome Research Laboratory, California Pacific Medical Center, 2340 Clay Street, San Francisco, CA 94115

INTRODUCTION

Over the past few years, rapid progress has been made in controlling the *in vivo* delivery and expression of therapeutic genes (Blaese et al., 1995). Presently, viral-based carriers of DNA are still the most common method of gene delivery, although there is a strong research effort for developing synthetic non-viral vectors, in particular, cationic liposome-DNA complexes (CLDC) (Gao and Huang, 1995; Felgner et al., 1995; Lasic and Templeton, 1996). When compared to viral vectors, liposomal gene delivery systems offer several advantages, including the lack of viral gene elements, low immunogenic and inflammatory responses, potential for transfer of expression units of essentially unlimited size, and the possibility for cell-specific targeting (Stewart et al, 1992, Crystal, 1995.). Such preparations appeared initially to be ineffective for many *in vivo* applications because of instability in serum; therefore, much effort has been devoted to optimizing the complexes. This included not only synthesizing more efficient cationic lipids (Crystal, 1995; Solodin et al., 1995; Stephan et al., 1996), alternative helper lipids (Hong et al., 1997; Liu et al., 1997), as well as plasmid expression vectors (Hartikka et al., 1996), but also stabilizing the whole CLDC by polyamines and poly (ethylene glycol)-phospholipid conjugates (Hong et al., 1997). With more than 1 μg protein expressed per gram of lung tissue (Hong et al., 1997; Liu et al., 1997, Templeton et al., 1997) trans-gene expression levels have recently approached those achievable with adenovirus.

Cationic liposome/DNA complexes (CLDC) display a variety of polymorphic and metastable structures, such as multilayered arrays of alternating lipid bilayers and DNA monolayers (Lasic et al., 1997, Rädler et al., 1997, Boukhnikachvili et al., 1997), invaginated liposomes (Templeton et al., 1997), fibrillar structures, free or attached to the complexes, among them *spaghetti*-like tubules (Sternberg et al., 1994, Sternberg, 1996) or small protrusions and "map-pin" structures (Sternberg et al., 1998), bundled and folded loops of strands (Dunlap et al., 1997), and last but not least hexagonal (H_{II})-lipid arrangements (honeycomb-like structure, Sternberg, 1996; Tarahovsky et al., 1996). CLDC are metastable with time and their structural instability is connected with the loss of *in vivo* transfection activity (Hong et al., 1997, Liu et al., 1997).

In order to optimize CLDC for *in vivo* applications we studied the influence of lipid components providing steric stabilization (poly[ethylene glycol]-phospholipid conjugates [PEG-PE], (Papahadjopoulos et al., 1991) on the morphology of complexes made of plasmid DNA, dimethyl-dioctadecylammonium bromide (DDAB), and cholesterol (Chol) in buffer of low ionic strength and in mouse serum by freeze-fracture electron microscopy. In parallel, we investigated their transfection activity *in vivo*, using luciferase as a marker gene, expressed in mouse lung following i.v. injection. Furthermore, in some control experiments, we looked for structural modifications of conventional liposomes made of hydrogenated soybean phosphatidylcholine (HSPC) and Chol, and sterically stabilized by monosialoganglioside (GM1); (Gabizon and Papahadjopoulos, 1988) or PEG-PE (Papahadjopoulos et al., 1991) by freeze-fracture as well as freeze-etch electron microscopy (Sternberg, 1992).

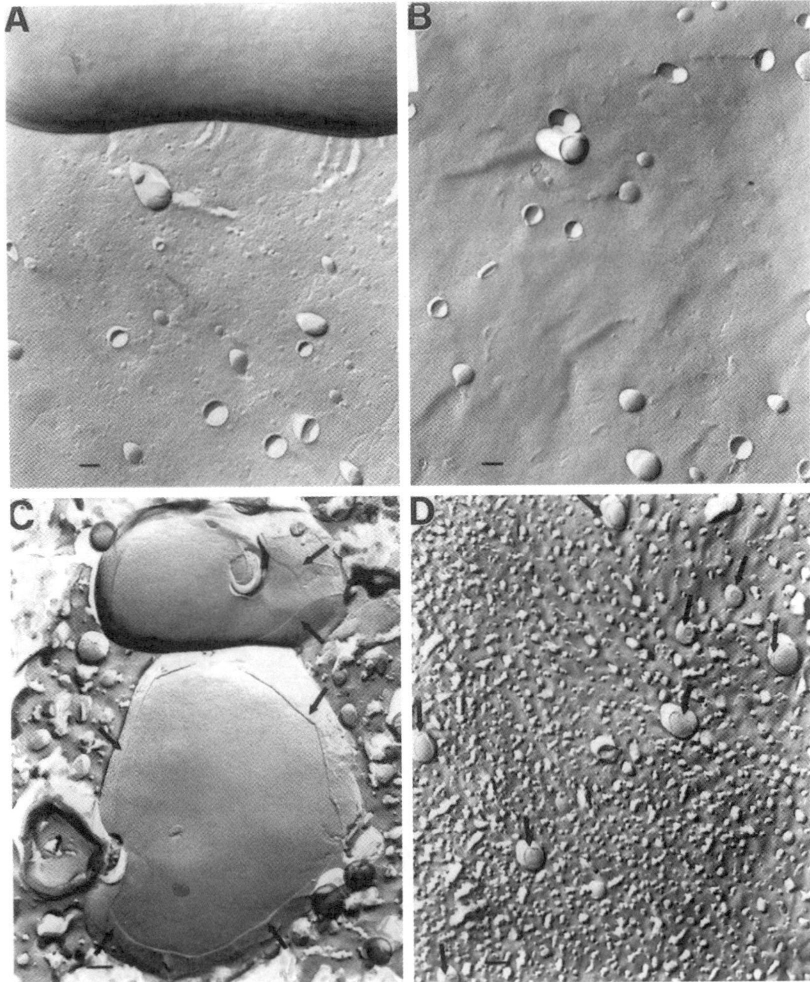

Figure 1. Conventional liposomes made of HSPC:Chol (2:1 mole ratio) and stabilized by GM1 (7%; A and C) or PEG-PE (10%; B and D) on freeze-fracture (A and B) or freeze-etch electron micrographs (C and D). Some of the true outer surfaces , revealed by freeze-etching, are marked by arrows in (C and D). Bars on all electron micrographs represent 100 nm and the shadow direction is running from bottom to top.

MORPHOLOGY OF STERICALLY STABILIZED LIPOSOMES

Control experiments on conventional liposomes (MLV`s and SUV`s) made of HSPC:Chol (2:1 molar ratio) and sterically stabilized by 7 mol% GM1 (Fig.1A and 1C) or 10 mol% PEG (Fig.1B and 1D) revealed totally unobtrusive liposomes showing smooth fracture faces exposed by freeze-fracture technique (Fig.1A and 1B) as well as true outer and inner surfaces exposed by freeze-etching (Fig. 1C and 1D, some of the true outer surfaces are marked by arrows), i.e., lowering of the ice level of the frozen liposome preparations in high vacuum by sublimation. Neither domain formation initiated by GM1 or PEG-PE nor any distinct structure formed by the PEG-chains are detectable although both types of sterically stabilized liposomes have a prolonged circulation time in blood (Gabizon and Papahadjopoulos, 1988 and Papahadjopoulos et al., 1991). Obviously the resolution limit of freeze-fracture as well as freeze-etching technique, limited to 2 nm for periodical structures, excludes the depicting of the highly flexible PEG-chains. This is certainly true for the pancake conformation of the PEG-chains which was measured and calculated to be ~1.5 nm for PEG-2000 at 5 mol% (Baekmark et al., 1995) but also for the mushroom as well as brush conformation with a length of ~3.5 or ~5 nm respectively (Kuhl et al., 1994).

MORPHOLOGY OF STABILIZED CATIONIC LIPOSOME/DNA COMPLEXES

Preparations of CLDC are metastable with time and since structural instability may be connected with the loss of transfection activity, it is highly desirable for *in vivo* applications to develop well-defined formulations which are stable in buffer and in serum over a reasonably long period of time. Stabilization of cationic-plasmid DNA complexes was achieved by using Chol instead of DOPE, by addition of PEG, and also by pre-condensation of the DNA by polyamines (Hong et al., 1997). In order to study the influence of PEG as a stabilizing factor on the morphology of CLDC, we investigated by freeze-fracture electron microscopy the structure of DDAB/Chol/DNA complexes (1:1 molar ratio of the lipids and 5-12 nmol DDAB per μg DNA as control) in buffer at low ionic strength as well as in mouse serum, and compared it with the structure of the same complexes but with 1 mole % PEG-PE of DDAB added shortly after complex formation.

Morphology of Stabilized CLDC in Low Ionic Strength Buffer

Freeze-fracture electron micrographs of DNA/DDAB/Chol complexes (control) display small spherical particles (average diameter in the range of 100-250 nm) with few protrusions (marked by an arrow in Figure 2A). This is true in buffer of low ionic strength (1.0 mM MES; pH 5.5) and after 3 days of storage time at 4°C (Figure 2A). Within 6 days of storage time, however, the spherical particles obviously adhere, fuse, and form larger complexes with diameters between 300 and 600 nm (Figure 2B). In contrast, the morphology of similar complexes but additionally stabilized by PEG-PE is very similar to the 3-day-old controls even when the electron micrographs were taken from 6-day-old samples, still displaying small spherical particles up to 300 nm in diameter and also showing few protrusions (marked by an arrow in Figure 2C).

CORRELATION BETWEEN MORPHOLOGY IN SERUM AND TRANSFECTION ACTIVITY *IN VIVO*

We investigated the transfection activity *in vivo* of plasmid DDAB/Chol/DNA complexes

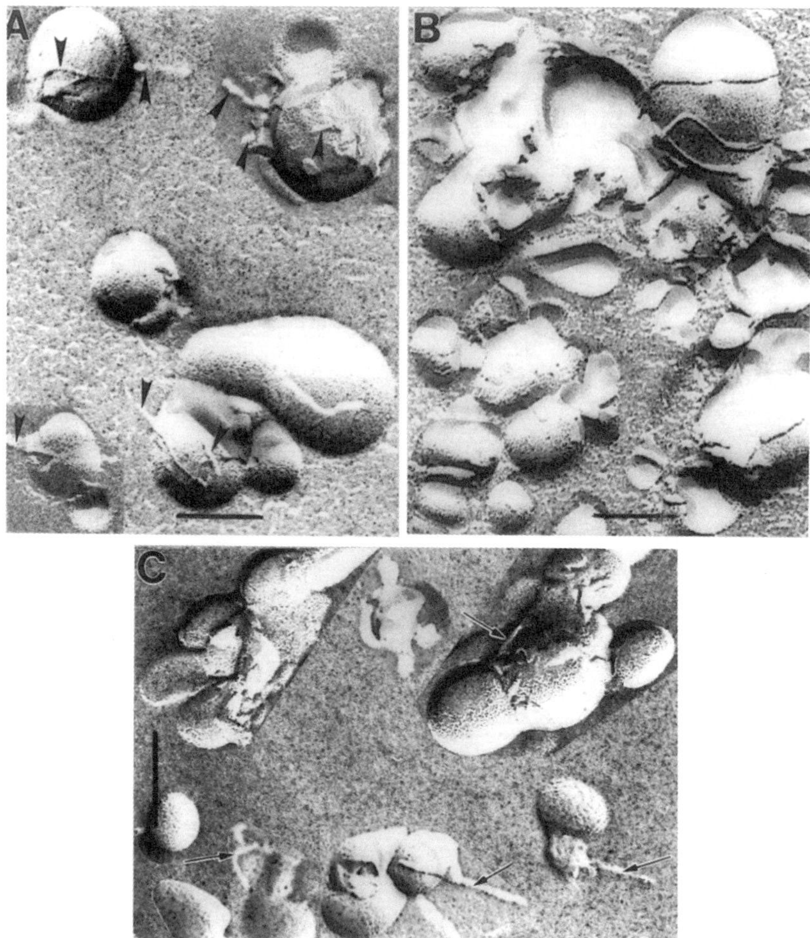

Figure 2. Freeze-fracture electron micrographs of DDAB/Chol/DNA complexes in buffer of low ionic strength. A: DDAB/Chol/DNA (control), 3 days old (some of the protrusions are marked by arrow heads); B: control, 6 days old; C: DDAB/Chol/DNA stabilized by 1% PEG-PE, 6 days old. All samples were prepared and kept in low-ionic strength dextrose buffer and diluted 50% v/v before final incubation with 1.0 mM MES (pH 5.5). Bars on all electron micrographs represent 100 nm and the shadow direction is running from bottom to top.

as a control, and of similar CLDC but additionally stabilized by 1% PEG-PE. In parallel, we have studied their morphology in serum by freeze-fracture electron microscopy. The *in vivo* studies were carried out in mice following i.v. injection and therefore the morphology of the CLDC was investigated in mouse serum.

When examined in 50% mouse serum (10 min incubation time), three-day-old DDAB/Chol/DNA complexes are as small as they are in buffer at low ionic strength (100-250 nm) but show very few protrusions (Fig. 3A). Such complexes show high transfection activity in mouse lungs after i. v. injection (expression of luciferase protein up to 3 ng per mg tissue protein; Hong et al., 1997) and we take this level of activity as 100 % for further comparisons. After the same incubation time in mouse serum, CLDC of the same composition but 6 days old, appear as densely packed aggregates of spherical particles where the number of attached particles in each aggregate is quite high (approx. 10-14; Fig. 3B). Such formulations have lost

Figure 3. Freeze-fracture electron micrographs of DDAB/Chol/DNA complexes in mouse serum. DDAB/Chol/DNA (control), 3 days old; B: control, 6 days old; C: DDAB/Chol/DNA stabilized by 1% PEG-PE, 6 days old. All samples were prepared and stored in low-ionic strength dextrose buffer and diluted 50% v/v with mouse serum before final incubation (10 min). Bars on all electron micrographs represent 100 nm and the shadow direction is running from bottom to top.

all their *in vivo* transfection activity within 4 days (Hong et al., 1997). Residual fibrillar protrusions were not observed.

DDAB/Chol/DNA complexes stabilized additionally by PEG-PE and incubated with mouse serum, are quite small (approx. 4-6 attached particles) even when 6 days old (Fig. 3C). These stabilized complexes, freshly prepared, show about one half of the *in vivo* lung transfection activity (60%) observed with similar, freshly prepared complexes but not stabilized by PEG (100%). Interestingly, similar preparations but PEG-stabilized do not lose activity within a storage period of 6 days. It is remarkable that among the stabilized, one month old samples there were much higher expression rates (about 130% for PEG-PE stabilized complexes) than for the control non-stabilized complexes which, after the same storage time, showed less than 20% of the standard value.

CONCLUSIONS

Our structural investigations on conventional liposomes, using freeze-fracture as well as freeze-etch electron microscopy, show that steric stabilization by GM1 (Fig. 1A and 1C) or PEG (Fig.1B and 1D) in the concentrations applied does not alter their morphology at the resolution limit of both cryofixation techniques.

Our structural investigations on CLDC, using freeze-fracture electron microscopy, reveal that stabilization by adding 1% of PEG-PE shortly after complex formation does not change much the morphology of freshly prepared complexes kept in buffer at low ionic strength, as seen by comparison of Fig. 2A with Fig. 2C, or in serum as seen by comparison of Fig. 3A with Fig. 3C. However, stabilization of the morphology of CLDC is being achieved, by the addition of poly(ethylene glycol)-phospholipid conjugates such as PEG-PE. Whereas complexes non-stabilized by PEG grow much larger during 6 days of storage time even in buffer at low ionic strength (as seen from Fig. 2B compared with Fig. 2A), the complexes stabilized by PEG-PE maintain the same morphology, displaying small-size complexes (not larger than 300 nm) and some relatively small protrusions connected with the CLDC (marked by an arrow in Fig. 2C).

Adding PEG-PE shortly after complex formation is an effective method to stabilize both the morphology and the *in vivo* transfection activity of CLDC made of DDAB/Chol/DNA. Whereas the control complexes, one day old and incubated with 50% mouse serum for 10 min, are still small (100-200 nm; Fig. 3A) and show 100% *in vivo* transfection activity (standard value) they form much larger and tighter packed aggregates with time (tight aggregates of 10-14 spherical particles after 6 days, Fig. 3B). Significantly, they lose totally their *in vivo* transfection activity within 4 days (Hong et al., 1997). Similar complexes when stabilized by the methods described above are still quite small (aggregates of 4-6 particles) even when 6 days old and when incubated with 50% serum for 10 min (Fig. 3C). These preparations were functionally stable for a month at 4°C and gave reproducibly high transfection activity *in vivo* (lungs) after i.v. injection in mice (Hong et al., 1997).

It should be noted that the morphology of the CLDC is different in buffer of low ionic strength from that found at high ionic strength such as cell medium (Sternberg et al., 1998) or serum. Structural features such as small protrusions, observed in MES-buffer (Fig. 2A and 2C) do not usually survive incubation with cell medium or serum (Figs. 3A and 3C). Structural transitions from bilayer to non-bilayer structures are also observed especially at DOPE-containing CLDC by changing the aqueous milieu (Sternberg et al, 1998). Thus, in order to define the active structure in terms of transfection it is important to investigate the morphology of the complexes in the appropriate aqueous milieu.

It appears from our studies that small, stabilized, spherical particles (100-300 nm), forming only small aggregates of a few particles in serum, even after storage in buffer at 4°C for 6 days, are possibly responsible for the high luciferase expression observed in mouse lungs. Although much smaller, these stabilized particles might have some similarity to recently described heterogenous particles in the shape of flat, concentric, bent or amorphous stacks of bilayers in the size range of 200-500 nm (Lasic et al., 1997) or birefringent liquid-crystalline condensed globules with sizes in the order of 1000 nm (Rädler et al., 1997).

Our investigations on DDAB/Chol/DNA complexes show clearly that their *in vivo* transfection activity is primarily related to small complexes, stabilized with respect to serum and time. It is possible that this stabilization allows the complexed DNA to travel within the blood stream avoiding immediate clearance by the RES, and eventually to be adsorbed on endothelial cells in lung capillaries where transfection can occur. Based on these observations we believe that the stabilization of formulation and morphology of the CLDC in biological fluids is at least equally important for high transfection activity *in vivo* as the specific chemical nature of cationic amphiphiles.

Acknowledgements

Brigitte Sternberg wishes to thank Dr T. Allen for the sterically stabilized liposome preparations, Mrs. I.-M. Herrmann and Mrs. R. Kaiser for technical assistance in freeze-fracture work and Mrs. G. Engelhardt and Mrs. G. Vöckler for their phototechnical work.

REFERENCES

Baekmark, T.R., Elander, G., Lasic, D.D., and Sackmann, E., 1995. Conformational transitions in monolayers of phospholipid-polyethyleneoxide lipo-polymers and interaction forces with solid surfaces. Langmuir 11:3975.

Blaese, M., Blankenstein, T., Brenner, M., Cohen-Haguenauer, O., Gansbacher, B., Russel, S., Sorrentino, B., and Velu, T., 1995. Vectors in cancer therapy: how will they deliver? Cancer Gene Ther. 2:291.

Boukhnikachvili, T., Aguerre-Chariol, O., Airiau, M., Lesieur, S., Ollivon, M., and Vacus, J., 1997. Structure of in-serum transfecting DNA-cationic lipid complexes. FEBS Lett. 409:188.

Crystal, R.G., 1995. Transfer of genes to humans: early lessons and obstacles to success. Science 270:404.

Dunlap, D.D., Maggi, A., Soria, M.R., and Monaco, L., 1997. Nanoscopic structure of DNA condensed for gene delivery. Nucl. Acids Res. 25:3095.

Felgner, P.L., Tsai, Y.J., Sukhu, L., Manthrope, M., Marshall, J. et al., 1995. Improved cationic lipid formulations for gene therapy. Ann. N.Y. Acad. Sci. 772:126.

Gabizon, A. and Papahadjopoulos, D., 1988. Liposome formulations with prolonged circulation time in blood and enhanced uptake in tumors. Proc. Natl. Acad. Sci. U.S.A. 85:6949.

Gao, X., and Huang, L., 1995. Cationic liposome-mediated gene transfer. Gene Therapy 2:710.

Hartikka, J., Sawdey, M., Cornefert-Jensen, F., Mangalith, M., Barnhart, K., Nolasco, M., Norman, J., and Manthorpe, M., 1996. An improved plasmid DNA expression vector for direct injection into skeletal muscle. Hum. Gene Ther. 7:1205.

Hong, K., Zheng, W., Baker, A., and Papahadjopoulos, D., 1997. Stabilization of cationic liposome-plasmid DNA complexes by polyamines and poly(ethylene glycol)-phospholipid conjugates for efficient *in vivo* gene delivery. FEBS Lett 400:233.

Kuhl, T., Leckband, D.E., Lasic, D.D. and Israelachvili, J.N., 1994. Modulation of interaction forces between lipid bilayers exposing short-chained ethylene oxide headgroups. Biophys. J. 66:1479.

Lasic, D.D., and Templeton N.S., 1996. Liposomes in gene therapy. Advanced Drug Deliv. Rev. 20:221-266.

Lasic, D.D., Strey, H., Stuart, M.C.A., Podgornik, R., and Frederik, P.M., 1997. The structure of DNA-liposome complexes. J. Amer. Chem. Soc. 119:832.

Liu, Y., Mounkes, L.C., Liggitt, H.D., Brown, C.S., Solodin, I., Heath, T.D., and Debs, R.J., 1997. Factors influencing the efficiency of cationic liposome-mediÿated intravenous gene delivery. Nature Biotechnology 15:167.

Papahadjopoulos, D., Allen, T.M., Gabizon, A., Mayhew, E., Huang, S.K., Lee, K.-D., Woodle, M.C., Lasic, D.D., Redemann, C., and Martin, F.J., 1991. Sterically stabilized liposomes: improvements in pharmacokinetics, and anti-tumor therapeutic efficacy. Proc. Natl. Acad. Sci. U.S.A. 88:11460.

Rädler, J.O., Koltover, I., Salditt, T., and Safinya, C.R., 1997. Structure of DNA-cationic liposome complexes: DNA intercalation in multilamellar membranes in distinct interhelical packing regimes. Science 275:810.

Solodin, I., Brown, C.S., Bruno, M.S., Chow, C.-Y., Jang, E.-H., Debs, R.J., and Heath, T.D., 1995. High efficiency *in vivo* gene delivery with a novel series of amphilic imidazolinium compounds. Biochemistry 34:13537.

Stephan, D.J., Yang, Z.-Y., Simari, R.D., San, H., Wheeler, C.J., Felgner, P.L., Gordon, D., Nabel, G.J., and Nabel, E.G., 1996). A novel cationic liposome DNA complex enhances the efficiency of arterial gene transfer *in vivo*. Hum. Gene Ther. 7:1803.

Sternberg, B., 1992. Freeze-fracture electron microscopy of liposomes. in: Liposome Technology, G. Gregoriadis, Ed; 2nd Ed, CRC Press, Inc., Boca Raton 21:363.

Sternberg, B., Sorgi, F.L., and Huang, L., 1994. New structures in complex formation between DNA and cationic liposomes visualized by freeze-fracture electron microscopy. FEBS Lett. 356:361.

Sternberg, B., 1996. Morphology of cationic liposome/DNA complexes in relation to their chemical composition. J. Liposome Res. 6:515.

Sternberg, B., Hong, K., Zheng, W., and Papahadjopoulos, D., 1998. Optimization of cationic liposome-DNA complexes *in vivo*: correlation between transfection activity and morphology. submitted.

Sterwart, M.J., Plautz, G.E., Del Buono, L., Yang, Z.Y., Xu, L., Gao, X., Huang, L., Nabel, E.G., and Nabel, G.J., 1992. Gene transfer *in vivo* with DNA-liposome complexes: safety and acute toxicity in mice. Hum. Gene Ther. 3:267.

Tarahovsky, Y.S., Khusainova, R.S., Gorelov, A.V., Nicolaeva, T.I., Deev, A.A., Dawson,A.K., and Ivanitsky, G.R., 1996. DNA initiates polymorphic structural transitions in lecithin. FEBS Lett. 390:133.

Templeton, N.S., Lasic, D.D., Frederik, P.M., Streey, H., Roberts, D.D., and Pavlakis, G.N., 1997. Improved DNA:liposome complexes for increased systemic delivery and gene expression. Nature Biotechnology 15:647.

POLYSIALIC ACIDS: POTENTIAL FOR LONG CIRCULATING DRUG, PROTEIN, LIPOSOME AND OTHER MICROPARTICLE CONSTRUCTS

Gregory Gregoriadis, Ana Fernandes, Brenda McCormack,
Malini Mital and Xiaoqin Zhang

Centre for Drug Delivery Research, School of Pharmacy,
University of London, 29-39 Brunswick Square, London, UK

INTRODUCTION

Optimal use of drugs often requires their extended presence within the vascular system or in extravascular areas (Gregoriadis et al, 1994). For instance, some antibiotics and cytostatics and a variety of peptides and proteins including hormones, cytokines, enzymes, antibodies and haemoglobin (as a blood surrogate) are excreted or removed from the circulation by tissues rapidly and before therapeutic concentrations in target areas can be achieved. Such drugs could be more effective, less toxic and also used in smaller quantities if their presence in the blood circulation (and hence interaction with corresponding receptors or substrates intravascularly or extravascularly) could be prolonged (Lee et al., 1995). In the same way, prolonged circulation of drug delivery systems such as liposomes (Gregoriadis, 1995), other colloidal systems (Davis et al., 1984) and polymers (Domb et al., 1997) would facilitate targeting of drugs to cells other than those (e.g. the reticuloendothelial system; RES) by which many of these systems are normally intercepted (Gregoriadis, 1995; Lee et al., 1995)

The half-lives of a number of short-lived proteins have been successfully augmented. (Nucci et al., 1991) by coating these with low molecular weight (750-5,000) monomethoxypoly(ethyleneglycol) (mPEG). Liposomes and polystyrene microspheres coated with mPEG or poloxamers are also known to exhibit increased half-lives (see chapters in this book). It is thought that mPEG action in prolonging the circulation time of proteins and particles is due to the formation of a shell of mPEG molecules around their surface which sterically hinders interaction with factors responsible for their clearance (Torchilin and Papisov, 1994).

POLYSIALIC ACIDS

We have previously reported (Gregoriadis et al., 1993) on an alternative type of

Targeting of Drugs 6: Strategies for Stealth Therapeutic Systems
Edited by Gregoriadis and McCormack, Plenum Press, New York, 1998

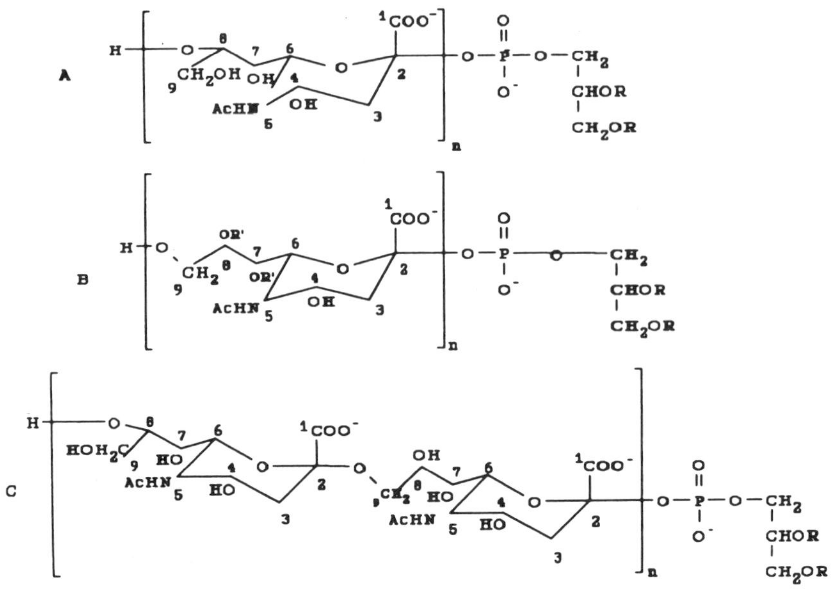

Figure 1. Structures of polysialic acids. (A) Serogroup B capsular polysialic acid B (PSB) from *N. meningitidis* or *E. coli* K1 is a homopolymer ($n = 199$) of α-(2-8)-linked *N*-acetyl neuraminic acid. (B) Serogroup C capsular polysialic acid (PSQ from *N. meningitidis* C is a homopolymer ($n = 74$) of α-(2-9)-linked *N*-acetyl neuraminic acid; R′: H or Ac. (C) Polysialic acid (PSK92) from *E. coli* K92 is a heteropolymer ($n = 78$) of alternate units of α-(2-8)-α-(2-9)-linked *N*-acetyl neuraminic acid. All three polysialic acids contain a phospholipid molecule covalently linked to the reducing end of the polymers. (From Gregoriadis et al., 1993, with permission)

macromolecules which may serve to increase the half-life not only of small peptides and conventional drugs (a role which, incidently, low molecular weight mPEG, known to be excreted rapidly through the kidneys, is unlikely to fulfill) but also of larger proteins, other biopolymers, liposomes and other microparticles. Such macromolecules are the naturally occurring polymers of *N*-acetyl neuraminic acid (NeuNAc) (polysialic acids) and include the serogroup B capsular polysaccharide (or polysialic acid) from *Neisseria meningitidis* B and *Escherichia coli* KI, the serogroup C capsular polysaccharide C from *N. meningitidis C,* and the polysaccharide K92 from *E. coli* K92 (Fig. 1) as well as their shorter chain derivatives.

Polysialic acid clearance from the circulation

The highly hydrophilic nature of polysialic acids (see Fig. 1) and the absence of a known receptor in the body for NeuNAc suggested to us (Gregoriadis et al., 1993) that polysialic acids may circulate in the blood for prolonged periods after intravenous injection and, therefore, serve as carriers of small drugs or peptides which are normally removed rapidly from the circulation. Thus, experiments were carried out where mice were injected with a variety of polysialic acids. NeuNAc levels in blood plasma were measured by an assay modified (Gregoriadis et al., 1993) in this laboratory to exclude the NeuNAc of plasma glycoproteins.

The clearance pattern of polysaccharide B (PSB) (see structure in Fig. 1) from the blood circulation was biphasic with 50% of the dose removed 3 min after injection (Fig. 2). The

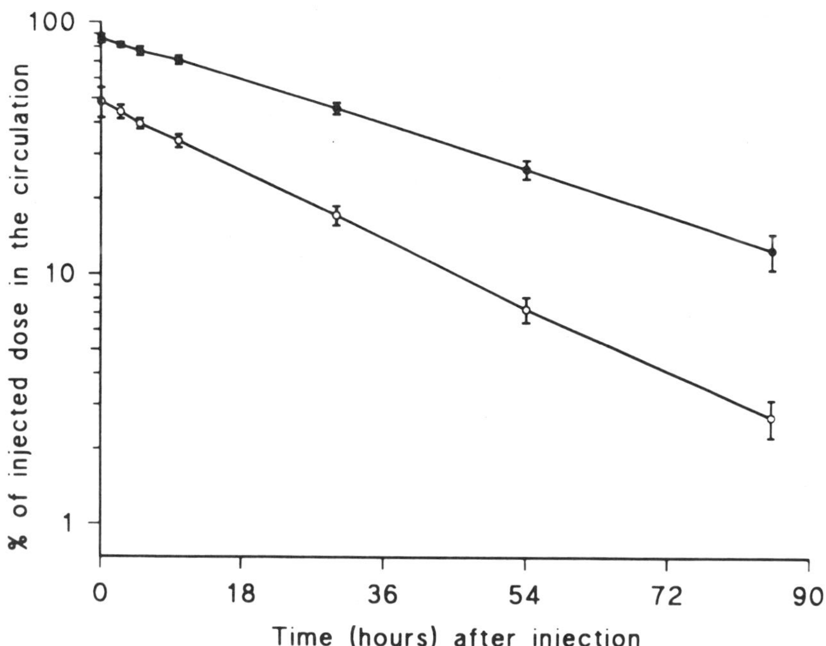

Figure 2. Clearance of PSB from the blood circulation. In six separate experiments, mice in groups of 3-4 animals were injected intravenously with 1.1-2.0 mg of intact (○) or deacylated (●) PSB and bled at time intervals. NeuNAc in the blood plasma samples was assayed as described (Gregoriadis et al., 1003) and expressed as % ± S.D. of the dose in total blood. (Values from all groups treated with intact and deacylated PSB, respectively, were pooled). Blood volume was estimated as 7% of the body weight (From Gregoriadis et al., 1993, with permission)

remainder assumed a linear rate of clearance with a half-life of 20 h. PSB, as well as polysaccharide C (PSC) and polysaccharide K92 (PSK92) in solution, have a phospholipid moiety covalently attached through its phosphate group to the reducing end (Fig. 1). As a result, polysialic acids exhibit micellar behaviour and form aggregates (Gotschlich et al., 1981). The PSB used here was probably partially deacylated (Gregoriadis et al., 1993) because of long-term storage, and only the acylated remainder would be expected to form aggregates. If this were so, it would explain the rapid partial loss of PSB from the circulation: hydrolysis of the acyl groups (e.g. by alkali treatment) abolishes the slightly hydrophobic nature of the acylated molecules, leads to their deaggregation and, perhaps, a slower rate of clearance. Results suggest that this is indeed the case; only 5-10% of the fully deacylated PSB was removed rapidly, the remainder exhibiting a linear rate of clearance with a half-life of 30 h (Fig. 2). On the other hand, there was no apparent difference in the clearance patterns of PSK92 before and after full deacylation (Fig. 3). Following a relatively slow clearance during the first 6 h, patterns became linear, with half-lives of 40 h.

It thus appears that the rate of removal of a given polysialic acid from the circulation may be dependent on the presence or absence of phospholipid acyl groups and, since the α-(2-8)-linked PSB is cleared more rapidly than the α-(2-8)-α-(2-9)-linked PSK92 (Figs 2 and 3), also its structure. Further, as the chain length of polysialic acids is an average, the preparations are polydisperse. Consequently, lower molecular weight moieties may contribute to the early rapid removal of some of the polysialic acid from the circulation (as observed for PSB and, to a lesser extent, for PSK92) (Figs 2 and 3 respectively). In this respect, experiments with a PSB of short chain length (15 NeuNAc units) have revealed (data not shown) that over 90% of the injected dose is removed from the circulation within 30 min.

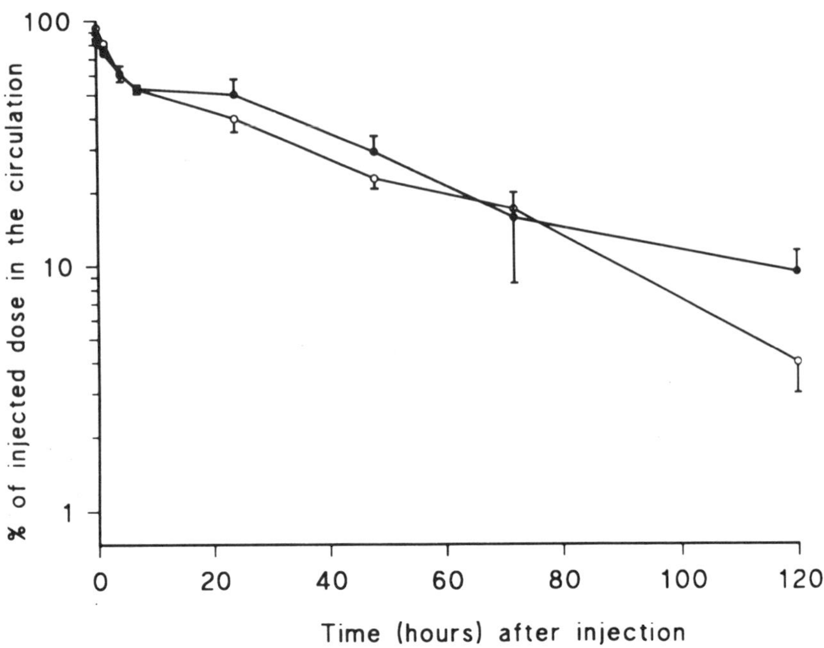

Figure 3. Clearance of PSK92 from the blood circulation. Mice in two groups of 4 were injected intravenously with 1.8 mg of intact (○) or deacylated (●) PSK92. For other details see legend to Fig. 2. (From Gregoriadis et al., 1993, with permission)

Clearance of a model drug bound to polysialic acid

The observation of prolonged circulation of the polysialic acids was encouraging in terms of using these as a means to extend the half-life of small drugs. To that end, studies were carried out with a model drug (fluorescein) coupled to a low molecular weight deacylated PSB (82 NeuNAc units). Data in Fig. 4 show that clearance of coupled fluorescein (which, as such, is removed from the circulation very rapidly; Fig. 4), is independent of the dose of injected PSB for the amounts tested. Following the removal of about 80% of the dose (fluorescein) within 2.5 h, the remainder exhibited a half-life of 5 h, presumably that of the conjugate.

Our results so far indicate that large molecular weight polysialic acids such as those described here, could potentially retain rapidly cleared drugs and small peptides within the vascular and extravascular areas for prolonged periods of time. Moreover, because of the dependence of polysialic acid clearance not only on the type used and the state of the phospholipid (intact or deacylated) but also the molecular size (Gregoriadis et al., 1993), it would be possible to tailor clearance rates of drugs and peptides to satisfy specific needs. It is envisaged that large molecular weight polysialic acids would be suitable for the delivery of one or more molecules (per molecule of polysialic acid) of small molecular weight drugs and peptides. Shorter chain polysialic acids, on the other hand (derived by autohydrolysis of long-chain molecules), could be used to coat large proteins as well as drug delivery systems such as liposomes (Fig. 5). Conjugation of drugs and liposomes to polysialic acids could be carried out by a variety of methods, depending on the reactive groups available on the interacting entities. Possible sites (Fig. 1) of conjugation in polysialic acids include the non-reducing end which, on periodate oxidation, would generate a reactive aldehyde, the carboxyl and hydroxyl groups, and the amino groups becoming available on deacetylation. However, caution is required as coupling reactions could potentially damage the tertiary structure of the longer chain polysialic acids and alter their clearance patterns.

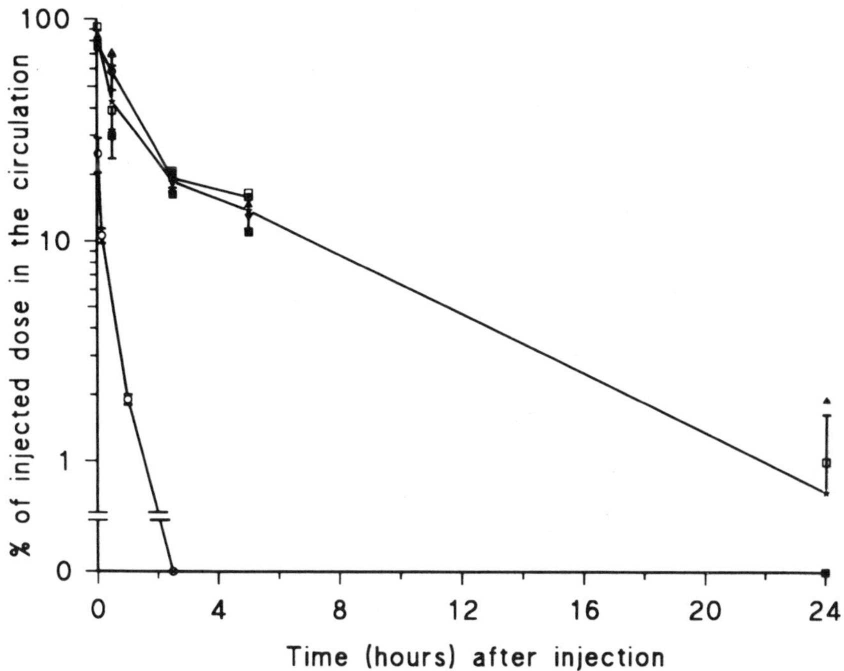

Figure 4. Clearance of low molecular weight fluorescein-PSB conjugate from the blood circulation. Mice in groups of three were injected intravenously with 28 (▾), 102 (■), 510 (▴) and 1,528 μg (●) of PSB conjugated to [^{125}I]fluorescein or with 40 μg fluorescein only (○). Values are means ± S.D. of ^{125}I-radioactivity (closed symbols). NeuNAc (□) or fluorescein. Stars denote the mean of ^{125}I mean values for all doses at each time interval. For other details see legend to Fig. 2. (From Gregoriadis et al., 1993, with permission)

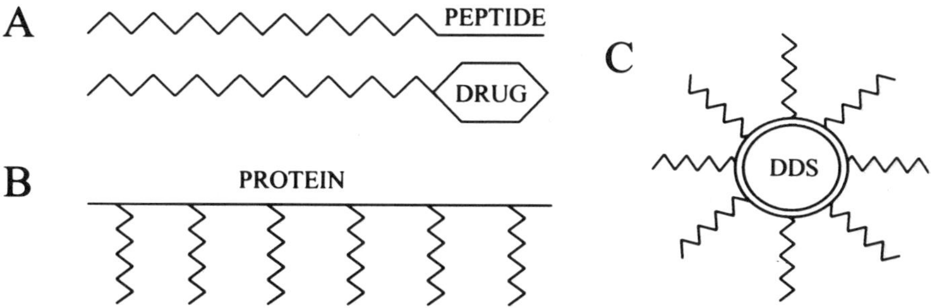

Figure 5. Schemes of polysialic acid use in drug delivery. Long polysialic acids can be used to prolong the circulation time of small drugs and peptides. Shorter polysialic acids bound to the surface of proteins or liposomes will render them more hydrophilic and extend their half-lives

POLYSIALYLATED ENZYMES

Use of proteins in intravenous therapy can be severely hampered by short half-lives in the circulation and proteolytic degradation (Nucci et al., 1991). Moreover, administration of large doses of poteins required for the maintenance of therapeutic efficacy can often provoke toxicity and also increase the possibility of adverse immune responses. As discussed elsewhere (Fernandes and Gregoriadis, 1996; Nucci et al., 1991), circumvention of such problems can be achieved with various degrees of success by covalent coupling of proteins to hydrophilic macromolecules such as albumin, dextrans and mPEG. Of these, mPEG is probably the most successful and comprehensively studied and its use has now been extended to particulate drug delivery systems such as liposomes (this book).

In our work with polysialylated enzymes (catalase and asparaginase), we have employed a low molecular weight polysialic acid, colominic acid (CA) (Fig. 6), as a means to render catalase and asparaginase more hydrophilic. This could, in turn, not only improve the pharmacokinetics and stability of the enzymes but also reduce their immunogenicity. Catalase was chosen as a model therapeutic protein because of its increasing use as an oxygen radical scavenger or in enzyme replacement therapy (Scott et al., 1991).Asparaginase on the other hand, catalyses the hydrolysis of the non-essential amino-acid L-asparagine to L-aspartic acid and ammonia and is currently in clinical use for the treatment of acute lymphoblastic leukaemia (Keating et al., 1993). It is also active against non-Hodgkin's lymphoma and pancreatic carcinoma (Yunis et al., 1977). The antitumour activity of the enzyme is based on the dependance of certain tumour cells that are deficient in L-asparagine synthetase (Haskell et al., 1969) on the external supply of L-asparagine. Normal cells produce the synthetase and are therefore not affected by the treatment. Thus, unlike conventional cancer therapy, asparaginase therapy is highly selective.

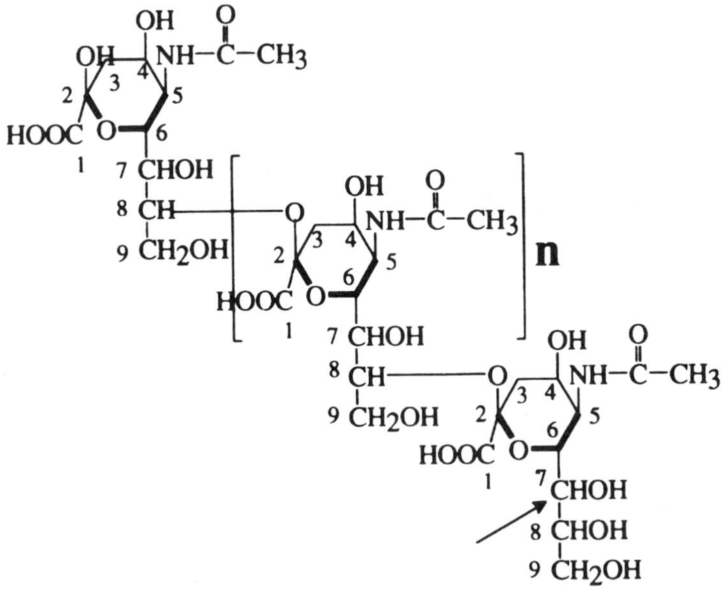

Figure 6. Structure of colominic acid. *N*-acetylneuraminic acid units are linked via α-(2→8) glycosidic linkages. Arrow indicates the carbon atom (C_7) at the non-reducing end of the sugar where periodate oxidation introduces an aldehyde group. (From Fernandes and Gregoriadis, 1996, with permission)

Polysialylated catalase

Results (Fernandes and Gregoriadis, 1996) on catalase polysialylated with colominic acid after periodate oxidation of the latter at the non-reducing end (carbon 7) and subsequent coupling to the enzyme by reductive amination in the presence of sodium cyanoborohydride, showed that the extent of polysialylation was modest (3.8±0.4 moles of CA per mole of catalase). Polysialylated catalase, however, retained 70% of its initial activity at the end of the coupling reaction compared with values of 29-39% for the enzyme controls incubated in the absence of CA and reagents. Formation of sialylated catalse was confirmed by ammonium sulphate or trichloroacetic acid precipitation, molecular sieve chromatography and SDS-PAGE electrophoresis (Fernandes and Gregoriadis, 1996). Compared to native enzyme, sialylated catalase was much more stable in the presence of specific proteinases, especially chymotrypsin (Fernandes and Gregoriadis, 1996).

Polysialylated asparaginase

Attempts made previously to increase the half-life of the asparaginase in the blood circulation, include entrapment into liposomes (Neerunjun and Gregoriadis, 1976) or erythrocytes (Kravtzoff et al., 1990), and covalent coupling to mPEG (Cao et al., 1990). This, can lead, however, to substantial loss of enzyme activity during the coupling reaction.

Activation of colominic acid: CA was oxidized (activated) with 0.1M sodium periodate (10 mg CA/ml periodate solution) at 20°C in the dark and in the presence of ethylene glycol to expend excess periodate. The solution was then dialysed extensively at 4°C against ammonium carbonate solution, freeze-dried and kept at -40°C until further use (Fernandes and Gregoriadis, 1996, 1997).

Preparation of asparaginase-colominic acid conjugates: Activated CA was covalently coupled to asparaginase (previously dialyzed to remove dextrose monohydrate) by reductive amination in the presence of $NaCNBH_3$ as previously applied for catalase (Fernandes and Gregoriadis, 1996). In brief, activated CA was added to asparaginase dissolved in 0.75 M K_2HPO_4, pH 9.0 and the mixture stirred for 48 h at 35-40°C in the presence of $NaCNBH_3$. Three different molar ratios of CA:asparaginase (50:1, 100:1 and 250:1) were used in the coupling reaction. The neoglycoprotein formed was then isolated by ammonium sulphate (70% saturation) precipitation followed by ultracentrifugation. The pellets obtained were washed with an ammonium sulphate solution of the same concentration to remove non-conjugated CA. Pellets containing the conjugates were then dissolved in 0.15 M sodium phosphate buffered saline, pH 7.4 (PBS) and extensively dialysed at 4°C against the same buffer. The dialysed samples were filtered through a low protein binding filter to remove insoluble material, and asparaginase activity and protein concentration in the filtrates determined spectrophotometrically. CA bound to the enzyme was measured (Fernandes and Gregoriadis, 1997) and values expressed as moles of CA per mole of asparaginase or as percentage of polysialylated lysine residues. Solutions containing a known amount of polysialylated asparaginase (450-475 U/mg protein) were freeze-dried and kept at 4°C until further use.

The coupling reaction between asparaginase and CA relies on the ability of $NaCNBH_3$ to selectively reduce the Schiff base formed between the aldehyde group introduced at the non-reducing end of the saccharide and the free ε-amino groups of the enzyme. Results (Fernandes and Gregoriadis, 1997) on the extent of CA coupling to asparaginase and activity retention by the polysialylated enzyme indicate that polysialylation is directly dependent on the molar ratio of CA and enzyme used in the coupling reaction, with the highest degree of polysialylation (8.1 ± 1.7 moles of CA/mole asparaginase) obtained by the use of a 250 fold excess CA in the reaction mixture. (However, as CA is polydisperse, values of degree of polysialylation are only average). This value (8.1) corresponds to an average of 11% of the available lysine ϵ-amino groups (Fernandes and Gregoriadis, 1997).

Chemical modification of asparaginase by other methods is known to lead to severe loss of enzyme activity, for instance as much as 70-90% for pegylated asparaginase (Cao et al., 1990). In contrast, the coupling procedure used here leads to only a modest loss (14-18%) of initial asparaginase activity in the polysialylated enzyme (Fernandes and Gregoriadis, 1997). It appears that polysialylation (or at least the presence of CA in the reaction mixture during the coupling procedure), protects the enzyme from inactivation: only 17% of the activity was retained by asparaginase subjected to identical reaction conditions in the absence of CA. Our results (Fernandes and Gregoriadis, 1997) also indicate that, retention of activity is independent of the amount of coupled CA, suggesting that the enzyme may tolerate an even greater degree of polysialylation.

Radiolabelling of asparaginase: Native asparaginase was tritiated as described elsewhere (Fernandes and Gregoriadis, 1997). Following isolation of the labelled enzyme with ammonium sulphate and precipitation with trichloroacetic acid, more than 90% of the radioactivity was recovered with the enzyme pellet. Moreover, polyacrylamide gel electrophoresis of the labelled enzyme confirmed (results not shown) the absence of higher molecular mass enzyme species that could have formed through formaldehyde-induced methylene bridges. In other experiments, asparaginase was simultaneously radiolabelled and coupled to CA as described (Fernandes and Gregoriadis, 1997), by employing $NaCNB[^3H]_3$ in the reaction mixture. The enzyme conjugate was isolated and excess label removed by

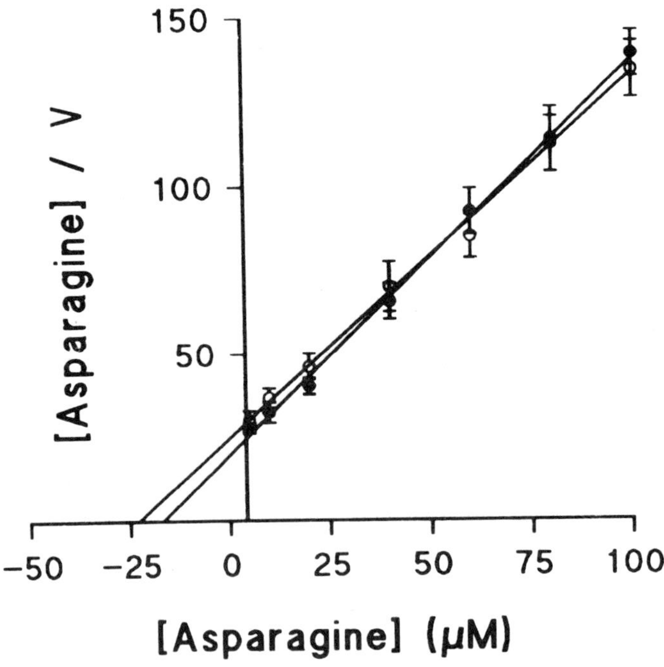

Figure 7. Hanes Woolf plots for native (●) and polysialylated asparaginase (○). Values denote means ± S.D. (3 different experiments); V denotes velocity, expressed as μmol of liberated ammonia per min per unit (U) of enzyme. K_m values were obtained by extrapolation to the abscissa. (From Fernandes and Gregoriadis, 1997, with permission)

ammonium sulphate precipitation followed by extensive dialysis. Typically, more than 90% of the radioactivity in the polysialylated asparaginase preparations could be precipitated by trichloroacetic acid, indicating that, because of the obligatory participation of ^3H in the structure of the polysialylated enzyme (reductive methylation), most of the radioactivity was associated with it.

Enzyme kinetics: Fig. 7 shows the effect of substrate concentration on the activity of native and polysialylated asparaginase, with the plots suggesting a modest (but not significant; P>0.05) increase in the enzyme's Km value after polysialylation. Km values estimated according to Hanes-Woolf were 1.68 x 10^{-5} M (native asparaginase) and 1.90 x 10^{-5} M, 2.15 x 10^{-5} M and 2.29 x 10^{-5} M respectively for the polysialylated constructs made with different CA to enzyme ratios (see Fig. 8). These values are of the same order of magnitude as those reported (Howard and Carpentier, 1972) for the clinically useful asparaginases (e.g. 10^{-5} M). From the same plots of Fig. 7 and in the same order, calculated V_{max} values were 0.847, 0.901, 0.910 and 0.919 μmole min^{-1} U^{-1} (results not shown). Thus, covalent coupling of CA to asparaginase does not affect significantly the enzyme's action on asparagine, regardless of its degree of polysialylation.

Stability of asparaginase in plasma: Polysialylated asparaginase appears (Fig. 8) to be more stable in the presence of (mouse) plasma at 37°C than the native (non-sialylated) enzyme. For instance, whereas polysialylated asparaginase retained most (65-83%) of its initial activity after 6 h of incubation, that of the native enzyme decreased to 13.5%. Moreover, activity retention by the polysialylated asparaginase was significantly higher for the preparation with

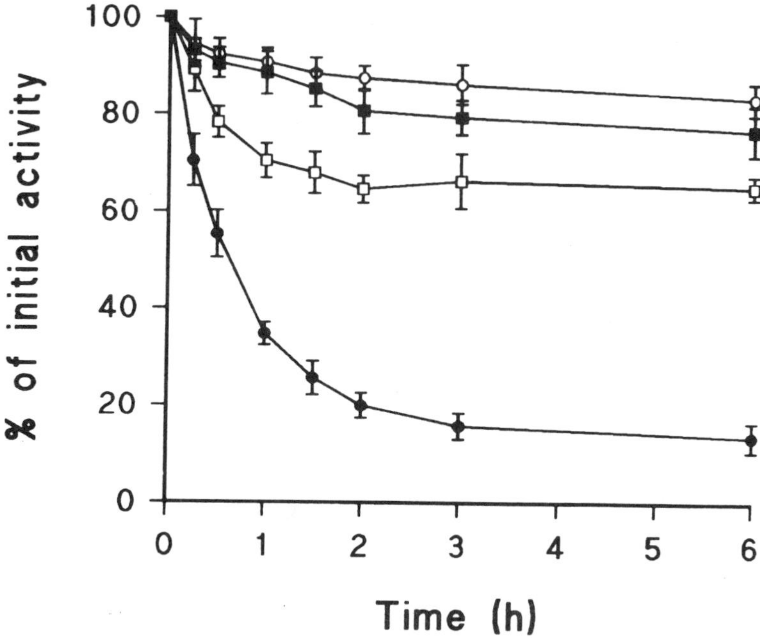

Figure 8. Retention of asparaginase activity in the presence of plasma. Native (●) and polysialylated asparaginase preparations A (□), B (■), C (○) were incubated in the presence of mouse plasma at 37°C; values denote means ± S.D. (3 different preparations). *Statistics*: results obtained at 6 h, were compared by ANOVA and *P* values corrected by the Bonferroni test. Native vs. A, B and C, *P*<0.001; A vs. B, n.s.; B vs. C, n.s.; A vs. C, *P*<0.05; n.s. = non-significant (From Fernandes and Gregoriadis, 1997)

the greatest number of CA molecules (preparation C; Fig. 8), suggesting that a greater degree of polysialylation than that achieved here, may further reduce the plasma-induced destabilization of the enzyme. The influence of polysialylation on the stability of asparaginase, also observed for polysialylated catalase (Fernandes and Gregoriadis, 1996), has been attributed to changes in the microenvironment of the enzyme which must have occurred by the presence of the highly hydrophilic, negatively charged CA molecules. It is conceivable that a combination of a hydrophilic environment (promoted by polysialylation) and a shielding effect of the CA chains, contributes to the inaccessibility of the enzyme to plasma proteases.

Clearance of native and polysialylated asparaginase from the circulation: Preliminary work (results not shown) had indicated that tritiated asparaginase does not exhibit an altered pattern of clearance (in terms of enzyme activity) from the circulation. We therefore compared the clearance of tritiated asparaginase with that of the tritiated polysialylated constructs. As alluded to earlier, radioactivity in the latter was expected to represent polysialylated asparaginase rather than non-sialylated enzyme possibly present in the preparation. Moreover, the two (radiolabelled) constructs (see above) were prepared under conditions (i.e. CA:enzyme molar ratio used in the coupling reaction) that were identical to those used for the unlabelled conjugates, except that $NaCNB[^3H]_3$ was added during the reaction.

Results (Fig. 9) show that all three constructs of polysialylated asparaginase are removed from the circulation at slower rates than the native enzyme, both in terms of radioactivity (Fig. 9A) and enzyme activity (Fig. 9B). However, as also observed with other modified enzymes (eg. Nucci et al., 1991), the bulk of the administered dose (about 75% and 60-65% for native and polysialylated asparaginase respectively) was cleared within 2 h after injection, the remainder exhibiting slower, linear clearance rates (Fig. 9A,B). Moreover terminal half-lives ($t_{1/2\beta}$) on the basis of enzyme activity for native (about 15 h) and polysialylated asparaginase (about 38 h) (estimated from the linear portions of asparaginase activity and ^3H radioactivity clearance patterns), were found to be independent of the dose injected, at least for the range of doses tested (0.5- 2.0 mg) (Fernandes and Gregoriadis, 1997).It was of interest to note (Fernandes and Gregoriadis, 1997) that for each of the polysialylated conjugates there was no difference between terminal half-lives derived from radioactivity and enzyme activity values, confirming that radioactivity measurements accurately reflect the presence of active enzyme.

Clearance of injected proteins occurs either through non-specific uptake by the reticuloendothelial system (often as a result of their association with blood components) or through receptor-mediated endocytosis (Delgado et al., 1992) by cells of the liver and other tissues where protein degradation eventually occurs. Moreover, molecular mass, shape and charge influence protein clearance (Bocci, 1987; Benbough et al., 1979) by determining the extent to which proteins undergo transcapillary passage or renal filtration (Bocci, 1987). In the latter case, the molecular mass cut-off is 66-70 kDa (Delgado et al., 1992), i.e. well below the molecular mass (135 kDa) of asparaginase. Judging from data in Fig. 8, it is likely that, at least in part, increases in the half-life of polysialylated asparaginase are due to a greater resistance to plasma proteases. It is also feasible that, upon polysialylation and as a result of the loss of some of the free ϵ-amino groups of asparaginase through the coupling of the CA molecules, the modified protein is intrinsically more negatively charged. This, together with a shielding effect of the CA chains discussed elsewhere (Fernandes and Gregoriadis, 1996), could interfere with the interaction of the enzyme with blood and tissue components and thus curtail its recognition by tissues and removal from the circulation.

The improvement in the pharmacokinetics of the polysialylated asparaginase observed in the present study is not as great as that claimed for the pegylated enzyme (eg. Cao et al., 1990). However, the following should be taken into consideration: (a) administered amounts of both

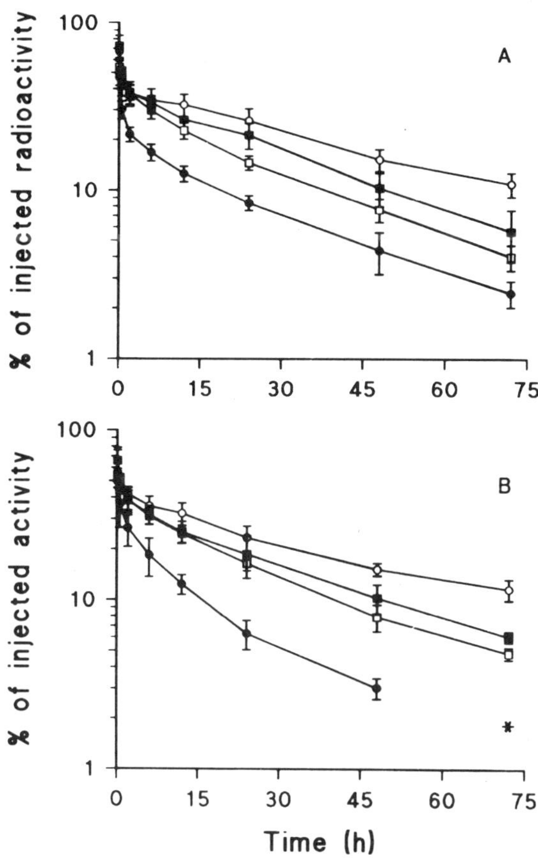

Figure 9. Clearance of asparaginase from the circulation. Mice were injected intravenously with 1 mg (550 U) tritiated native (●) and 1 mg (450-475 U) tritiated polysialylated asparaginase prepared by the use of 50:1 (□), 100:1 (■) and 250:1 (○) molar ratios of CA: asparaginase in the coupling reaction. Blood plasma obtained at time intervals was assayed for ^{3}H (A) and asparaginase activity (B). The pharmacokinetics profiles demonstrate biphasic patterns of clearance which are consistent with a two-compartment model. Values denote means ± S.D.; n=4 animals. *Native asparaginase activity was not detectable at 72 h (From Fernandes and Gregoriadis, 1997)

native and pegylated enzyme in these studies were too low (up to 40 U per animal) compared to those (450-550 U) used in the present work. Therefore, measurements of enzyme activity may not have been as accurate, especially over extended periods of time, in turn contributing to an overestimation of improvement in the half-lives for the pegylated asparaginase; (b) as already mentioned, pegylation of asparaginase leads to quantitative loss (Cao er al., 1990; Uren and Ragin, 1979) of enzyme activity. Present results show that enzyme activity is largely maintained for the polysialylated enzyme (Fernandes and Gregoriadis, 1996, 1997). Therefore, polysialylation as a means to extend the half-lives of asparaginase and other therapeutic proteins, may be a prefered alternative as it would limit wastage of proteins; (c) to the present authors' knowledge, there is no information on the fate and effect of the mPEG moiety of pegylated proteins subsequent to their uptake by tissues. mPEG is non-biodegradable and its accumulation intracelullarly, especially on chronic use, may prove undesirable.

POLYSIALIC ACIDS: IMMUNOGENICITY AND ANTIGENICITY

Polysialic acids have the advantage of being biodegradable and catabolic products (eg. NeuNAc) are not known to be toxic. Furthermore, polysialic acids, like other polysaccharides, are T-independent antigens and do not induce immunological memory. For instance, PSB is non-immunogenic in animals and humans which has hampered attempts to produce a vaccine against *N.meningitidis* group B or *E.coli* K1 (Moreno et al., 1985). Although PSC and PSK92 have been shown to be immunogenic in humans, it is necessary to use polysaccharides with molecular weights in excess of 50.000 Da (average chain length greater than 170 NeuNAc units). This, however, is not true when considering polysialic acids coupled to proteins. Here, polysialic acids can become T cell dependent antigens with induction of memory, and no restriction on the size of the polymer applies. Nonetheless, immune responses are difficult to achieve (especially for PSB) although immunogenic vaccines have been manufactured by coupling polysialic acids to some protein carriers. Another, perhaps more important, consideration in selecting a polysialic acid as a drug carrier, is antigenicity (i.e. binding of the antigen to its antibodies). Although it has been shown that antibodies against some of the polysialic acid structures examined here exist at low levels in circulation, they are generally of low affinity, especially against α-(2-8) linked PSB (Mandrell and Zollinger, 1982). Indeed, α-(2-8)-linked sialic acid structures are known to be present on host cell surfaces, thereby limiting any immunological response (Finne, 1982). Finally, taking into account the pathogenicity of *N.meningitidis* it would be easier, from the practical point of view, to produce polysialic acids from slightly or non-pathogenic bacteria. Since PSB (deacylated) and PSK92 exhibit the longest half-lives (Figs. 2 and 3) and can be derived from the slightly pathogenic *E.coli* K1 (PSB) or the non-pathogenic *E.coli* K92 (PSK92) bacteria, these materials and their lower molecular weight products should be adopted for conjugation to drugs and drug delivery systems.

Acknowledgements

The authors thank Mrs Concha Perring for excellent secretarial assistance in the short time required to complete this book.

REFERENCES

Benbough, J.E., Wiblin, C.N., Rafter, T.N.-A., Lee, J., 1979, The effect of chemical modification of L-asparaginase on its persistence in circulating blood of animals, *Biochem.Pharmacol.* 28:833.

Bocci, V., 1987, Metabolism of protein anticancer agents, *Pharmacol.Ther.* 34:1.

Cao, S., Zhao, Q., Ding, Z., Ma, L., Yu, T., Wang, J., Feng, Y., and Cheng, Y., 1990, Chemical modifications of enzyme molecules to improve their characteristics, *Ann. New York Acad.Sci.* 613:460.

Davis, S.S., Illum, L., McVie, J.-G., and Tomlison, E., eds., 1984, *Microspheres and Drug Therapy*, Elsevier, Amsterdam.

Delgado, C., Francis, G.E., and Fister, D., 1992, The uses and properties of PEG-linked proteins, *Crit.Rev.Ther. Drug Carrier Syst.* 9:249.

Domb, A.J., Kost, J., and Wiseman, D.M., eds., 1997, *Handbook of Biodegradable Polymers*, Harwood Academic Publishers, Amsterdam.

Fernandes, A., and Gregoriadis, G., 1996, Synthesis, characterization and properties of sialylated catalase, *Biochim.Biophys.Acta*, 1293:92.

Fernandes, A., and Gregoriadis, G., 1997, Polysialylated asparaginase: preparation, activity and pharmacokinetics, *Biochim.Biophys.Acta*, 1341:26.

Finne, J., 1982, Occurrence of unique polysialosyl carbohydrate units in glycoproteins of developing brain, *J.Biol.Chem.* 257:11966

Gotschlich, E.C., Fraser, B.A., Nischimura, O., Robbins, J.B., and Liu, T.-Y., 1981, Lipid on capsular polysaccharides of gram-negative bacteria, *J.Biol.Chem.* 256:8915.

Gregoriadis, G., 1995, Engineering liposomes for drug delivery: progress and problems, *Trends in Biotechnology*, 13:527.

Gregoriadis, G., McCormack, B., Wang, Z., and Lifely, R., 1993, Polysialic acids: Potential in drug delivery, *FEBS Lett.*, 315:271.

Gregoriadis, G., McCormack, B., and Poste, G., eds, 1994, Liposomes *in vivo*: Control of behaviour, in: *Targeting of Drugs: Advances in System Constructs*, Plenum, New York.

Haskell, C.M., Canellos, G.P., Leventhal, B.G., Carbone, P.P., and Black, J.B., 1969, L-asparaginase resistance in human leukemia-asparaginase synthetase, *Biochem.Pharmacol.* 18:2578.

Howard, J.B., and Carpentier, F.H., 1972, L-asparaginase from *Erwinia carotovora*, *J.Biol.Chem.* 247:1020.

Keating, M.J., Holmes, R., Lerner, S., Ho, D.H., 1993, L-asparaginase and PEG asparaginas -past, present and future, *Leukemia Lymphoma*, 10:153.

Kravtzoff, R., Ropers, C., Laguerre, M., Muh, J.P., and Chassaigne, M., 1990, Erythrocytes as carriers for L-asparaginase. Methodological , *J.Pharm.Pharmacol.*42:473.

Lee, V.H.L., Hashida, M., and Mizushina, Y., eds., 1995, *Trends and Future Perspectives in Peptide and Protein Drug Delivery*, Harwood Academic Publishers, Chur.

Mandrell, R.E., and Zollinger, W.D., 1982, Measurement of antibodies to meningococcal group B polysaccharide: low avidity binding and equilibrium binding constants, *J.Immunol.* 129:2172.

Moreno, C., Lifely, M.R., and Esdaile, J., 1985, Immunity and protection of mice against *Neisseria meningitidis* group B by vaccination using polysaccharide complexed with outer membrane proteins: a comparison with purified B polysaccharide, *Infect.Immun.* 47:527.

Neerunjun, D.E., and Gregoriadis, G., 1976, Tumour regression with liposome-entrapped asparaginase. Some immunological advantages, *Biochem.Soc.Trans.* 4:133.

Nucci, M.L., Shorr, R., and Abuchowski, A., 1991, The therapeutic value of poly(ethylene-glycol)-modified proteins, *Adv. Drug Del.Rev.* 6:133.

Scott, M.D., Lubin, B.H., Zuo, L., and Kuypers, F.A., 1991, Erythocyte defence against hydrogen peroxide: preeminent importance of catalase. *J.Lab.Clin.Med.* 118:7.

Torchilin, V.P., and Papisov, M.I., 1994, Why do polyethylene glycol-coated liposomes circulate so long?, *J.Liposome Res.* 4:725.

Uren, J.R., and Ragin, R.C., 1979, Improvement in the therapeutic, immunological and clearance properties of *Escherichia coli* and *Erwinia carotovora* L-asparaginase by attachment of poly-DL-alanyl peptides, *Cancer Res.* 39:1927.

Yunis, A.A., Arimura, G.K., and Russin, D.J., 1977, Human pancreatic carcinoma (MIA PaCa-2) in continuous culture: sensitivity to asparaginase, *Int.J.Cancer*, 19:128.

TAILOR-MADE SOLUBLE POLYMER CARRIERS

Paolo Ferruti[1], Ruth Duncan[2] and Simon Richardson[2]

[1]Dipartimento di Chimica Organica e Industriale,
Università di Milano, Via Venezian 21, 20133 Milano, Italy and
[2]The School of Pharmacy, University of London,
Center for Polymer Therapeutics, Brunswick Square, London, UK

INTRODUCTION

Poly(amido-amines) (PAAs) are synthetic polymers characterized by the presence of amido and tertiary amino groups regularly arranged along their macromolecular chain. Therefore, they can be classified as polycations.

The potential of PAAs for medical applications as heparin-complexing agents was first discovered by Marchisio in the early seventies (Marchisio et al., 1973) and PAAs are currently being employed as components of de-heparinizing filters for hemodialysis and, in the form of block and graft copolymers, for imparting to conventional polymers the capability of stably retaining heparin on their surface thus improving their hemocompatibility. However, the potential of PAAs as soluble drug carriers has been only recently recognized (Duncan et al., 1994). The aim of this presentation is precisely to report on the potential of of poly(amido-amine)s as soluble drug carriers.

Preparation and properties of poly(amido-amine)s

Preparation: Linear PAAs are obtained by polyaddition of primary monoamines or bis-(secondary)amines, to bis-acrylamides:

a) $$CH_2{=}CH{-}\overset{\overset{\displaystyle O}{\|}}{C}{-}\underset{\underset{\displaystyle R_1}{|}}{N}{-}R_2{-}\underset{\underset{\displaystyle R_1}{|}}{N}{-}\overset{\overset{\displaystyle O}{\|}}{C}{-}CH{=}CH_2 \; + \; H{-}\underset{\underset{\displaystyle R_3}{|}}{N}{-}H \longrightarrow$$

$$\left[CH_2{-}CH_2{-}\overset{\overset{\displaystyle O}{\|}}{C}{-}\underset{\underset{\displaystyle R_1}{|}}{N}{-}R_2{-}\underset{\underset{\displaystyle R_1}{|}}{N}{-}\overset{\overset{\displaystyle O}{\|}}{C}{-}CH_2{-}CH_2{-}\underset{\underset{\displaystyle R_3}{|}}{N}{-} \right]_n$$

b)

$$CH_2{=}CH{-}\overset{O}{\overset{\|}{C}}{-}\underset{R_1}{N}{-}R_2{-}\underset{R_1}{N}{-}\overset{O}{\overset{\|}{C}}{-}CH{=}CH_2 \; + \; H{-}\underset{R_3}{N}{-}R_4{-}\underset{R_3}{N}{-}H$$

$$\downarrow$$

$$\left[-CH_2{-}CH_2{-}\overset{O}{\overset{\|}{C}}{-}\underset{R_1}{N}{-}R_2{-}\underset{R_1}{N}{-}\overset{O}{\overset{\|}{C}}{-}CH_2{-}CH_2{-}\underset{R_3}{N}{-}R_4{-}\underset{R_3}{N}- \right]_n$$

The polymerization reaction usually takes place very smoothly in aqueous or alcoholic media, at room temperature, and without added catalysts, with only a moderate evolution of heat in the very first stages. Amines react under these conditions only as free bases, but not in their protonated form, and therefore the pH of the reaction medium should be sufficiently high to ensure that a substantial amount of the aminic monomer is present as free base. Aprotic solvents are unsuitable as reaction media, as only low molecular weight products are obtained. As regards concentration, the sum of the monomers in the reacting mixture should be kept as high as possible. A total solids content of 30-60% by weight will usually perform well (Ferruti, 1985) and if one of the monomers is not soluble to this extent in the reaction medium, but the resulting PAA is, the former will gradually dissolve as the polymerization proceeds.

The above method is a general one for aliphatic amines and for all bisacrylamides. The presence of additional functions in the monomers does not usually bias the polyaddition reaction, with the obvious exception of additional primary or secondary amino groups and of thiol groups, which undergo the addition reaction, and of carboxyl groups. In fact, if aminoacids are used as monomers, the reaction conditions must be adjusted by adding a strong base in order to meet the requirement of having the amino groups in their unprotonated form. Even so, however, natural α-aminoacids other than glycine give as a rule only low molecular weight products, but the polyaddition proceeds reasonably well with all aminoacids bearing no substituents on the carbon atom in α position, as well as with peptides whose first aminoacid residue fulfills the same condition. It may be concluded that PAAs carrying many additional functions as side substituents can be easily obtained starting from the appropriate monomers.

Oligomeric compounds carrying either two secondary- or one primary end-amino group, such as for instance α,ω-bis(*sec*-amino)polyoxyethylenes, or α-methyl-ω-aminopoly-oxyethylenes, can be used as monomers in the same way as simple amines, giving, respectively, block(Ranucci and Ferruti, 1991) - or graft (Vansteenkiste et al., 1992) copolymers based on PAA. The structures of a number of PAAs described in the literature (Ferruti et al., 1985; Ferruti, 1996) are given in Table 1.

Under non selective conditions, bis-primary amines usually give crosslinked products on reaction with bis-acrylamides, since they react tetrafunctionally. However, it has been reported that under special conditions including low reactant concentrations, low initial temperatures, and an excess, on a molar basis, of bis-amines, soluble PAAs carrying secondary amino groups in their main chain can be obtained. These PAAs have been found to provide suitable soluble carriers for drug attachement (Caldwell et al., 1993).

PAAs in which the amido and ter-amino groups are not regularly distributed along the macromolecular chain have been synthesized by poly(acylation-addition) of piperazine and other bis-secondary amines, with various activated derivatives of acrylic acid (Rusconi et al., 1982a; Tanzi et al., 1984a). "Alternating" PAAs, in which one amino-group is strictly followed by one amido-group all along the macromolecular chain have also been synthesized (Rusconi et al., 1982b).

Table 1. Some examples of poly(amino-amines)

No.	Structure of the repeating unit	[η] (dl/g)
1	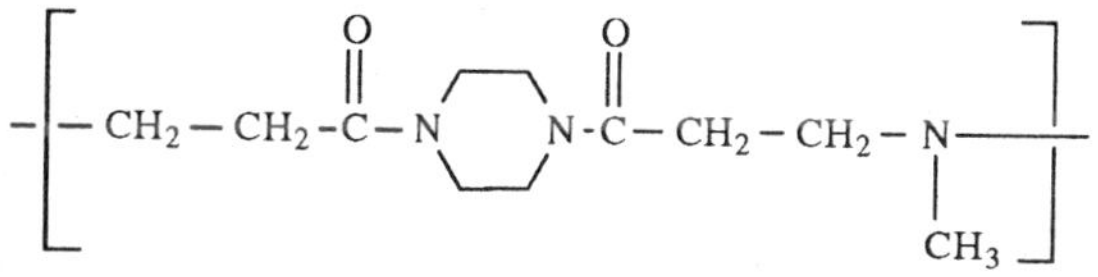 $-[-CH_2-CH_2-C(=O)-N(CH_2CH_2)_2N-C(=O)-CH_2-CH_2-N(CH_3)-]-$	0.40
2	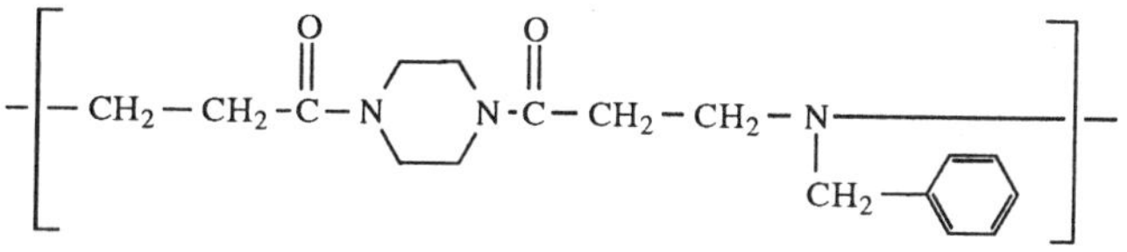 $-[-CH_2-CH_2-C(=O)-N(CH_2CH_2)_2N-C(=O)-CH_2-CH_2-N(CH_2-C_6H_5)-]-$	0.15
3	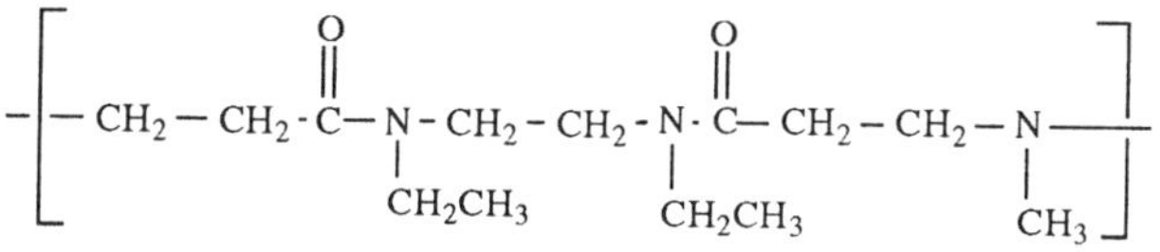 $-[-CH_2-CH_2-C(=O)-N(CH_2CH_3)-CH_2-CH_2-N(CH_2CH_3)-C(=O)-CH_2-CH_2-N(CH_3)-]-$	0.20
4	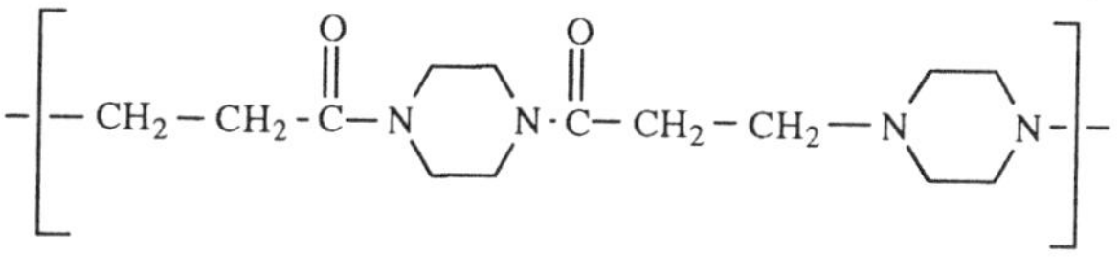 $-[-CH_2-CH_2-C(=O)-N(CH_2CH_2)_2N-C(=O)-CH_2-CH_2-N(CH_2CH_2)_2N-]-$	0.46
5	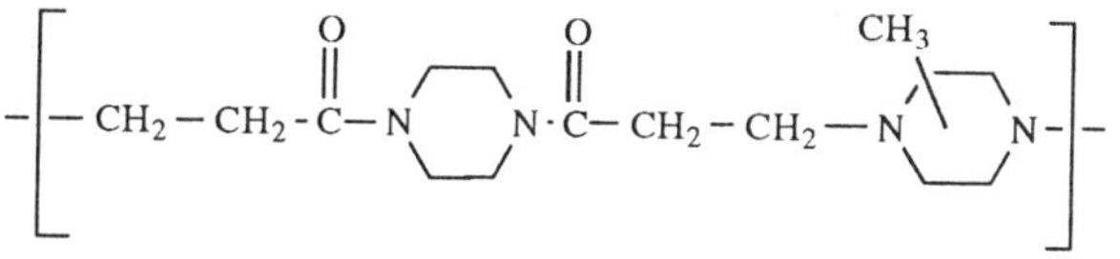 $-[-CH_2-CH_2-C(=O)-N(CH_2CH_2)_2N-C(=O)-CH_2-CH_2-N(CH(CH_3)CH_2)(CH_2CH_2)N-]-$	0.81
6	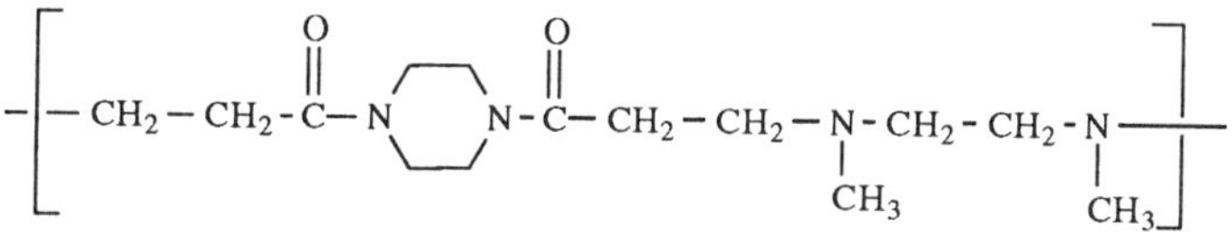 $-[-CH_2-CH_2-C(=O)-N(CH_2CH_2)_2N-C(=O)-CH_2-CH_2-N(CH_3)-CH_2-CH_2-N(CH_3)-]-$	0.41
7	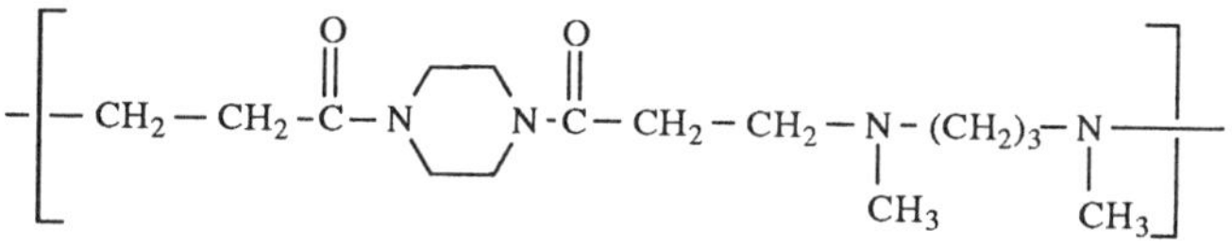 $-[-CH_2-CH_2-C(=O)-N(CH_2CH_2)_2N-C(=O)-CH_2-CH_2-N(CH_3)-(CH_2)_3-N(CH_3)-]-$	0.43
8	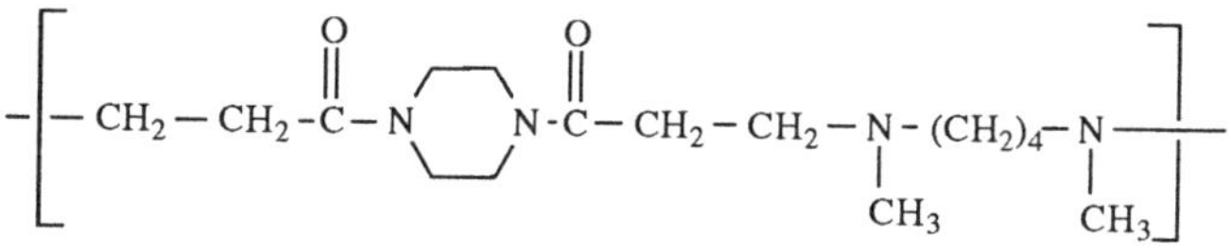 $-[-CH_2-CH_2-C(=O)-N(CH_2CH_2)_2N-C(=O)-CH_2-CH_2-N(CH_3)-(CH_2)_4-N(CH_3)-]-$	0.40
9	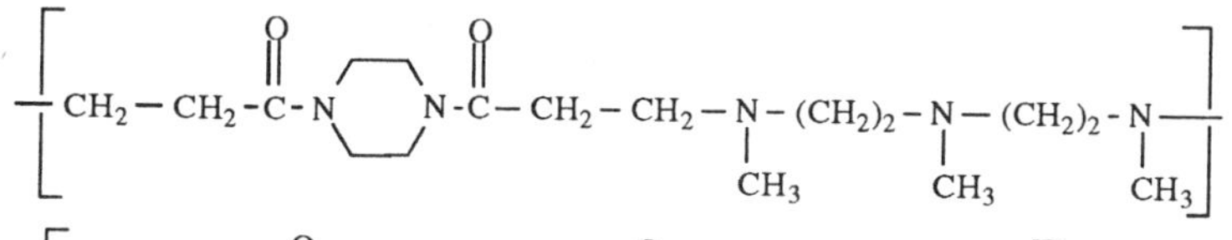 $-[-CH_2-CH_2-C(=O)-N(CH_2CH_2)_2N-C(=O)-CH_2-CH_2-N(CH_3)-(CH_2)_2-N(CH_3)-(CH_2)_2-N(CH_3)-]-$	0.27
10	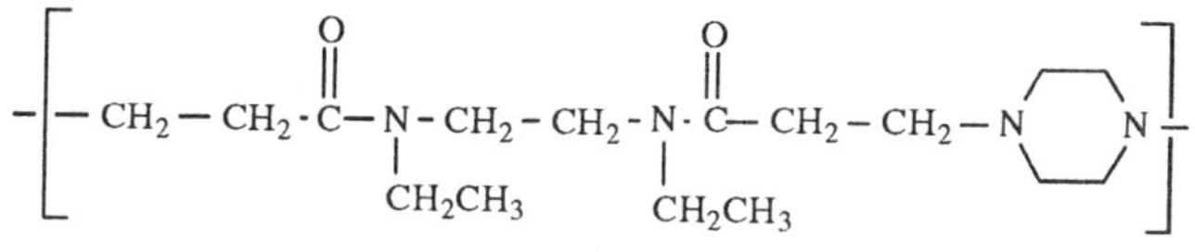 $-[-CH_2-CH_2-C(=O)-N(CH_2CH_3)-CH_2-CH_2-N(CH_2CH_3)-C(=O)-CH_2-CH_2-N(CH_2CH_2)_2N-]-$	0.20

Table 1. Some examples of poly(amino-amines) (cont/ed)

No.	Structure of the repeating unit	[η] (dl/g)
11	$-[-CH_2-CH_2-\overset{O}{\overset{\Vert}{C}}-NH-CH_2-CH_2-HN-\overset{O}{\overset{\Vert}{C}}-CH_2-CH_2-N\langle\ \rangle N-]-$	0.12
12	$-[-CH_2-CH_2-\overset{O}{\overset{\Vert}{C}}-NH-CH_2-HN-\overset{O}{\overset{\Vert}{C}}-CH_2-CH_2-N\langle\ \rangle N-]-$	1.0
13	$-[-CH_2-CH_2-\overset{O}{\overset{\Vert}{C}}-NH-CH_2-HN-\overset{O}{\overset{\Vert}{C}}-CH_2-CH_2-N\langle^{CH_3}\ \rangle N-]-$	0.82
14	$-[-CH_2-CH_2-\overset{O}{\overset{\Vert}{C}}-NH-CH_2-HN-\overset{O}{\overset{\Vert}{C}}-CH_2-CH_2-\underset{CH_3}{\underset{\vert}{N}}-CH_2-CH_2-\underset{CH_3}{\underset{\vert}{N}}-]$	0.37
15	$-[-CH_2-CH_2-\overset{O}{\overset{\Vert}{C}}-NH-CH_2-HN-\overset{O}{\overset{\Vert}{C}}-CH_2-CH_2-\underset{CH_3}{\underset{\vert}{N}}-(CH_2)_6-\underset{CH_3}{\underset{\vert}{N}}-]$	0.54
16	$-[-CH_2-CH_2-\overset{O}{\overset{\Vert}{C}}-NH-CH_2-HN-\overset{O}{\overset{\Vert}{C}}-CH_2-CH_2-\underset{N(CH_3)_2}{\underset{\vert}{\underset{(CH_2)}{\underset{\vert}{N}}}}-]$	0.25
17	$-[-CH_2-CH_2-\overset{O}{\overset{\Vert}{C}}-N\langle\ \rangle N-\overset{O}{\overset{\Vert}{C}}-CH_2-CH_2-\underset{CH_2CH=CH_2}{\underset{\vert}{N}}-]-$	0.35
18	$-[-CH_2-CH_2-\overset{O}{\overset{\Vert}{C}}-N\langle\ \rangle N-\overset{O}{\overset{\Vert}{C}}-CH_2-CH_2-\underset{CH_2CH_2OH}{\underset{\vert}{N}}-CH_2-CH_2-\underset{CH_2CH_2OH}{\underset{\vert}{N}}-]$	0.37
19	$-[-CH_2-CH_2-\overset{O}{\overset{\Vert}{C}}-N\langle\ \rangle N-\overset{O}{\overset{\Vert}{C}}-CH_2-CH_2-\underset{N(CH_3)_2}{\underset{\vert}{\underset{(CH_2)_2}{\underset{\vert}{N}}}}-]-$	0.19
20	$-[-CH_2-CH_2-\overset{O}{\overset{\Vert}{C}}-N\langle\ \rangle N-\overset{O}{\overset{\Vert}{C}}-CH_2-CH_2-\underset{N(CH_3)_2}{\underset{\vert}{\underset{(CH_2)_2}{\underset{\vert}{N}}}}-(CH_2)_2-\underset{N(CH_3)_2}{\underset{\vert}{\underset{(CH_2)_2}{\underset{\vert}{N}}}}-]$	0.25

Table 1. Some examples of poly(amino-amines) (cont/ed)

No.	Structure of the repeating unit	$[\eta]$ (dl/g)
21	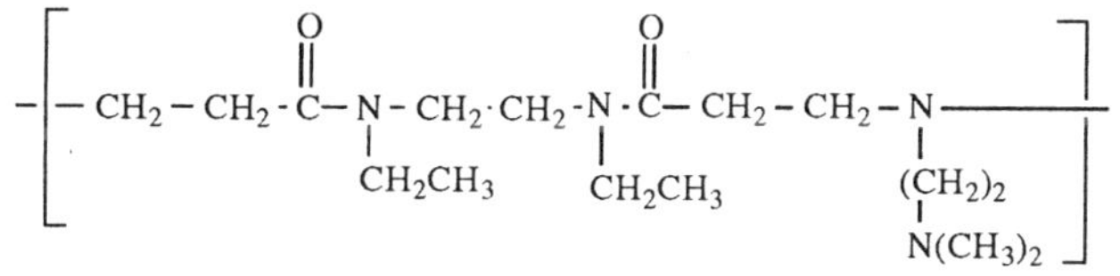 $-[-CH_2-CH_2-C(=O)-N(CH_2CH_3)-CH_2\cdot CH_2-N(CH_2CH_3)\cdot C(=O)-CH_2-CH_2-N((CH_2)_2N(CH_3)_2)-]-$	0.12
22	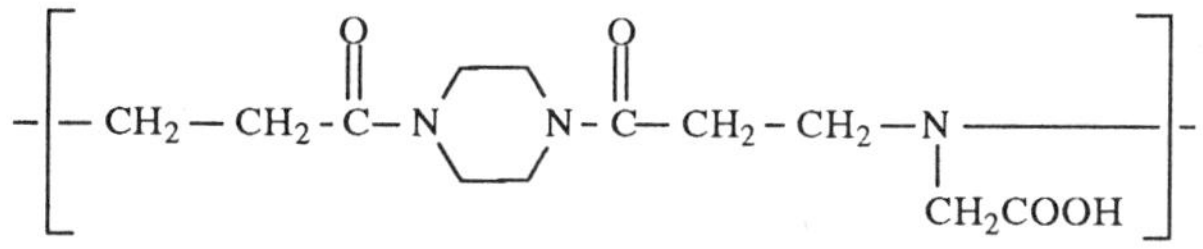 $-[-CH_2-CH_2-C(=O)-N\langle piperazine \rangle N-C(=O)-CH_2-CH_2-N(CH_2COOH)-]-$	0.16
23	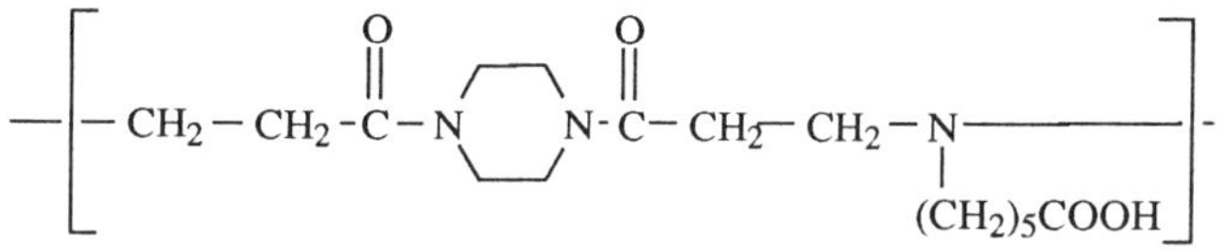 $-[-CH_2-CH_2-C(=O)-N\langle piperazine \rangle N-C(=O)-CH_2-CH_2-N((CH_2)_5COOH)-]-$	0.23
24	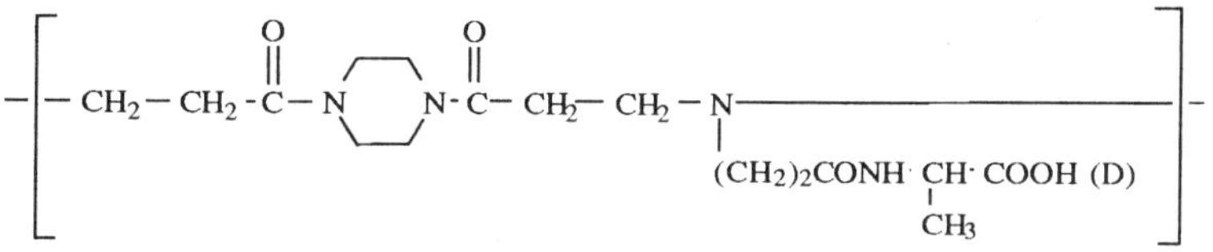 $-[-CH_2-CH_2-C(=O)-N\langle piperazine \rangle N-C(=O)-CH_2-CH_2-N((CH_2)_2CONH\cdot CH(CH_3)\cdot COOH\ (D))-]-$	0.25

General properties

Most PAAs synthesized so far are soluble in, or at least swollen by water. Most water-insoluble PAAs are soluble in aqueous acids. Besides water, PAAs are usually soluble in chloroform, lower alcohols, and highly polar organic solvents. PAAs bearing carboxy groups as side substituents, however, tend to dissolve only in water.

The intrinsic viscosities $[\eta]$ of PAAs in organic solvents or aqueous media range from about 0.15 to 1 dl/g. The number-average molecular weights of some PAAs with intrinsic viscosities lower than 0.4 dl/g were found (Tanzi et al., 1984b; Tanzi et al., 1992a) to range between 3×10^3 and 1×10^4. The number-average molecular weights of polymers having intrinsic viscosities higher than 0.4 dl/g were not determined.

Molecular weight distributions of water-soluble PAAs can be conveniently studied by SEC in aqueous buffers at slightly alkaline pH's. A convenient system is, for istance, TSK-GEL G3000 PW and G4000 PW columns connected in series, using 0.1 M TRIS buffer pH 8.1 in NaCl 0.2M as mobile phase. A calibration curve was obtained with a model PAA, namely N° 6 of Table 1, by synthesizing a set of PAA samples having different molecular weights, and carefully characterizing them by means of ^1H, ^{13}C, and bidimensional homo- and heterocorrelated NMR spectrometry. Particular attention was devoted to the quantitative determination of functionalities, which enabled to calculate the number average molecular weight of PAA samples. The GPC results were combined with NMR data and a linear calibration curve of log M versus retention time was obtained. From viscometric measurements the Mark-Houwink constants were also determined. For the particular PAA considered, the following relation was obtained (Bignotti et al., 1994):

$$[\eta] = 1.76 \times 10^{-5} \times M^{1.14}$$

Owing to their regular structure, many PAAs are partially crystalline in the solid state (Danusso and Ferruti, 1970). In some instances, crystallization can be induced by solvent treatment. Partially crystalline PAAs usually have melting points in the range 80°-120°, but when cyclic structures are present melting points as high as 270° (with decomposition) have been measured.

The thermal stability of PAAs is not very high. It may be safely assumed that in most instances decomposition begins at about 140° in air and 170-180° under inert atmosphere (Danusso and Ferruti, 1970). Shelf stability in air is good for PAA salts but is rather poor for PAAs as free bases which should be stored under nitrogen to avoid discoloration. Precautions must always be taken to exclude moisture, as both PAAs and their salts are hygroscopic.

PAAs as Polyelectrolytes

All PAAs apart form the presence of additional ionizable functions as side substituents, contain ter-aminic nitrogens in their main chain and therefore can be classified as polyelectrolytes. However, in all the PAAs studied, except those deriving from amino acids (see below), the results of the potentiometric titrations clearly show that the basicity of the aminic nitrogens N-R of each repeating unit does not depend on the degree of protonation of the whole macromolecule (Ferruti et al., 1981; Ferruti and Barbucci, 1984; Barbucci and Ferruti, 1979; Barbucci et al., 1979; Barbucci et al., 1980a; Barbucci et al., 1980b; Barbucci et al., 1981a; Barbucci et al., 1981b; Barbucci et al., 1982a; Barbucci et al., 1982b). The substantial independence of the repeating units is possibly due to the presence of the bis-acrylamido groups sheltering the positive charges of the protonated nitrogens present in different units. Consequently, "real" basicity constants can be determined. The basicity constants determined for each PAA are always as many as the aminic nitrogens present in its repeating unit, and their values are very close to those determined for its non-macromolecular model. It may be noticed that this behaviour is quite unusual in the polyelectrolytes domain (Mandel, 1992). The basicity constants of some representative PAAs are reported in Table 2.

Viscometric titrations in 0.1M NaCl (Barbucci et al., 1981a; Barbucci et al., 1982b) proved that the conformational freedom of PAAs in aqueous media is reduced on protonation. All the polymers in their neutral form are most likely in a random coil structure due to the great number of accessible conformations of about the same energy. For PAA 7 of Table 1, for instance, the first protonation leads to the formation of a strong hydrogen bond between "onium" ions and carbonyl groups belonging to the same monomeric unit. When the first protonation of all the monomeric units is complete, the above effect strongly reduces the conformational freedom of the whole polymer, which tends to assume a rigid structure. This is cleary shown by a pronounced jump of its reduced viscosity at $\alpha = 0.5$ (α = degree of protonation). The second protonation leads to an electrostatic repulsion between the positively charged onium ions belonging to the same monomeric unit, and this effect further compels the polymer to adopt a more rigid structure. Consequently, a second jump is observed at $\alpha = 1$.

PAAs deriving from amino acids, and therefore carrying carboxylate groups as side substituents (for instance PAAs 22 and 23 of Table 1), behave in a different way with respect to all other PAAs (Barbucci et al., 1986). In fact, their logK values depend on the degree of protonation of the whole macromolecule, *i.e.* they follow the modified Henderson-Hasselbach equation:

$$\log K_i = \log K_i^\circ + (n-1)\log[(1-\alpha)/\alpha]$$

and therefore they exhibit a typical polyelectrolyte behaviour.

Table 2 . Basicity constants of some PAAs, and their non-macromolecular models, at 25°C in 0.1 M NaCl

Structure[a)]	Polymer		Model	
	pK_{a1}	pK_{a2}	pK_{a1}	pK_{a2}
6	8.09	4.54	8.25	4.80
19	8.80	4.11	9.05	4.35
4	7.01	2.98	7.12	2.39
1	7.79	-	8.07	-
20	9.02	7.91[b)]	9.05	8.34[c)]

a): The numbers refer to Table 1
b): $pKa_3 = 4.46$; $pKa_4 = 2.12$.
c): $pKa_3 = 4.42$; $pKa_4 = 2.43$

Degradation behaviour of poly(amido-amines) in aqueous media

All PAAs, bearing hydrolyzable amidic bonds in their main chain, are in principle degradable in the presence of water. This property is obvously advantageous when PAAs are considered as soluble drug carriers, since it ultimately permits their elimination through the usual excretion routes.

The degradation of several PAAs in solution was studied at 37°C in 0.2 M phosphate buffer pH 7.5 by means of viscometric and chromatographic techniques (Ranucci et al., 1991;

Table 3. Typical PAAs selected for degradation studies

No.	Structure	Monomers[a)]	η_{sp}/C[b)] (dl/g)
PAA-1	[$CH_2CH_2C(=O)$-N(piperazine ring)N-$C(=O)CH_2CH_2$ - N(2-methylpiperazine ring, CH_3)N—]	BP + 2MeP	0.46
PAA-2	[$CH_2CH_2C(=O)$-$NHCH_2NHC(=O)$ CH_2CH_2N(2-methylpiperazine ring, CH_3)N—]	MBA + 2-MeP	0.31
PAA-3	[$\cdot CH_2CH_2C(=O)$-NHCH(COOH)NH-$C(=O)CH_2CH_2$ - N(2-methylpiperazine ring, CH_3)N—]	BAC + 2-MeP	0.27

a): BP= 1,4-bis(acryloyl)piperazine, MBA = N,N'-methylenebis(acrylamide), BAC= 2,2-bis(acrylamido acetic acid), 2-MeP= 2-methylpiperazine.
b): C= 0.2% in 0.2 M phosphate buffer pH 7.5.

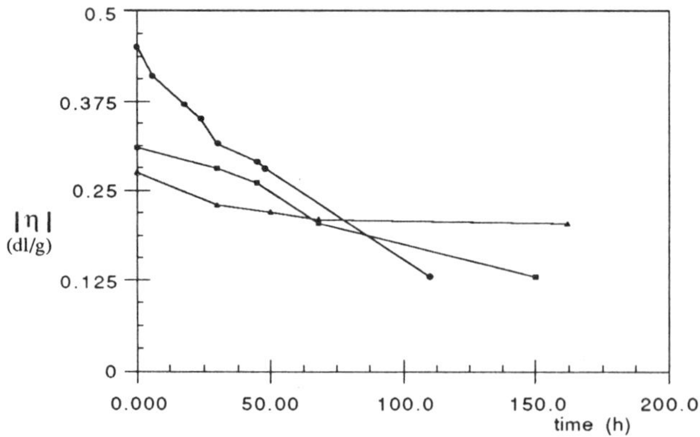

Figure 1. Degradation behaviour of PAA-1, PAA-2 and PAA-3 of Table 2, at 37°C and in 0.2M phosphate buffer (pH 7.5). Viscometric measurements.

Ferruti et al., 1994a; Ferruti et al., 1994b). The results of viscometric studies on some model PAAs whose structures are presented in Table 3 are reported in Fig. 1.

It was found that the structure of both the aminic and amidic moieties has an influence on the degradation rate. From the data of Fig. 1 it is apparent that the rates at which viscosities decrease can be arranged in the order PAA1>PAA2>PAA3, *i.e.*, the aminic portion being the same, the PAA carrying a carboxyl groups in the amidic moiety is the most stable one. This trend has been confirmed by our studies on several other PAAs having different aminic portions coupled with the same three amidic portions. The viscometric results have been strictly confirmed by GPC experiments. As regards the degradation mechanism, it seems to be purely hydrolytic, as no vinyl groups, such as those which would have derived from a β-elimination reaction, could be determined by NMR analysis (Bignotti et. al., 1994). This was confirmed by the finding that the degradation pattern of PAA-3 of Table 3 was not affected by the presence of a 20-fold excess, on a molar basis, of 2-mercapto- ethanol which would have acted as scavenger of the activated double bonds resulting from a hypothetical elimination reaction (Ferruti et al., 1994a; Ferruti et al., 1994b).

At any rate, it is relevant for a use of PAAs as drug carriers to observe that under conditions mimicking the physiological conditions all the PAAs considered were reduced to oligomeric level after relatively short periods of time. Moreover, it was demonstrated that the degradation rate is strongly influenced by temperature and pH but does not seem affected by the presence of tritosomal enzymes at pH 5.5 (Ranucci et al., 1991).

POLY(AMIDO-AMINE)S AS SOLUBLE DRUG CARRIERS

General Considerations and Toxicological Studies

Early in the 1970's several PAAs were found to display inherent antitumour activity. They displayed the ability to reduce the number and average weight of metastases in mice injected with several different types of model tumour (Ferruti et al., 1973). A number of other polycations including poly L-lysine have since been explored as drug carriers for a use in cancer chemotherapy (Chu et al., 1981, Citro et al., 1994), and more recently have been proposed as vehicles for gene and oligonucleotide delivery, for example in liver-directed gene therapy and antisense nucleotide delivery (Wu et al., 1993). It has been suggested that polycations may display a degree of selectivity towards cancer cells, whose membranes usually possess a more anionic character than those of normal cells. Although this may be true *in vitro,* the fact that polycations interact very strongly will all cell membranes has made it impossible to demonstrate selective antitumour activity *in vivo*.

Polycations are in general very toxic. For instance, the IC_{50} of poly-L-lysine *in vitro* is the range 1-60 μg/ml, depending on the cell type and incubation conditions (Sgouras, 1990). Tert-amino polymers obtained by ring-opening polymerization of dialkylamino-substiduted episulphides:

$$\left[-S-\underset{\underset{NR_2}{|}}{\underset{CH_2}{\underset{|}{CH}}}-CH_2- \right]_n$$

rendered water-soluble by partial quaternarization were also found to be very toxic in vivo (Vert, 1986), a disappointing result as these polymers showed interesting pH-mediated conformational change in a narrow pH range (Huguet, 1985).

For more than a decade we have sought to identify PAA structures that are neither cytotoxic nor heamolytic *in vitro*. For example, the cytotoxicity of PAAs ISA 22 and 23 was evaluated using various cancer cell lines (Hep G2- human liver; B16F10 - mouse melanoma and Mewo - human melanoma) in combination with the MTT assay to assess cell viability after exposure to PAAs and dextran and poly-L-lysine as reference control polymers. It can be seen that both PAAs are much less toxic than poly-L-lysine, their IC_{50} being higher by two or three orders of magnitude (Table 4 and Fig. 2). This is probably attributable to their low degree of protonation at neutral pH and indicates that such polymers have definite potential as drug carriers suitable for *in vivo* use.

PAAS AS CARRIERS OF ANTICANCER AGENTS

Soluble polymeric carriers are already showing considerable potential as carriers of antitumour agents (Duncan R. et al,1996). N-2-(hydroxypropyl)-methacrylamide (HPMA) copolymers have been developed that carry anthracyclines such as doxorubicin (Duncan, 1992) and taxol and facilitate both drug targeting to solid tumours and controlled release of drug intratumourally. Three HPMA copolymer antitumour conjugates have entered Phase I/II clinical testing. It is already clear that HPMA copolymer-doxorubicin (PK1) is considerably less toxic

Table 4. Summary of the *in vitro* cytotoxicity of PAAs ISA 22 and ISA 23 when incubated with various different cell lines

	IC_{50} (mg/ml; mean ± SD)		
Polymers	Hep G2	B16 F10	MeWo
Poly(L-Lysine)	0.05 ± 0.01	0.05 ± 0.01	0.01 ± 0.001
Dextran	>5	>5	>5
ISA 22	>5	4.00 ± 1.41	4.63 ± 0.53
ISA 23	>5	>5	4.23 ± 1.10

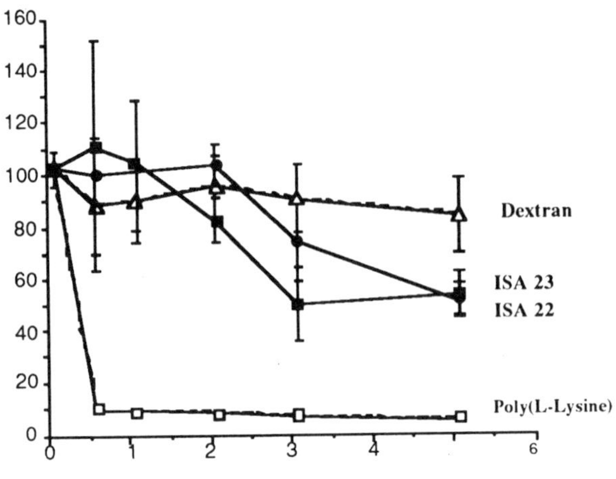

Figure 2. Cytotoxicity of various polymers against B16F10 melanoma cells after a 72h incubation *in vitro*. Cells were incubated with either the PAAs ISA 22 and ISA 23 or dextran and poly-L-lysine as reference materials. Viability was assessed by the MMT assay and results are expressed as percentage of viability seen in cells not exposed to polymer.

free drug. The maximum tolerated dose in man was found to be 320mg/m^2 dox-equivalent for PK1 compared to 60-80 mg/m^2 for free doxorubicin during the Phase I study , and this compound is currently under Phase II evaluation.

PAAs have also been investigated as carriers of antitumour agents. They bring the potential advantages in comparison with other polymeric carriers in that they are biodegradable in the polymeric main chain and thus more versatile in terms of the molecular weight range that can be used.

PAA-Mitomycin Conjugates

Soluble covalent adducts between mitomycin C (MMC), and PAAs carrying pendant hydroxy or amino groups have been patented (Schacht et al., 1997). The synthesis of the

Table 5. Effect of MMC and PAA-MMC on a panel of cell lines in vitro

Cell line	MMC$^{a)}$	PAA-MMC$^{a)}$	PAA alone$^{b)}$
L1210	0.4	16.0	>>100
P388	0.3	2.1	35.0
LoVo	2.4	22.0	>>100
Res LoVo	38.0	50.0	>>100

a): Results are expressed as the concentration of MMC (μg/ml) required to reduce call viability (MMT assay) by 50 % (IC_{50})

b): Results are expressed as the concentration of polymer required to reduce call viability (MMT assay) by 50 % (IC_{50})

poly(amido-amine)/MMC conjugate was performed in three steps. Firstly, a PAA copolymer was prepared by polyaddition of a 2:1 mixture (on a molar basis) of 2-methylpiperazine and N,N'-bis(2-hydroxyethyl)- ethylenediamine to 1,4-bis(acryloyl)piperazine. The two PAA homopolymers deriving from the polyaddition of 2-methylpiperazine and N,N'-bis(2-hydroxyethyl)- ethylenediamine to 1,4-bis(acryloyl)piperazine, besides having a low cell toxicity (Ranucci et al.,1991) are both highly water-soluble. The hydroxy groups present as side-substituent in the repeating unit of the latter provide a convenient way for attaching drug moieties. The hydrophilicity of the same unit can be expected to be sharply reduced by coupling with MMC. Therefore, a copolymer was selected, in which the units deriving from 2-methylpiperazine, remaining unsubstituted, act as solubilizing unit. Furthermore, by rarifying the hydroxy groups along the macromolecular chain, the onset of side reactions, such as cyclization or crosslinking reactions, during the subsequent steps are substantially avoided. The second step was the activation of the hydroxy groups and finally MMC coupled to the activated PAA carrier and the product had a MMC loading of approximately 25%.

Covalent binding of an antitumour agent to a macromolecular carrier changes the route of cellular entry from rapid transmembrane passage for the low molecular weight compound to much slower uptake by endocytosis for the macromolecular adduct. It is clear that *in vitro* screening of antitumour activity of polymeric antitumour agents has limited value in terms of their probable *in vivo* efficacy (Duncan et al., 1996; Duncan, 1992). Nontheless as a prelude to *in vivo* studies, the antitumour activity of PAA-MMC was investigated *in vitro* using a panel of normal and multidrug resistant tumour cell lines and the activity observed compared with that see for free MMC. Cytotoxicity of PAA-MMC was considerable lower than that seen for free MMC (Table 5). The mouse leukaemia cell lines P388 and L1210 were relatively sensitive to free MMC (IC_{50} values of 0.3-0.4 μg/ml) and in this case the PAA conjugates were 7-40 times less active. The human colon cancer cell line lovo was inherently resistant to MMC, as was a multidrug resistant variant, res. lovo. In this case the IC50 values observed for free MMC were 2.4 and 28 μg/ml respectively. Although the PAA-MMC conjugates were again less active than free drug it was interesting to note that the activity was only 1.3 - 9.2 fold that seen for free drug in this case. It is noteworthy that the PAA polymers used as carriers in these studies did not show any significant cytotoxicity in these experiments (Table 5).

Preliminary experiments were carried out to investigate the efficacy and toxicity of these PAA-MMC conjugates *in vivo*. DBA2 mice bearing L1210 tumour cells intraperitoneally (i.p) were treated with free MMC or PAA-MMC conjugates by a single i.p. dose on day 1 after tumour inoculation.

It can been seen from Fig.3 that the PAA-MMC conjugates were equi-active with MMC via the i.p. route and indeed in this pilot experiment we did observe a long term survivor in each group treated with conjugate. Obviously this experiment requires repetition to determine the real therapeutic potential of PAA-MMC. It was clear (Fig 3c) that the polymer conjugates were substantially less toxic than free MMC and no animal weight loss was observed when the conjugates were administered at a MMC equivalent dose of 5 mg/kg. This is the maximum tolerated dose of free MMC that can be given to animals and even so it obviously causes a substantial reduction in weight (Fig.3c).

DEVELOPMENT OF PAAS FOR USE IN INTRACYTOPLASMIC DELIVERY SYSTEMS

As PAAs have amido and tertiary amino groups arranged along a linear back bone this can give rise to changes in polymer tertiary conformation in a pH-responsive manner. Therefore we have proposed that these molecules may make good candidates as vehicles for intracytoplasmic drug delivery following systemic administration. We have pointed out

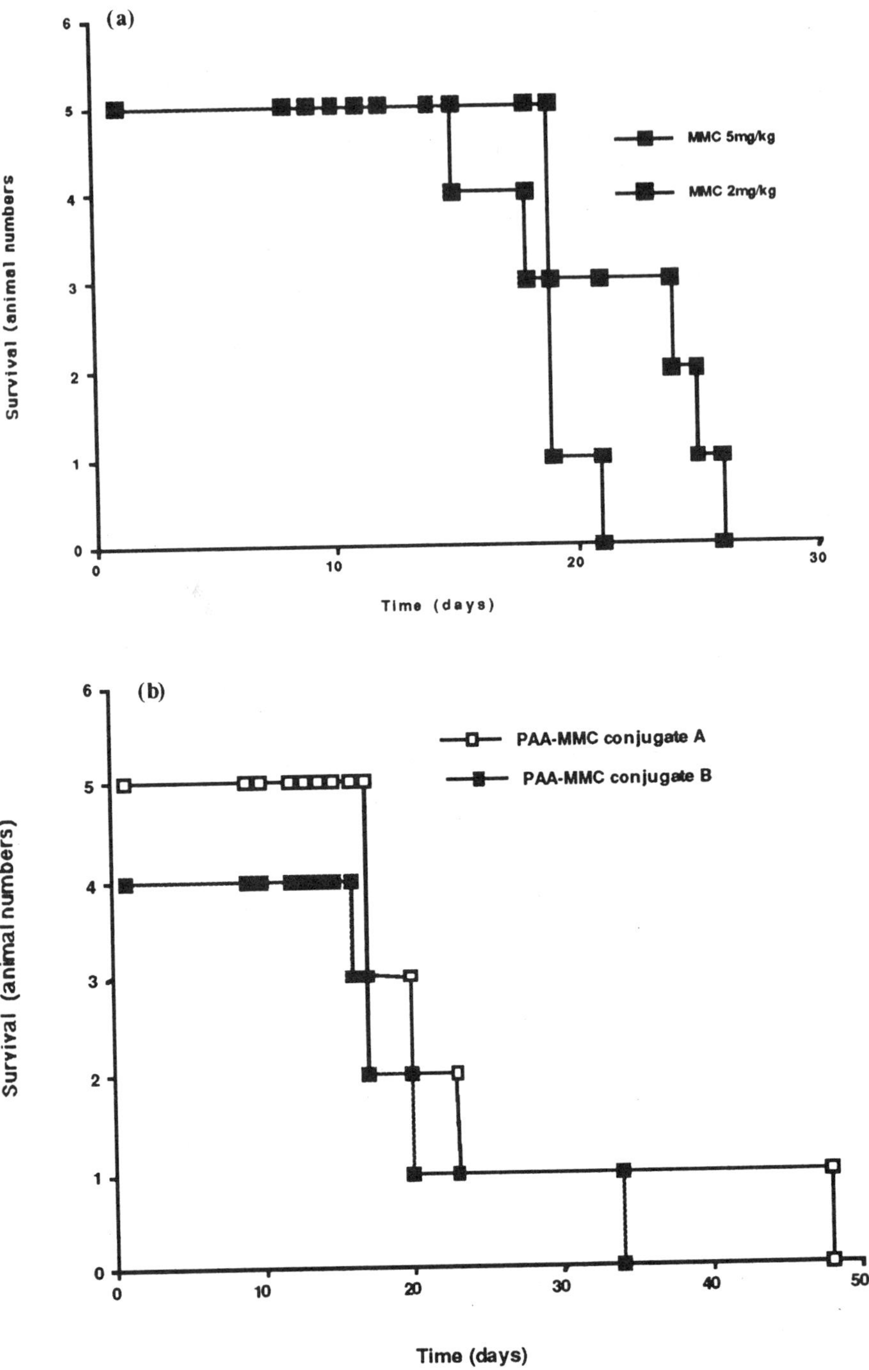

Figure 3. Antitumour activity and toxicity of PAA-MMC conjugates (5 mg/kg MMC equivalent) administered intraperitoneally on day one in DBA2 mice inoculated intraperitoneally with 10^6 L1210 leukaemia cells

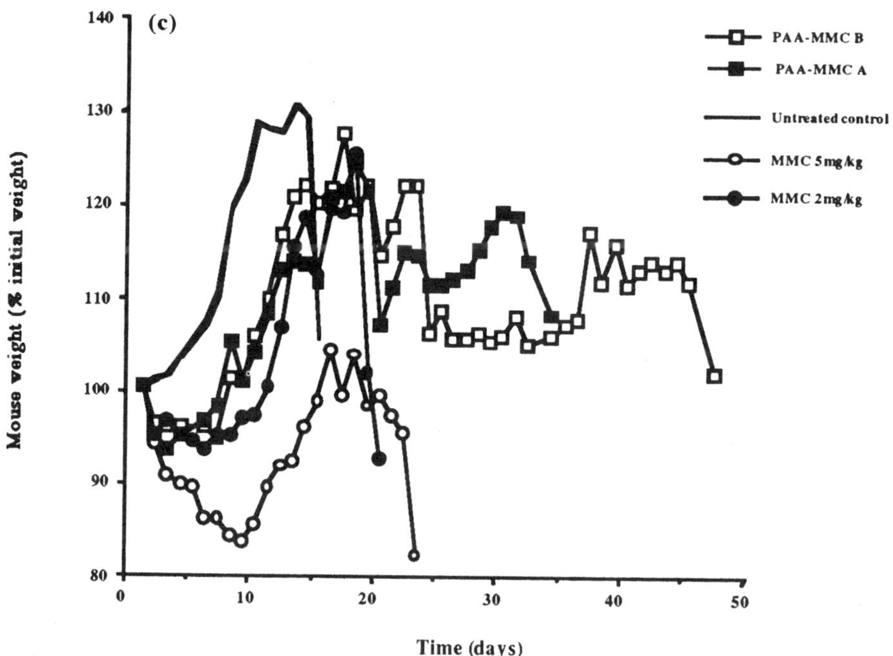

Figure 3. Continued.

previously that PAAs bearing two aminic nitrogens in their repeating unit show a marked conformational change during movement from a neutral pH to acidic one as a result of the modification of their average charge, this effect being more pronounced when the aminic nitrogens were separated by only two methylene groups (Barbucci et al., 1981a; Barbucci et al., 1982b). This property provides, in principle, the possibility of designing polymer-drug conjugates that are, following intravenous administration, relatively compacted and thus protect a drug payload in the circulation, where the pH is 7.4, but following pinocytic internalization into acidic (pH 5.5) intracellular compartments unfold, permitting pH-triggered intracellular drug delivery (Fig. 4). If the unfolding PAA also is able to carry membrane breaching agents the contruct would also have the ability to pass across the endosomal membrane and enter the cytoplasm of the cell.

To study the feasibility of hiding a biologically active molecule within a PAA polymer coil, a covalent conjugate was prepared comprised of a hydroxylated PAA bound to the non-ionic detergent, reduced Triton X-100 (Duncan et al., 1994). The PAA chosen for this study, prepared by polyaddition of N,N'-bis(2-hydroxy-ethyl)-ethylenediamine to 1,4-bis(acryloyl)-piperazine (N° 22 of Table 1), was found to be non-toxic *in vitro* when incubated with liver human cells ($ID_{50} > 2000$ µg/g) (Sgouras. and Duncan, 1990). About the detergent, it is well known that reduced Triton X-100, as well as its aromatic parent Triton X-100, are very toxic to cells causing cytolysis induced by detergent binding to cell membrane lipid bilayer (Tiller, 1984). Triton X-100 was linked to the PAA main chain by means of carbonate bonds (Fig. 4).

The degree of substitution (65% of the available hydroxyl groups) was the highest possible still ensuring water solubility. The ability of the conjugate to lyse red blood cells *in vitro* was used as an indicator of conjugate conformation at different pHs. It was found that free Triton X-100 was highly lytic at pH 5.5, 7.4, and 8.0, while the parent PAA was not lytic under

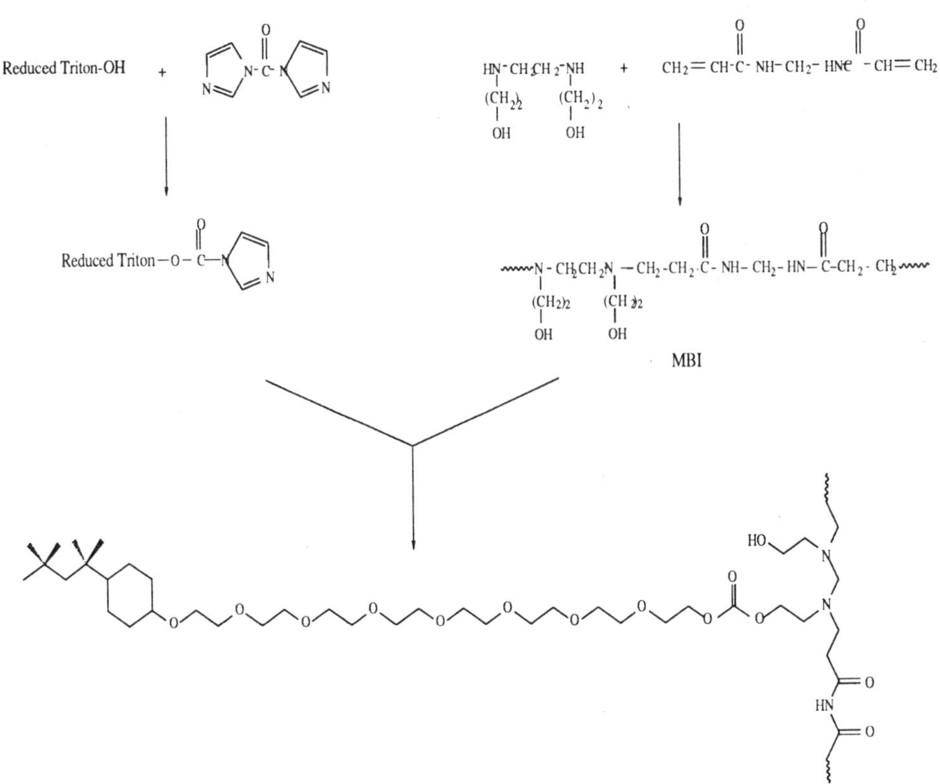

Figure 4. Synthesis of PAA-Triton X-100 conjugates

any condition. In contrast, the PAA-Triton conjugate showed concentration-dependent red blood cell lysis at pH 5.5, but was inactive at pH 7.4 and higher (Fig. 5).

Moreover, cell toxicity studies employing human leukaemic cells showed that the conjugate though 6-fold more toxic than the parent PAA, is about 140-fold less toxic than Triton X-100, and 30-fold less toxic than poly-L-lysine (IC_{50} values of 100, 600, 0.7, and 23.9

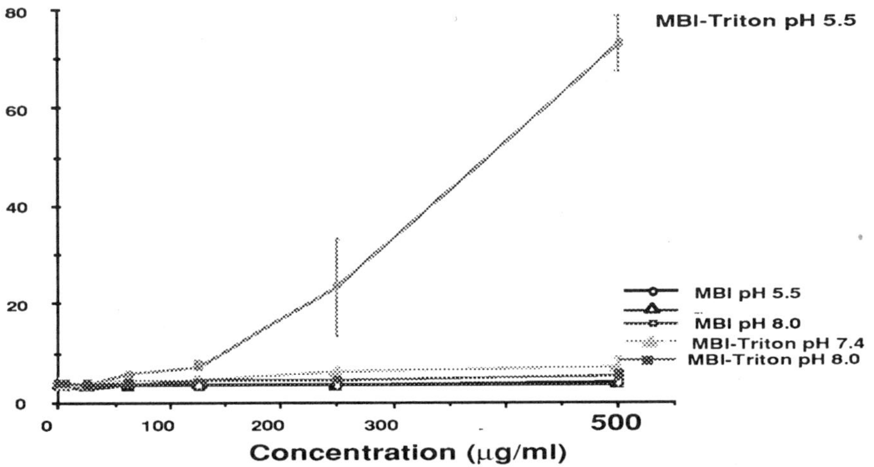

Figure 5. pH-Dependent rat red blood cell lysis following 1h incubation with MBI-Triton X-100 conjugate

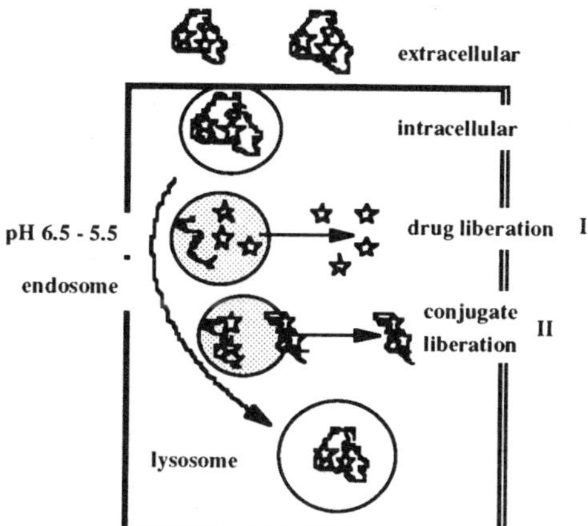

Figure 6. Schematic representation of the changes in conformation of PAAs during cellular internalisation by endocytosis.

μg/ml in that order) (Duncan et al. 1994). Although Triton X-100 was used as a model compound in these studies, the data presented show that a polymer-bound detergent can rupture the red cell membrane under acidic conditions. It remains to be confirmed whether Triton X-100, or for that matter a less potent penetration enhancer, could be used to microinject macromolecules into the cytoplasm via intracellular acidic compartments with maintenance of cell viability. Future experiments will evaluate whether such an effect might usefully be used to open the endosomal membrane to facilitate microinjection of macromolecules, without concomitant decrease in cell viability. *In vitro* cytotoxicity was assessed in B16 F10 and Mewo cell lines. IC_{50} values for each polymer were attained and seen to be >4 mg/ml in each instance.

A schematic representation of the behaviour expected from these kind of polymers is reported in Fig. 6. On entering the endosomes, which have a pH of approximately 6.5 the polymer chain would become prontonated and as it unfolds this conformational change could be used to either (I) liberate drug which would then pass across the endosomal membrane or (II) the polymer could itself become membrane active allowing penetration of the whole conjugate across the endosomal membrane into the cytoplasm of the cell.

Obviously it would be advantageous if PAAs could be designed that display inherent ability to transiently breach the endosomal membrane. In this context a famility of PAAs have been studied to determine whether or not they can display pH-dependent membrane perturbation. (Richardson et al., 1996, 1997) During incubation of the PAA s with isolated rat red blood cells it was shown that ISA 23 deponstrated pH-depent red blood cell lysis (Richardson et al., 1996). Whereas red blood cells showed minimal lysis over a 24 h incubation at all pHs studied, ISA 23 displyed approximately 10% lysis at pH 7.4, approximately 30% lysis at pH 6.5 and 90% lysis at pH 5.5. A marked change in membrane interaction at the different pHs. Studies are currently underway to detyermine the body distribution of these PAAs after intravenous administration with a view to using them *in vivo* as targetable carriers (Richardson et al., 1998).

CONCLUSIONS

In summary, PAAs are polycations that can be designed to show little general cytotoxicity. They are water soluble and can be easily functionalized in order to provide sites of attachement for drug moieties. The polymers have the distinct advantage that they are degradable in aqueous media allowing the possibility to tailor their molecular weight over a wide range according to the specific needs of a particular drug delivery system. In addition their ability to show sharp conformational changes within short pH intervals provides a unique family of 'smart' materials that can can show remarkable responses to local changes in pH.

The theoretical opportunities for future development of pH-dependent PAAs as drug targeting and delivery systems might be the following. Preparation of covalent conjugates, or complexes with anionic drugs including oligonucleotides. This could afford protection against degradation during transport in the circulation, but following pinocytic internalisation might allow intra-cytoplasmic delivery if the polymer conjugate is able to breach the endosomal membrane barrier. In the case of conjugates containing agents suitable for lysosomotropic delivery, it would be relatively simple to synthesise polymers containing covalent linkers designed for intracellular cleavage (we have previously described pH-sensitive linkers, or peptidyl spacers) as well as residues selected to facilitate receptor-mediated pinocytic uptake, thus enabling drug targeting (Duncan, 1992). Galactose (Wu. and Wu , 1993), transferrin (Citro G., unpublished), and folic acid (Citro et al., 1994) have been used to target poly-L-lysine constructs containing oligo-and polynucleotides. These moieties can be easily introduced as side substituents into PAAs by simple modifications of the synthetic process.

REFERENCES

Barbucci, R., and Ferruti, P., 1979, Protonation and complex formation of some novel poly(amido-amines), *Polymer*, 20:1061.

Barbucci, R., Ferruti, P., Micheloni, M., Delfini, M., Segre, A.L., and Conti, F., 1980a, Thermodynamics and C^{13} n.m.r. data on the protonation of polymeric bases whose repeating units behave independently towards protonation, *Polymer,* 21:81.

Barbucci, R., Barone, V., Ferruti, P., and Delfini, M., 1980b, Macro-inorganics. Part 3. Chelation of Copper (II) ion with some polymers having a poly(amido-amine) structure and their non-macromolecular models", *J. Chem. Soc. Dalton Transact.*, 253.

Barbucci, R., Casolaro, M., Ferruti, P., Barbone, V., Lelj, F., and Oliva, L., 1981a, Macroinorganics. 7. Property-structure relationships for polymeric bases whose monomeric units behave independently toward protonation, *Macromolecules*, 14:1203.

Barbucci, R., Barone, V., Ferruti, P., and Oliva, L., 1981b, Thermodynamics of protonation of polymeric bases whose repeating units behave independentely, *J. Polymer Sci., Polymer Symp.*, 69:49.

Barbucci, R., Casolaro, M., Ferruti, P., and Barone, V., 1982a, Macroinorganics. 8. Chelation of copper(II) ion with some new poly(amido-amines), *Polymer*, 23:148.

Barbucci, R., Casolaro, M., Danzo, N., Beni, M.C., Barone, V., and Ferruti, P., 1982b, Thermodynamics of protonation and complex formation of multifunctional polymers, *Gazz. Chim. It.,* 112:105.

Barbucci, R., Casolaro, M., Nocentini, M., Corezzi, S., Ferruti, P.. and Barone, V., 1986, Acid-base and metal ion complex formation properties of polymers containing amino acid residues, *Macromolecules*, 19:37.

Barbucci, R., Ferruti, P., Improta, C., La Torraca, M., Oliva, L., and Tanzi, M.C., 1979, Macroinorganics - IV - Thermodynamic functions relative to the protonation of poly(amido-amine)s with repeating unit containing amino groups, *Polymer,* 20,:1298.

Bignotti, F., Sozzani, P., Ranucci, E., and Ferruti, P., 1994, NMR studies, molecular characterization, and degradation behavior of poly(amido amine)s. 1. Poly(amido amine) deriving from the polyaddition of 2-methylpiperazine to 1,4-bis(acryloyl) piperazine, *Macromolecules*, 27:171.

Caldwell, G., Neuse, E., and Stephanou, A., 1993, Synthesis of water-soluble poly(amidoamine)s for biomedical applications. II. Polymers possessing interchain-type secondary amino groups suitable for side-chain attachment, *J. Appl. Polym. Sci.*, 50:393.

Chu, B.C.F., and Howell, S.B., 1981, Pharmacological and therapeutic properties of carrier bound methotrexate against tumor confined to third space body compartment, *J. Pharmacol. Exp. Ther.* 219:389.

Citro, G., Szczylik, C., Ginobbi, P., Zupi, G., and Calabretta, B., 1994, Inhibition of leukaemia cell proliferation by folic acid-polylysine-mediated introduction of c-myb antisense oligodeoxynucleotides into HL-60 cells, *Br. J. Cancer*, 69:463.

Danusso, F., and Ferruti, P., 1970, Synthesis of tertiary amino polymers, *Polymer,* 11:88.

Duncan, R., 1992, Drug polymer conjugates potential for improved chemotherapy, *Anti-Cancer Drugs,* 3:175.

Duncan, R., Dimitrijevic, S., and Evagorou, E., 1996, The role of polymer conjugates in the diagnosis and treatment of cancer", *S.T.P. Pharma Sci.* 6:237.
Duncan, R., Ferruti, P., Sgouras, D., Tuboku-Metzger, A., Ranucci, E., and Bignotti, F., 1994, A polymer-Triton X-100 conjugate capable of pH -dependent red blood cell lysis: a model system illustrating the possibility of drug delivery within acidic intracellular compartments, *J. Drug Targeting,* 2:341.
Ferruti, P., and Barbucci, R., 1984, Linear amino polymers: synthesis, protonation, and complex formation, *Advanc. Polym. Sci.*, 58:57.
Ferruti, P., Danusso, F., Franchi, G., Polentarutti, N., and Garattini, S., 1973, Effects of a series of new synthetic high polymers of cancer metastases, *J. Medicinal Chem.* 16:497.
Ferruti P., 1996, Ion chelating polymers, in: The polymeric materials encyclopedia, J.C. Salamone,. ed., CRC Press, Boca Raton.
Ferruti, P., 1985, Poly(amido-amines), *Macromol. Syntheses*, 9:25.
Ferruti, P., Marchisio, M.A., and Barbucci, R., 1985m Synthesis physico-chemical properties and biomedical applications of poly(amido-amine)s, *Polymer*, 26:1336.
Ferruti, P., Danzo, N., Oliva, L., Barbucci, R., and Barone, V., 1981, Macro-inorganics. Part 6. Protonation and complex formation of a new series of polymers whose repeating units behave independently, *J.Chem. Soc. Dalton,* 539.
Ferruti, P., Ranucci, E., Sartore, L., Bignotti, F., Marchisio, M.A., Bianciardi, P., and Veronese, F.M., 1994a, Recent results on functional polymers and macromonomers of interest as biomaterials or for biomaterial modification, *Biomaterials*, 15:1235.
Ferruti, P., Ranucci, E., Bignotti, F., Sartore, L., Bianciardi, P., and Marchisio, M.A., 1994b, Degradation behaviour of ionic stepwise polyaddition polymers of medical interest, *J. Biomat. Sci., Polym. Ed.,* 6:833.
Huguet, J. and Vert, M., 1985, Partially quaternised poly(tertiary amines) as pH-dependent drug carrier for solubilisation and temporary trapping of liopophilic drug in aqueous media, *J. Controlled Rel.*, 1:217.
Mandel M., 1992, Polyelectrolytes in *Encyclopedia of Polymer Science and Technology*, H.F. Mark, N.G. Gaylord and N.M. Bikales, eds., Interscience, New York..
Marchisio, M.A., Longo, T., and Ferruti, P., 1973, A selective de-heparinizer filter made of new cross-linked polymers of a poly-amido-amine structure, *Experientia,* 29:93.
Ranucci, E., and Ferruti, P., 1991, Block copolymers containing poly(ethylene glycol) and poly(amido-amine) or poly(amido-thioether-amine) segments, *Macromolecules*, 24:3747.
Ranucci, E., Spagnoli, G., Ferruti, P., Sgouras, D., and Duncan, R., 1991, Poly(amidoamine)s with potential as drug carriers: degradation and cellular toxicity, *J. Biomat. Sci. Polym. Ed.,* 2:303 .
Richardson, S., Bignotti, F., Ferruti, P., and Duncan, R., 1997, Poly(amido-amines) as potential delivery systems for oligonucleotides, *Proceedings of the 2nd International Symposium on Polymer Therapeutics*, Kumamoto, April 18-19.
Richardson, S., Bignotti, F., Ferruti, P., and Duncan, R., 1996, Poly(amidoamine)s as pH-sensitive drug carriers: evaluation of biocompatibility and membrane interaction, *Proceedings of the 1st International Symposium on Polymer Therapeutics: Form the Laboratory to Clinical Practice*, London)
Richardson, S., Ferruti, P., and Duncan, R., 1998, The body distribution of poly(amido-amines) and their DNA complexes, *Proceedings of the 3rd International Symposium on Polymer Therapeutics*, London, Jan. 7-9.
Rusconi, L., Tanzi, M.C., Ferruti, P., Angiolini, L., Barbucci, R., and Casolaro, M., 1982a, Thermodinamics of protonation and complex formation of multifunctional polymers, *Polymer* 23:1223.
Rusconi, L., Tanzi, M.C., Garlaschelli, G., Ferruti, P., and Angiolini, L., 1982b, Synthesis and characterization of a poly(amido-amine) with amido and tertiary amino groups regulary alternating along the macromolecular chain, *Makromol. Chem., Rapid Commun.*, 3:909.
Schacht, E., Ferruti, P., and Duncan, R., 1997, Drug delivery agents incorporating mitomycin", *PCT Intern. Pat. Appl. N° PCT/IB94/00259,* Oct. 1.
Sgouras, D., and Duncan, R., 1990, Methods for the evaluation of biocompatibility of of soluble synthetic polymers which have potential for biomedical use: 1- Use of tetrazolium -based colorimetric assay (MTT) as a preliminary screen for evaluation of in vitro cytotoxicity, *Journal of Materials Science: Materials and Medicine*, 1:61.
Tanzi, M.C., Barzaghi, B., Anouchinsky, R., Bilenkis, S., Penhasi, A., and Cohn, D., 1992, Grafting reactions and heparin adsorption of poly(amidoamine)-grafted poly(urethane-amide)s, *Biomaterials,* 13:425.
Tanzi, M.C., Rusconi, L., Barozzi, C., Ferruti, P., Angiolini, L., Nocentini, M., Barone, V., and Barbucci, R., 1984a, Synthesis and characterization of piperazine-derived poly(amido-amine)s with different distribution of amido- and amino-froups along the macromolecular chain, *Polymer*, 25:863.
Tanzi, M.C., Tieghi, G, Botto, P., Barozzi, C., Cardillo, P., 1984b, Synthesis and characterization of poly(amido-amine)s belonging to two different homologous series, *Biomaterials*, 5:357.

Tiller, G.E., Mueller, T.J., Dockter, M.E., and Struve, W.G., 1984, Hydrogenation of Triton X-100 eliminates its fluorescence and ultraviolet light absorption while preserving its detergent properties, *Anal. Biochem.* 141:262.

Vansteenkiste, S., Schacht, E., Ranucci, E., and Ferruti, P., 1992, Synthesis of a new family of poly(amidoamine)s carrying poly(oxyethylene) side chains, *Makromol. Chem.*, 193:937.

Vert, M., 1986, Polyvalent polymeric drug carriers, *CRC Crit. Rev. Ther. Drug Carrier Syst.*, 2:291.

Wu, J.Y., and Wu, C.H., 1993, Liver directed gene delivery, *Adv. Drug. Del. Rev.*, 12:159.

DEVELOPMENT OF NOVEL TECHNOLOGIES FOR THE SYNTHESIS OF BIODEGRADABLE PEGYLATED NANOPARTICLES

Maria Teresa Peracchia, *Didier Desmaële, Christine Vauthier,
Denis Labarre, Elias Fattal, *Jean d'Angelo, and Patrick Couvreur

URA CNRS 1218 and *URA CNRS 1843
Faculty of Pharmacy, University of Paris-Sud XI, France

INTRODUCTION

Polymeric injectable carriers, able to deliver drugs or other compounds to specific sites of action for a prolonged time, represent a potential therapeutic approach for several diseases. However, therapeutic activity is compromised by particle recognition by the macrophages of the mononuclear phagocyte system (MPS) and their elimination within seconds, or minutes after intravenous injection. Blood proteins (opsonins) adsorb onto the particles surface, making them recognizable to the macrophages of the MPS.

It has been shown that body distribution of injectable nanoparticles is correlated with their surface properties and that modifying the surface of particulate carriers with more hydrophilic polymers prolongs nanoparticle half-life in the blood compared to unmodified particles; among the most investigated and promising polymers, are poly(ethylene glycol) (PEG) and its derivatives or copolymers. Biodegradable characteristics are another important issue for the development of drug carriers: at the present, PEG-coated slowly biodegradable polymeric particles have been developed from polyesters (Gref et al., 1994; Bazile et al., 1995; Celikkaya et al., 1996; Peracchia et al., 1997a), polyanhydrides (Peracchia et al., 1997a) and albumin (Müller and Kissel, 1993). A PEG coating covalently bound to the nanoparticles is desirable, in particular in the case of biodegradable nanoparticles, where an adsorbed coating layer is supposed to be very unstable.

The development of biodegradable long circulating nanoparticles composed of polyalkylcyanoacrylate (PACA) could represent a promising injectable drug carrier, due to the degradation properties of PACA and its rapid excretion. Our work proposed to develop novel technologies for the preparation of pegylated PACA nanoparticles. A first strategy used was to obtain PEG-PACA nanoparticles during the nanoparticle preparation step, by emulsion polymerization of alkylcyanoacrylate monomer in an aqueous medium in the presence of polyethyleneglycol (PEG). The second strategy to obtain pegylated cyanoacrylates nanoparticles was to use preformed pegylated copolymers, obtained by a single step condensation of Me-PEG cyanoacetate with formaldehyde.

Targeting of Drugs 6: Strategies for Stealth Therapeutic Systems
Edited by Gregoriadis and McCormack, Plenum Press, New York, 1998

PEG-COATED NANOPARTICLES BY CHEMICAL COUPLING OF PEG DURING THE EMULSION/POLYMERIZATION OF THE CYANOACRYLIC MONOMER

The first strategy used was to obtain PEG-PACA nanoparticles during the nanoparticle preparation step, by emulsion polymerization of alkylcyanoacrylate monomer in an aqueous medium in the presence of polyethyleneglycol (PEG). PEG initiated the polymerization of the cyanoacrylic monomer and was chemically coupled to the growing polyisobutylcyanoacrylate (PIBCA). The pH of the polymerization medium, as well as the concentration of PEG in the medium, were found to be key parameters for obtaining the PEG-PIBCA nanoparticles. NMR and GPC studies confirmed the structure of the obtained copolymer, and DSC analysis suggested that the PEG was covalently coupled to the PACA polymer. Hydrophobic interaction chromatography (HIC) revealed that the pegylated nanoparticles were much more hydrophilic than the native PIBCA nanoparticles.

The use of PEG, methoxy-PEG (Me-PEG) or dimethoxy-PEG (Me_2PEG) suggested interesting information concerning the role of PEG functional groups in the initiation of the polymerization of IBCA, and about the organization of the PEG chains at the surface of the particles (Fig. 1). Finally, PEG-coated PACA nanoparticles were found to decrease complement activation dramatically compared to native PACA non PEG-coated nanoparticles.

Influence of the pH of the polymerization medium

Nanoparticles size and yield: Nanoparticles were formed by emulsion/polymerization in the presence of PEG, methoxyPEG 5000 or dimethoxyPEG 2000 (Peracchia et al., 1997b; Peracchia et al., 1997c; Peracchia et al., 1997d). After completion of the polymerization reaction (3 hours), it has been observed that nanoparticles with an unimodal distribution and high yield can be produced at a pH between 0.75 and 1.5 (Fig. 2), with an effect of the pH on the size (240 nm at pH 0.75, 350 nm at pH 1.5); at pH<0.75 a polymodal distribution was observed, with the monomer still present after several hours of polymerization, whereas a pH value >2 led to the formation of microparticles >1.5 μm. The faster polymerization observed at pH 1.5 led to the formation of the highest amount of nanoparticles, whereas the yield was drastically reduced at pH <1 (Peracchia et al., 1997b).

The pH of the polymerization medium is known to be a key parameter in the control of the polymerization of alkylcyanoacrylates monomers (Leonard et al., 1966), for nanoparticle

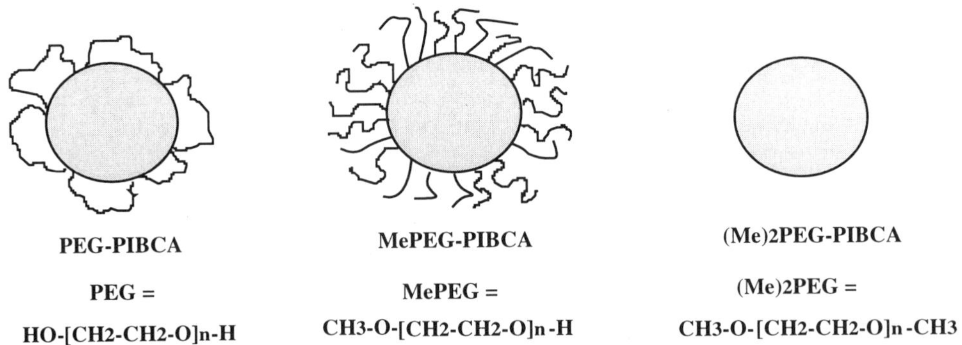

Figure 1. Structure of PIBCA nanoparticles prepared by emulsion/polymerization of IBCA in the presence of PEG, MePEG and $(Me)_2PEG$.

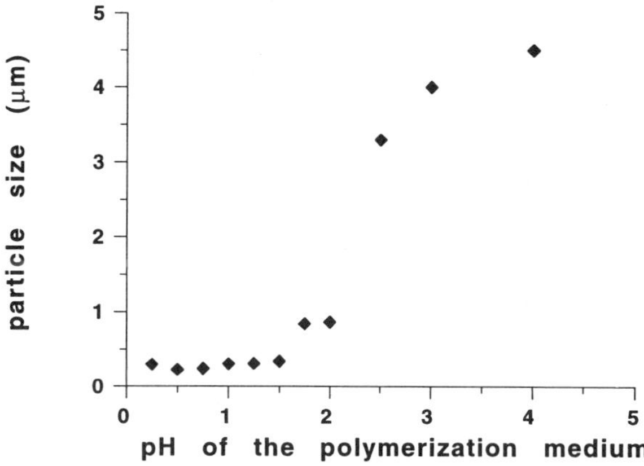

Figure 2. Effect of the pH of the polymerization medium on the nanoparticles size (from ref. 6, with permission).

formation. Currently, PIBCA nanoparticles are prepared in H_20 at pH 2.5, in the presence of dextran or other surfactants as stabilizers. It has been found that this pH value is the most appropriate for producing nanoparticles with a mean diameter around 200 nm and with a unimodal size distribution (Couvreur et al., 1979), thus suitable for intravenous administration. As one would expect, the presence of PEG in the polymerization medium affected polymerization, as well as nanoparticle formation, through an influence on the initiation of the polymerization reaction. A pH decrease slowed down polymerization kinetics, as clearly observable after adding IBCA in the PEG solution. Polymerization was apparently, at any pH value, completed in a minimum of 3 hours, but a pH <1.5 and, thus, a slower polymerization led to the formation of fewer particles (Peracchia et al., b).

The degree of polymerization of alkylcyanoacrylate results from a balance between the concentration of initiators (e.g. OH^-) and the concentration of terminating agents (H^+) (Peracchia et al., 1997b). In the case of emulsion/polymerization of IBCA in the presence of PEG, OH^- derived from water dissociation, but also OH moieties of PEG may also play the role of initiator of the reaction. As PEG concentration was maintained fixed at 2%, a pH decrease (from 1.5 to 0.75) resulted in a higher concentration of terminating agents and in a lower degree of polymerization, thus in the formation of fewer and smaller nanoparticles. A too acidic pH (<0.75) induced the protonation of PEG OH- groups, whereas the concentration of OH^- deriving from water dissociation was not enough for initiating the polymerization. On the other hand, the presence of an excess of H_3O^+ is supposed to end the propagation step of the polymerization rapidly. This explained the residual of monomer and the absence of formation of colloidal nanoparticles after IBCA polymerization at pH<0.75. On the contrary, a pH increase (>1.5) led to more OH^-, thus to a fast polymerization that did not allow the formation of colloidal nanoparticles, but only polymer aggregates. Indeed, OH^- deriving from water molecules is a stronger nucleophile than OH moieities of PEG, thus slight pH changes dramatically affected the process (Peracchia et al., 1997).

Nanoparticle PEG content: Another effect of the pH concerned the amount of PEG associated with the cyanoacrylic polymer. Interestingly, the highest PEG association was not obtained at pH 1.5, where the highest nanoparticle production was observed, but at pH 1, where PEG represented 5% of nanoparticle weight. As discussed above, the formation of colloidal

PACA nanoparticles and the amount produced depended on OH^- derived from either PEG or water molecules. Since PEG OH^- is a weaker initiator than water OH^-, PEG association could be higher at pH 1, because at this pH, the ratio between PEG OH and water OH^- was in favour of the former. At pH 1.5, the increased $[OH^-]$ led to a faster polymerization and to a faster formation of colloidal particles, limiting the PEG anchoring onto the growing nanoparticles (Peracchia et al., 1997).

Nanoparticle surface properties: Interestingly, data relative to surface charge can be compared to PEG association with nanoparticles as a function of the pH: at pH 0.75 a very negative surface charge (-30 mV) corresponded to a very low PEG association, whereas at pH 1 surface charge was less negative (-18 mV) and PEG content higher. This suggested at least a partial PEG disposition on the nanoparticle surface, even if the PEG total content was reduced in nanoparticles prepared at pH 1.25 and 1.5 when compared to nanoparticles prepared at pH 1 (Peracchia et al., 1997).

We also compared surface hydrophilicity with the surface charge behaviour: nanoparticles prepared at pH 1-1.5 were less negatively charged and exhibited in the meantime more hydrophilic properties, whereas nanoparticles prepared at pH 0.75 were hydrophobic and showed a highly negative surface charge. Thus, since a small amount of PEG was still present in nanoparticles prepared at pH 0.75, we hypothesized that in this case the whole amount of PEG associated with the nanoparticles was not concentrated on their surface or was not enough to ensure hydrophilic properties to the nanoparticles (Peracchia et al., 1997b).

Influence of PEG concentration in the polymerization medium

Nanoparticle formation: A dramatic effect of PEG concentration in the polymerization medium on the formation of nanoparticles was shown: at low concentration in the medium (<1% w/v), PEG was a weak initiator of the anionic polymerization of IBCA, and did not lead to the formation of colloidal particles. This result confirmed that the process is governed by the relative ratio of OH alcohol groups of PEG and OH^- ions generated from water molecules: since the pH was maintained fixed at 1.5, a too low PEG concentration was not able to control the polymerization of IBCA to form a colloidal suspension (Peracchia et al., 1997b,c).

Polymer molecular weight: We evaluated the MW of the forming copolymer as a function of the PEG initial concentration, showing the formation of a PEG-PIBCA copolymer at any PEG concentration, whereas an unimodal mass distribution was obtained only at PEG concentration >3% (Peracchia et al., 1997c). At PEG concentration in the polymerization medium <3%, a bimodal MW distribution was observed, with a first polymer fraction of 114-141 KDa and a second of approximately 1000-1200 Da (Peracchia et al., 1997c). Most likely, two competitive mechanisms occurred: the dispersion polymerization of IBCA and the copolymerization between IBCA and PEG, that led to the observed bimodal mass distribution. At higher PEG concentration (>3%), the process was better controlled, and only the peak corresponding to the PEG-PIBCA copolymer was detected.

Polymer thermal behavior: After thermal analysis by DSC of PEG-PIBCA nanoparticles prepared at different PEG concentrations, as shown in Figure 3, the Tg varied between 41.82°C and 51.81°C, depending on PEG concentration in the polymerization medium (Peracchia et al., 1997c). This important shift could be explained by the presence of an increasing PEG fraction in the formed copolymer, depending on the PEG initial concentration in the polymerization medium.

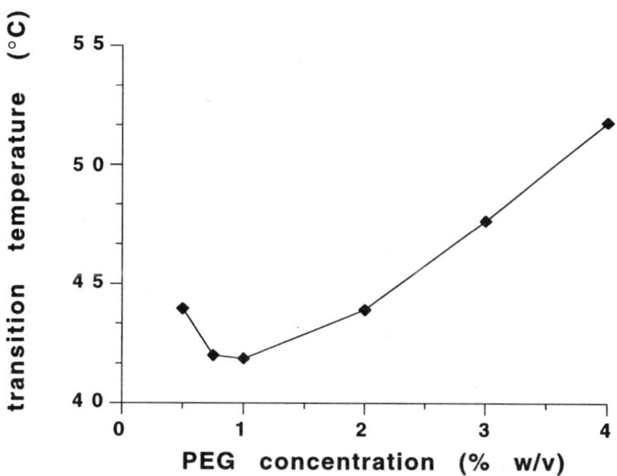

Figure 3. Effect of the PEG concentration on polymer thermal behavior (from ref. 7, with permission).

Influence of PEG chemical structure on the formation of PEG-PIBCA nanoparticles

The polymerization reaction of IBCA was followed in the presence of PEG and Me_2PEG in order to highlight the role of PEG and of its functional groups as initiators of the polymerization, since, theoretically, either PEG OH terminal groups or PEG -O- ether groups could act as nucleophiles (Peracchia et al., 1997c). Me_2PEG (without OH terminal groups) was not able to initiate the polymerization of isobutylcyanoacrylate through its -O- group (Peracchia et al., 1997c), and is supposed to act simply as a colloidal stabilizer during the polymerization reaction initiated by OH- ions deriving from water dissociation (Fig. 4).

Indeed, the optical density of the nanoparticle suspension during the polymerization reaction in the presence of Me_2PEG was of the same order of magnitude as when the polymerization proceeded in water, in the absence of PEG. This result suggests that only the PEG OH terminal group was needed for the initiation of the polymerization process. Thus, Me_2PEG was clearly not able to participate in the chemical reaction (Peracchia et al., 1997c).

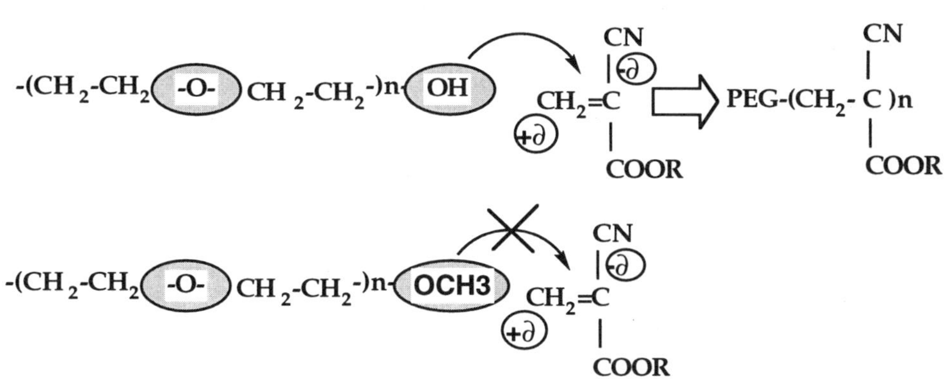

$-(CH_2-CH_2$ -O- $CH_2-CH_2-)n$ OH ; CN, ∂, $CH_2=C$, +∂, COOR ⟹ $PEG-(CH_2-C)n$ (CN, COOR)

$-(CH_2-CH_2$ -O- $CH_2-CH_2-)n$ OCH3 ✕ ; CN, ∂, $CH_2=C$, +∂, COOR

Figure 4. Role of PEG functional groups on the initiation of the polymerization of cyanoacylic monomers behavior

Furthermore, no PEG was found after degradation of the nanoparticles prepared in the presence of Me_2PEG: the absence of PEG suggested that PEG can remain associated with the nanoparticles only if chemically coupled after its involvement in the polymerization reaction (Peracchia et al., 1997c).

Interestingly, data relating to the surface charge reflect the results related to PEG association to nanoparticles (Peracchia et al., 1997c). The surface charge of PIBCA nanoparticles prepared in the presence of PEG or Me_2PEG was evaluated, and confirmed the absence of a PEG layer in the case of Me_2PEG. PEG-PIBCA nanoparticles presented, in PBS, a mean Z potential of -6.5 mV, whereas, in the presence of Me2PEG in the polymerization medium, the surface charge was much more negative (-26.8 mV), and similar to the zeta potential value of PIBCA nanoparticles prepared in the presence of dextran. Thus, the non-ionic PEG, that induced a zeta potential close to neutral in PEG-PIBCA nanoparticles, was not detectable.

Polymer chemistry investigations

NMR analysis. ^{1}H-NMR and ^{13}C-NMR clearly identified the chemical structure of PEG-PIBCA nanoparticles, and suggested that the PEG block contained in the nanoparticles was attached to PIBCA via an etheroxide bond (Peracchia et al., 1997b; Peracchia et al., 1997c).

The ^{1}H-NMR spectra of PEG-PIBCA nanoparticles showed the presence of both PEG (at 3.57 ppm, due to the methylene protons of PEG ethylene units) and PIBCA in the nanoparticle structure, and confirmed the 5% PEG:PIBCA ratio as determined through the colorimetric detection of PEG (Peracchia et al., 1997b; Peracchia et al., 1997c).

Other polymer characteristics. We carried out some investigations on the nature of PEG binding to PIBCA nanoparticles. If a covalent binding between PEG and PIBCA occurred, PEG m.p., detectable by DSC, should disappear, as described for pegylated copolymers. PEG-PIBCA nanoparticles did not reveal the presence of a PEG m.p. peak, whereas only PIBCA Tg was observable (Peracchia et al., 1997b), even if PEG was still associated with the nanoparticles. Thus, these results contributed to show that the part of PEG associated with the nanoparticles, was copolymerized with PIBCA (Peracchia et al., 1997b).

Finally, solubility of PEG-PIBCA nanoparticles in different solvents confirmed the formation of a copolymer (Peracchia et al., 1997b): PEG-PIBCA amphiphilic characteristics, derived from the modification of their physico-chemical properties, determined the solubility of the nanoparticles in solvents where PEG is currently insoluble. Toluene was found to be able to dissolve PEG-PIBCA nanoparticles and not PEG. Thus, this change of solubility profile was an additional proof that PEG was covalently linked to PIBCA nanoparticles (Peracchia et al., 1997b).

Comparative study on the complement consumption by PEG in different conformations chemically coupled to PIBCA nanoparticles

Complement consumption was evaluated as the residual lytic capacity of the serum after contact with PEG-PIBCA and MePEG-PIBCA nanoparticles (Peracchia et al., 1997d). Nanoparticles with the desired structure were formed by chemical coupling of PEG or MePEG during emulsion/polymerization of isobutylcyanoacrylate (IBCA). The methoxy-terminated PEG was supposed to link with the forming PACA nanoparticle only through one OH end group, whereas in the case of PEG two OH groups were available for coupling to PIBCA.

The conformational mobility of PEG and MePEG chains has been evaluated by surface charge measurement. The determination of the zeta potential in different media was a direct proof of the lower compressibility of PEG chains (anchored to the nanoparticles through both ends) and the higher compressibility of MePEG chains (linked through only one end-group, thus fully

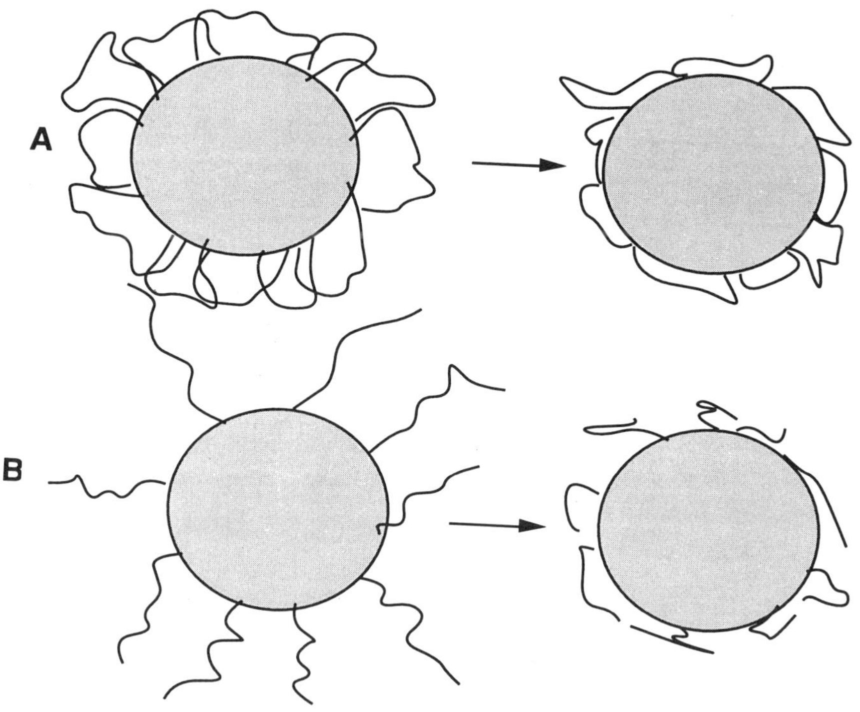

Figure 5. Possible mechanism of compression/extension of PEG (A) and MePEG (B) chains by increasing the ionic strength of the medium (from ref. 8, with permission).

extended in a low ionic strength medium). Zeta potential of PEG- and MePEG-coated nanoparticles in water was close to neutrality, due to the presence of the non-ionic PEG molecules on the particle surface (Peracchia et al., 1997d). Interestingly, zeta potential of PEG-PIBCA nanoparticles did not vary by increasing the ionic strength of the medium, whereas it did change in the case of MePEG-PIBCA nanoparticles. This result suggested a limited mobility of PEG chains when they are anchored through both ends to the nanoparticle core, and an incomplete PEG-coating in the case of MePEG-PIBCA. In Figure 5 we have schematically represented the compression/extension of PEG chains by increasing the ionic strength of the medium. On the contrary, there were no differences concerning the hydrophilicity of nanoparticles surface. Hydrophobicity surface studies showed that non-PEGylated PIBCA nanoparticles were hydrophobic, whereas, as expected, both MePEG- and PEG-PIBCA particles exhibited a more hydrophilic surface (Peracchia et al., 1997d).

Complement consumption was evaluated as the residual lytic capacity of the serum after contact with PEG-PIBCA and MePEG-PIBCA nanoparticles. In both cases, nanoparticles consumed complement in a dose-dependent manner (Peracchia et al., 1997d). Surprisingly, the PEG side-on configuration showed a lower complement consumption, better evidenced when CH50 data were represented versus the particles surface area (Fig. 6).

Indeed, in the case of MePEG-PIBCA particles, the number of CH50 units rose rapidly and almost linearly by increasing the nanoparticle concentration, suggesting a PEG conformation at the particle surface not effective enough to prevent complement consumption. On the contrary, PEG-PIBCA nanoparticles did not induce any response of the complement system up to a volume of 300 μl. At higher concentration, complement consumption took place, rising rapidly and in a parallel way with PEG-PIBCA nanoparticles. Thus, we hypothesized that the side-on conformation was able to prevent complement consumption better; in any case, when the number

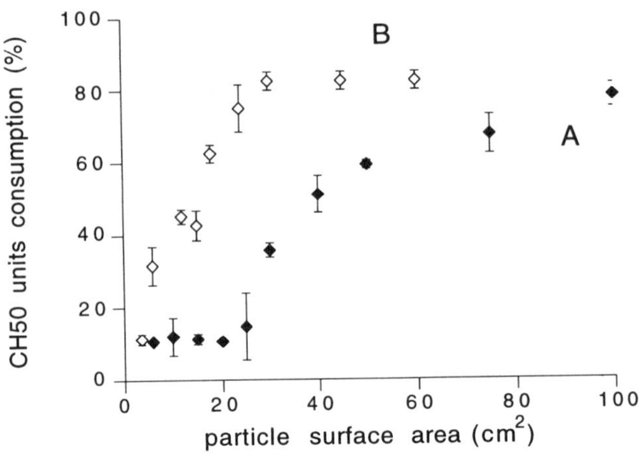

Figure 6. Complement consumption by MePEG-PIBCA (A) and PEG-PIBCA (B) nanoparticles (from ref. 8, with permission).

of nanoparticles in contact with a fixed volume of serum increased, complement consumption took place, probably amplified by the feedback loop.

Furthermore, if we consider the conformation of PEG and MePEG chains when a protein approaches a nanoparticle, our results suggested that a PEG chain anchored to both ends of the nanoparticle core could prevent opsonization more effectively because of a more effective conformational cloud: this seems to assure a better protection against complement consumption by hindering the approach of blood proteins, as represented in Figure 7.

Interestingly, the higher MePEG chains extendibility, even if it confers a higher chain mobility (as observed by zeta potential measurements), could allow protein adsorption, due to an easier penetrability by proteins. A PEG layer showed a better dysopsonic effect, confirming the importance of layer conformation in rejecting protein approach.

PEG-COATED NANOPARTICLES DESIGNED WITH A PREFORMED PEG-POLYCYANOACRYLATE COPOLYMER

The second strategy to obtain pegylated cyanoacrylates nanoparticles was to use preformed pegylated copolymers, obtained by a single step condensation of Me-PEG cyanoacetate with formaldehyde, a classical procedure for the formation of polyalkylcyanoacrylate oligomers via the formation of the amino derivative. This condensation procedure was chosen to allow a better control of the polymerization process than in the case of the anionic polymerization of alkylcyanoacrylates, which was more difficult to handle. This approach allowed the covalent linkage of PEG moieties to the cyanoacrylate polymer, thus avoiding the risk of desorption of PEG, as in the case of a PEG coating obtained by simple adsorption onto preformed nanoparticles. In addition, the condensation procedure allowed a better control of the polymerization process than in the case of anionic polymerization of alkyl cyanoacrylates (Peracchia et al., 1997e). Finally, we have shown that the hydrophobicity of the copolymer can be perfectly modulated by adjusting the MePEG/alkyl cyanoacetate ratio. Characterization of the copolymer has been performed by NMR, GPC and FTIR analysis (Peracchia et al., 1997e). Monodispersed nanoparticles were easily prepared by nanoprecipitation of this amphiphilic copolymer (Peracchia et al., 1997e; Peracchia et al., 1997f). Surface chemical analysis and characterization of surface properties allowed verification of the presence of PEG on nanoparticle surface (Peracchia et al., 1997f).

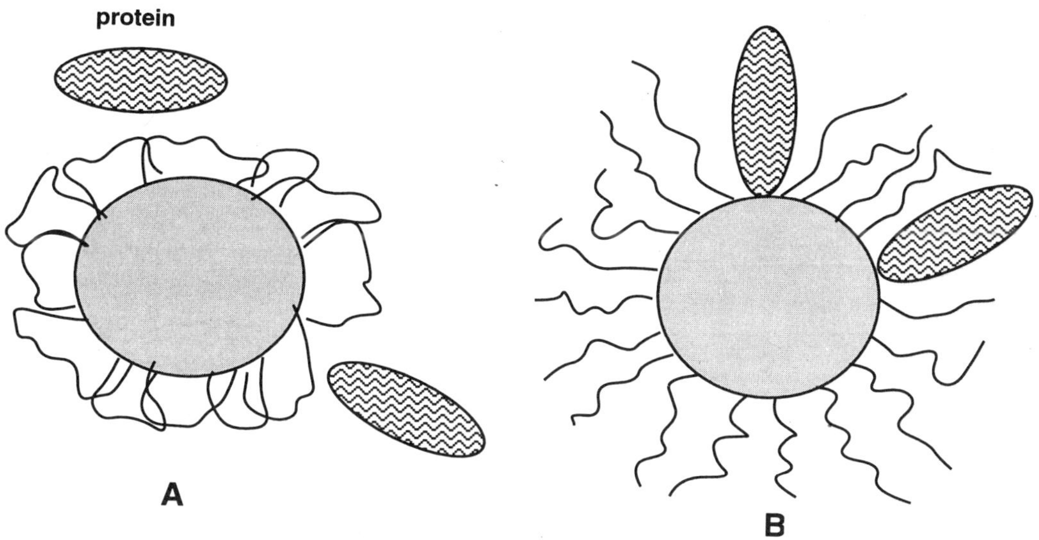

Figure 7. Probable mechanism of protein approach to PEG-PIBCA (A) and MePEG-PIBCA (B) nanoparticles (from ref. 8, with permission).

Synthesis and characterization of PEG-polycyanoacrylate copolymer

In order to obtain in a simple way a novel PEG-polycyanoacrylate copolymer, we proposed to use the base-catalyzed condensation of alkyl cyanoacetates with formaldehyde (Fig. 8), described in the literature for the synthesis of oligomers of poly(alkyl cyanoacrylates). The nature of oligomer formation in cyanoacetate-formaldehyde condensation involved a Michael addition as step-growth mechanism (Fig. 8), with chain lengths governed by steric hindrance.

The advantage of polyalkylcyanoacrylate synthesis from alkyl cyanoacetates instead of alkyl cyanoacrylate monomers is clearly that it is possible to control the polymerization process more easily, avoiding the extremely fast anionic polymerization of alkyl cyanoacrylates initiated by bases and other nucleophilic agents.

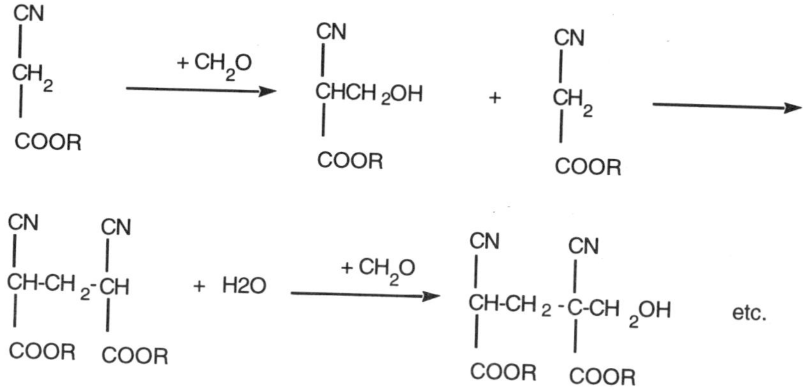

$$\mathrm{CN{-}CH_2{-}COOR} \xrightarrow{+\,CH_2O} \mathrm{CN{-}CHCH_2OH{-}COOR} \;+\; \mathrm{CN{-}CH_2{-}COOR} \longrightarrow$$

$$\mathrm{CH(CN)(COOR){-}CH_2{-}CH(CN)(COOR)} \;+\; \mathrm{H2O} \xrightarrow{+\,CH_2O} \mathrm{CH(CN)(COOR){-}CH_2{-}C(CN)(COOR){-}CH_2OH} \quad \text{etc.}$$

Figure 8. Condensation of alkyl cyanoacetates with formaldehyde.

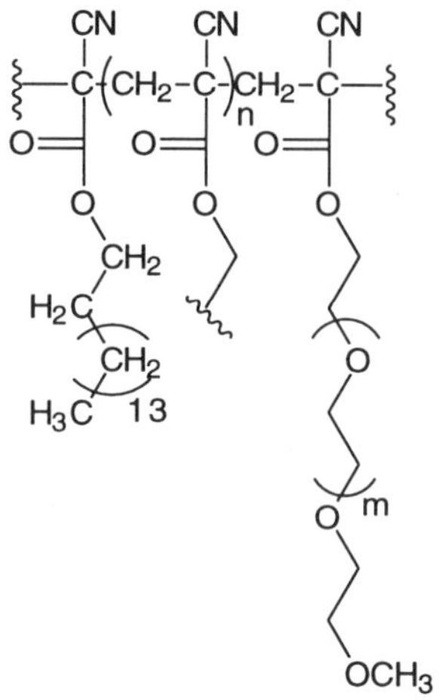

Figure 9. Structure of the synthetized copolymer.

The esterification of cyanoacetic acid with MePEG led to the formation of the corresponding esters in almost pure form and high yield (Peracchia et al., 1997e). The structure of the obtained ester was confirmed by ^{1}H-NMR (Peracchia et al., 1997e). The most striking feature was the singlet at 3.51 ppm, corresponding to the "active" methylene of cyanoacetate moiety, whereas the methylene groups of the PEG showed a large peak at 3.57 ppm.

In order to modulate the hydrophobicity of the polymer, we proposed to insert into the polymer structure an alkyl chain derived from a fatty alcohol (Fig. 9). To this end, the hexadecyl cyanoacetate was first prepared by using the same conditions to form MePEG cyanoacetate. The two esters were then combined at different ratios (1:1, 1:2, 1:3, 1:4 and 1:5, mole ratio between MePEG cyanoacetate and hexadecyl cyanoacetate) and condensed with formaldehyde. A thick suspension formed immediately. The use of formalin and dimethylamine in protic media allowed a smooth polymerization at room temperature. Unreacted MePEG cyanoacetate was eliminated by washing with water. Evaporation of the solvent left a yellow waxy material.

The ^{1}H-NMR spectrum is consistent with the structure of the expected copolymer. Figure 10 shows the spectrum related to a copolymer formed from MePEG cyanoacetate and hexadecyl cyanoacetate at the molar ratio 1:5. Integration data showed a good correlation between the initial molar ratios and the final copolymer composition. Since peaks "c" and "g" could be assigned to the resonance signals of methyl groups belonging to both ester moities, the ratio of their respective surfaces gave a direct access to the copolymere composition. Thus, for instance, in the case of the 1: 5 copolymer, this calculation was consistent with an average of 5.6 hexadecyl subunits to 1 PEG. Data related to the other molar ratios between the two cyanoacetates are reported in Table 1, which sumarizes the results for the copolymerization.

The ^{13}C-NMR spectrum is consistent with the proposed structure of the copolymer backbone. Ester carbonyl and cyano groups show broad positive peaks at 165.2 and 114.2 ppm respectively. The signal at 42.8 ppm was assigned to the quaternary carbon atoms of the polycyanoacrylate chain whereas the methylene carbons bearing the oxygen atoms of the side chains appear at 68 ppm. The terminal MeO group of the MePEG chain shows a negative peak at 55.4 ppm, whereas the methylene presented a large signal at 70.3 ppm. The hexadecyl chain exhibits several methylene groups between 34 and 25 ppm, and a methyl group at 18.13 ppm.

The hydrophobicity of the polymer was easily modulated by adjusting the MePEG/alkyl cyanoacetate ratio (molar ratio from 1:1 to 1:5). In this respect, it would be desiderable to use for nanoparticle preparation the copolymer with the highest MePEG/alkyl cyanoacetate ratio, in order to ensure a high PEG surface density. Indeed, it is expected that, due to the amphiphilic properties

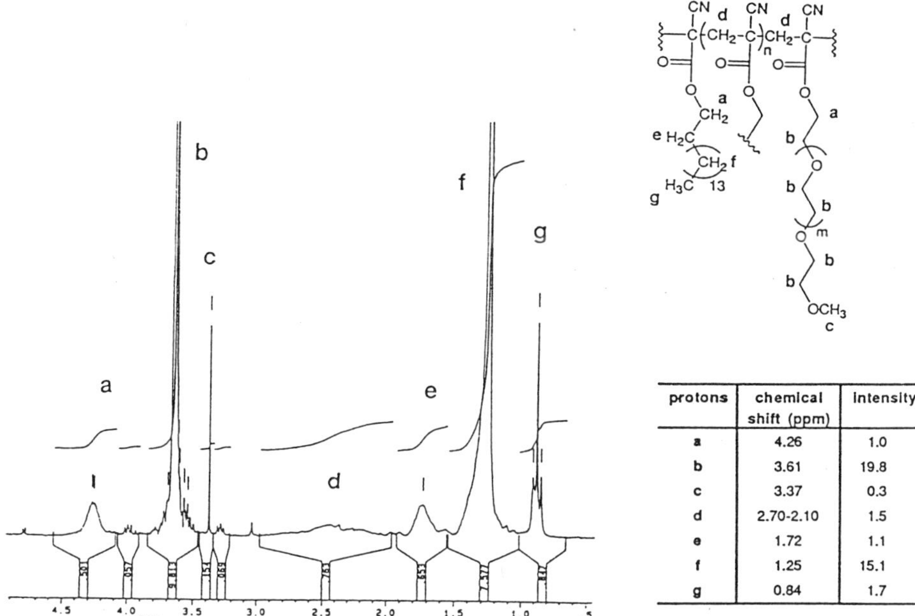

protons	chemical shift (ppm)	intensity
a	4.26	1.0
b	3.61	19.8
c	3.37	0.3
d	2.70-2.10	1.5
e	1.72	1.1
f	1.25	15.1
g	0.84	1.7

Figure 10. ^{1}HNMR spectrum related to the copolymer formed from MePEG cyanoacetate and hexadecyl cyanoacetate at the molar ratio 1:5 (from ref. 11, with permission).

of the obtained copolymer, the PEG chains migrate to the interface with the aqueous phase during nanoparticle preparation, disposing at the surface of the formed particles. However, it should be noted that the copolymer (MePEG/alkyl cyanoacetate at the ratio 1:1) was obtained in only low yield (Table 1), since most of it was lost during the washing with water. By increasing the molar ratio between the two cyanoacetates, it was possible to increase the yield up to 46% (Table 1).

Analysis of the product suggested that the condensation of the growing chain was governed by steric hindrance of MePEGcyanoacrylate monomer (Peracchia et al., 1997e).

Table 1. Results of the condensation/polymerization of MePEG cyanoacetate with hexadecyl cyanoacetate with formaldehyde. From ref. 11, with permission.

feed (ratio MePEGCA/HDCA)	polym. comp.[a]	yield (%)	weight-avg. GPC	MW calcd
1:5	1:5.6	46	3719	3780
1:4	1:4	40	3506	3441
1:3	1:2.5	35	3410	3102
1:2	1:2	30	2962	2763
1:1	1:0.6	3.5	2373	2424

[a] calculated from 1H-NMR spectra

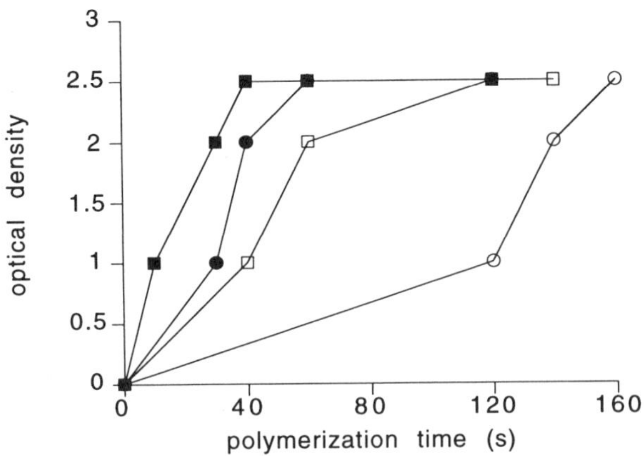

Figure 11. Turbidity during the polymerization of different initial molar ratios of MePEG cyanoacetate and hexadecyl cyanoacetate (1:2, 1:3, 1:4, 1:5) (from ref. 11, with permission).

We hypothesized that it is the sterically less demanding alkyl cyanoacetate which initiated the reaction, and that the polymerization stopped when one PEG cyanoacetate linked at the end of the growing material. To support this hypothesis, we condensed PEG cyanoacetate with increasing amounts of hexadecyl cyanoacetate. Indeed, the turbidity of the reaction medium showed an increment of the kinetics depending on the alkyl/PEG cyanoacetate ratio (Fig. 11), indicating that copolymerization rate is a function of the initial amount of alkylcyanoacetate moieties, and suggesting their major role in the polymerization process.

GPC allowed calculation of the MW of the obtained copolymer, as well as the index of polydispersity. The average MW is 3720 Da for an initial ratio of 1:5 between PEG cyanoacetate and hexadecylcyanoacetate), with a low index of polydispersity (1.3). The calculated MW would be 3780 Da, thus the method allowed a very good correlation between initial feed of the starting esters and final copolymer composition. Table 1 reports the data related to the experimental and theoretical MW for all the different molar ratios, showing in all cases a good approximation in the definition of the copolymer structure, according to NMR data. The unimodal mass distribution excluded the presence of poly(MePEGacrylate) or poly(hexadecylacrylate). The formation of low MW polyalkylcyanoacrylates was expected, since condensation of alkylcyanoacetates with formaldehyde is a method described for the synthesis of alkylcyanoacrylate monomers. Furthermore, the interest in using poly-alkylcyanoacrylate in controlled drug delivery is due to their rapid degradability by erosion, after hydrolysis of the ester lateral chain and the formation of water soluble polycyanoacrylic acid, eliminated by renal excretion. Thus, a low MW is an important issue for the degradability properties of the obtained material. Finally, the MW of the "hydrophobic" alkyl cyanoacrylate part of the copolymer was of the same order as that observed in nanoparticles prepared by emulsion/polymerization of IBCA, well known to degradade rapidly *in vivo*.

The FTIR spectrum of poly(MePEG-co-hexadecylcyanoacrylate) showed absorption bands related to CN stretching vibration at 2120.6 cm^{-1}, and ester-carbonyl at 1747.2 cm^{-1}. The C-O stretching of the PEG appeared at 1114.4 cm^{-1}.

Nanoparticle preparation and characterization

Monodispersed nanoparticles with a mean diameter between 98 and 162 nm were prepared in a single step by nanoprecipitation, depending on polymer concentration in the organic phase. Furthermore, due to the presence of PEG in their structure, no further stabilizer was needed during the preparation step. Nanoparticle suspension was shown to be very stable, since during 4 weeks storage no significant change in size was determined.

Surface charge analysis suggested the presence of PEG at the particle surface. Indeed, in stabilized suspensions, a zeta potential value close to neutral is due to non ionic materials on the particle surface.

Surface chemical analysis of poly(MePEG-co-HDCA) nanoparticles verified the presence of PEG on nanoparticle surface. As depicted in Fig. 12, the high resolution C(1S) peak related to PMePEGCA (A) fitted with a main component at 286.1 eV, corresponding to C-O, and a component at 285 eV (C-C and C-H). The same component at 286.1 eV is found for poly(MePEG-co-HDCA) nanoparticles (B), and it was not detectable for PHDCA (C), where only a main peak assigned to C-C and C-H was displayed.

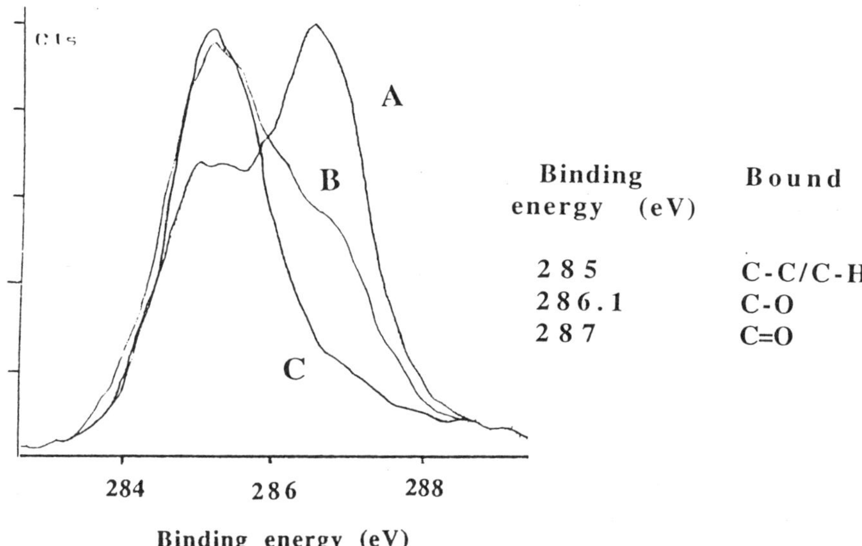

Figure 12. XPS high resolution C(1S) peaks related to PMePEGCA (A), poly(MePEG-co-HDCA) nanoparticles (B), and PHDCA (C) (from ref. 12, with permission).

CONCLUSIONS

Our work proposed to develop novel technologies for the preparation of pegylated polycyanoacrylate nanoparticles. A first strategy used was to obtain PEG-polyisobutylcyanoacrylate (PIBCA) nanoparticles during the nanoparticle preparation step, by emulsion polymerization of the alkylcyanoacrylate monomer in an aqueous medium in the presence of polyethyleneglycol (PEG).

The pH of the polymerization medium was a key-parameter for the preparation of PEG-coated PIBCA nanoparticles: the polymerization appeared to be governed by the relative ratio of OH moieties from PEG and OH^- and H_3O^+ ions generated from water molecules. PEG concentration in the polymerization medium was another important parameter: at low concentration, PEG was a weak initiator of the polymerization of IBCA, and did not lead to the formation of colloidal particles. Furthermore, a correlation between copolymerization and nanoparticle formation was shown.

The nucleophilic attack of the IBCA monomer occurred through the PEG OH end groups and not through PEG ether groups. Thus, when the emulsion/polymerization of IBCA was carried out in the presence of Me_2PEG, no PEG was associated with the nanoparticles.

^{1}H-NMR and ^{13}C-NMR allowed to clear identification of the chemical structure of PEG-PIBCA nanoparticles, and suggested that the PEG block contained in the nanoparticles was attached to PIBCA via an etheroxide bond. Other polymer chemistry investigations (DSC, NMR and the chemical determination of PEG) strongly suggested that the part of PEG contained in the nanoparticles was well copolymerized with PIBCA. This was an important issue, since a covalent PEG binding should ensure a stable coating , potentially able to prevent any interaction with blood components *in vivo*. Complement consumption was observed to be very sensitive to the number of particles in contact with human serum, as well as to the PEG conformation, suggesting PEG configuration could affect the particle exposed surface. In any case, pegylated nanoparticles exhibited a reduced complement activation compared to non-pegylated PIBCA nanoparticles.

The second strategy to obtain pegylated cyanoacrylate nanoparticles was to use preformed pegylated copolymers, obtained by a single step condensation of Me-PEG cyanoacetate with formaldehyde. This work proposed the synthesis of a novel amphiphilic PEG cyanoacrylate copolymer for the single step preparation of sterically-stabilized nanoparticles. The desired copolymer was obtained in a simple way by condensation/polymerization of PEG- and hexadecyl- cyanoacetate esters with formaldheyde. Moreover, the hydrophobicity of the copolymer could be perfectly modulated by adjusting the Me-PEG cyanoacetate/alkyl cyanoacetate ratio. Characterization of this novel copolymer has been performed by NMR, GPC and FTIR analysis.

Nanoparticles with unimodal size distribution and size of less than 100 nm were prepared by nanoprecipitation. The advantage of preparing PEG-coated injectable nanoparticles from PEG-R copolymers is clearly that a covalent PEG binding would ensure a stable coating, potentially able to prevent any interaction with blood components *in vivo*. Surface chemical analysis and characterization of surface properties allowed verification of the presence of PEG on nanoparticle surface. In conclusion, we designed two different pegylated particulate systems composed of polycyanoacrylates.

In conclusion, both the pegylated systems we proposed, represent potential stealth rapidly biodegradable carriers, possibly able to avoid MPS recognition *in vivo*.

Acknowledgments

This work was supported by a "Marie Curie" postdoctoral fellowship (TMR) from the European Community to M.T. Peracchia. The work was also supported by "Galileo 1995", joint program between Italy and France, sponsored by the French Ministery of Foreign Affairs and the Italian Ministery of University and Scientific and Technological Research (MURST), and by "Association pour la Recherche sur le Cancer" (ARC, Paris, France).

REFERENCES

Bazile, D., Prud'Homme, C., Bassoullet, M.T., Marlard, M., Spenlehauer, G., and Veillard, M., 1995, Stealth MePEG-PLA nanoparticles avoid uptake by the Mononuclear Phagocyte System, *J. Pharm. Sci.*, 84:493.

Celikkaya, E., Denkbas, E.B., and Piskin, E., 1996, PLA/PEG copolymer particles. I. Preparation and characterization, *J. Appl. Polym. Sci.*, 61:1439.

Couvreur, P., Kante, B., Roland, M., Guiot, P., Baudhuin, P., and Speiser, P., 1979, Poly(alkylcyanoacrylate) nanocapsules as potential lysosomotropic carriers: preparation, morphological and sorptive properties, *J. Pharm. Pharmacol.*, 31:331.

Gref, R., Minamitake, Y., Peracchia, M.T., Trubetskoy, V.,Torchilin, V., and Langer, R., 1994, Biodegradable long-circulating polymeric nanospheres, *Science*, 263:1600.

Leonard, F., Kulkarni, R.K., Brandes, G., Nelson, J., Cameron, J., 1966, Synthesis and degradation of poly(alkyl a-cyanoacrylates), *J. Appl. Polym. Sci.,* 10:259.

Müller, B.G., and Kissel, T., 1993, Camouflage nanospheres: a new approach to bypassing phagocytic blood clearance by surface modified particulate carriers, *Pharm. Pharmacol. Lett.*, 3:67.

Peracchia, M.T., Gref, R., Minamitake, Y., Domb, A., Lotan, N., and Langer, R., 1997a, PEG-coated injectable nanospheres prepared from amphiphilic diblock and multiblock copolymers: investigation of drug encapsulation and release characteristics, *J. Controlled Rel.*, 46:223.

Peracchia, M.T., Vauthier, C., Puisieux, F., and Couvreur, P., 1997b, Development of sterically stabilized poly(isobutylcyanoacrylate) nanoparticles by chemical coupling of polpoly(ethylene glycol), *J. Biomed. Mat. Res.*, 34:317.

Peracchia, M.T., Vauthier, C., Popa, M., Puisieux, F., and Couvreur, P., 1997c, An investigation on the formation of sterically stabilized PEG-PIBCA nanoparticles by chemical grafting of PEG during the polymerization of isobutyl cyanoacrylate, *S.T.P. Pharma Sci.,* 7:514.

Peracchia, M.T., Vauthier, C., Passirani, C., Couvreur, P. and Labarre, D., 1997d, Complement consumption by poly(ethylene glycol) in different configurations chemically coupled to poly(isobutyl 2-cyanoacrylate) nanoparticles, *Life Sci.*, 61:749.

Peracchia, M.T., Desmaële, D., Couvreur, P. and d'Angelo, J., 1997e, Synthesis of a novel poly(PEG-cyanoacrylate co-alkyl cyanoacrylate) amphiphilic copolymer for the development of "stealth" PEG-coated nanoparticles, *Macromolecules,* 30:846.

Peracchia, M.T., Desmaële, D., Appel, M., d'Angelo, J. and Couvreur, P., 1997f, Biodegradable PEG-coated nanoparticles from a novel poly(MePEGcyanoacrylate-co-alkyl cyanoacrylate) amphiphilic copolymer, *Proceed. Intern. Symp. on Control. Rel. of Bioact. Mater.,* 24:124.

DEVELOPMENT OF A NEW CLASS OF NANOPARTICLES WHICH AVOID PHAGOCYTOSIS BY INHIBITING COMPLEMENT ACTIVATION

Catherine Passirani, Gillian Barratt, Jean-Philippe Devissaguet and Denis Labarre

URA CNRS 1218, Centre d'Etudes Pharmaceutiques, Université Paris-Sud, Chatenay-Malabry, France

INTRODUCTION

The development of colloidal systems (liposomes, nanoparticles) as efficient drug carriers is limited by their distribution within the organism. It is now well established that after intravenous administration these particles are rapidly opsonized by plasma proteins. These adsorbed proteins promote recognition and uptake by cells of the mononuclear phagocyte system (MPS), particularly the Küpffer cells of the liver, but also spleen and bone-marrow macrophages (Puisieux et al., 1994). Activation of complement, especially by the alternative pathway, plays an important role in the processes of opsonization and phagocytosis. If colloidal drug carrier systems are to remain in the blood compartment for any length of time and deliver their contents to organs other than the MPS, strategies for avoiding opsonization must be developed.

STRATEGIES FOR PREPARING LONG-CIRCULATING DRUG CARRIER SYSTEMS

Two possible approaches to the problem of reducing opsonization of colloidal drug carrier systems can be envisaged. The first is to render their surface more hydrophilic by creating a steric barrier. The most effective way of doing so seems to be surface grafting of poly(ethylene glycol) (PEG) chains. Theoretical studies (Jeon et al., 1991) have shown that if the density of this barrier is sufficient, it not only prevents denaturation of the proteins on the surface but also repels proteins in solution, by a "brush" effect. This strategy has been extensively studied over the last ten years, particularly with reference to "Stealth" or "sterically-stabilized" liposomes (Woodle and Lasic, 1992), and is the subject of a number of chapters in this book. Recently, a similar strategy has been applied to nanoparticles prepared from biodegradable polymers, such as poly(d,l-lactide)(PLA) (Gref et al., 1994, Vittaz et al., 1996).

The second approach can be described as biomimetic because it seeks to imitate cells and pathogens which avoid phagocytosis by reducing or inhibiting the activation of complement. One example is the development of liposomes with a membrane composition imitating that of erythrocytes, by including the ganglioside GM_1 (Allen and Chonn, 1987). Some of these formulations were shown to have circulating half-lives as long as those of PEG-grafted liposomes (Allen et al., 1991). In the same way, Gregoriadis et al., (1993) suggested that the clearance of peptides and proteins from the blood might be delayed by coupling them to polysialic acids.

As far as polymeric nanoparticles are concerned, one method of introducing sialic acid groups onto the surface would be non-covalent adsorption of glycoproteins rich in this saccharide. This was tested with fetuin on PLA nanoparticles (Huve, 1994) and orosomucoid on poly(isobutylcyanoacrylate) nanoparticles (Olivier et al., 1996). However, the results were disappointing, since the half-life of these systems was not different from that of non-coated controls, probably because of desorption and exchange of the adsorbed proteins as soon as the particles were diluted in the plasma.

Another approach, this time involving a covalent linkage, was the use of a copolymer between PLA and poly(sialic acid) derived from the capsule of a bacteria, *Escherichia coli* strain K1, which owes its extreme virulence to the fact that it does not activate complement and is therefore not captured and destroyed by macrophages. Although the clearance of nanoparticles prepared from this copolymer was slightly reduced compared to PLA coated with fetuin, MPS capture remained rapid (Huve, 1994). Possible explanations are the water-solubility of the copolymer or a conformation of polysialic acid at the surface different to that found in the pathogen.

Many other compounds are able to modulate the action of complement, including heparin, an anionic, sulfated polysaccharide. As well as possessing anticoagulant properties, heparin has been shown to inhibit several steps in the complement cascade (Ekre et al., 1992). For these reasons, the grafting of heparin onto biomaterials destined to be in contact with blood, such as hemodialysis membranes has been proposed (Labarre, 1990). We reasoned that grafting of heparin onto colloidal systems would prevent or at least reduce the activation of complement and the deposition of opsonins on their surface, and thus create a long-circulating system. We therefore prepared copolymers of heparin and poly(methylmethacrylate) (PMMA), which, being amphiphilic, were able to self-assemble into nanoparticles. As a control, we also prepared nanoparticles from a copolymer with another polysaccharide, dextran, which is known to activate complement when rendered insoluble by crosslinking (Carreno et al., 1988a). Although the hydrophobic segment, PMMA, is not biodegradable, these particles represent a useful first model to test their interaction with the complement system and their clearance *in vivo*.

Before describing our experimental work, we will give a brief overview of the complement system and the interactions of heparin and dextran with this system.

THE COMPLEMENT SYSTEM

The complement system is one of the oldest defense systems of the organism. Although the activation of the complement cascade can lead to lysis of a number of bacteria species, its principal functions are to attract phagocytes (by chemotaxis) to sites of infection, to increase the expression of receptors for fragments of the complement protein C3 on the membranes of these cells and to promote phagocytosis of cells or particles which have these C3 fragments adsorbed or bound on their surface. The complement system consists of about 20 plasma proteins as well as a number of cell-membrane proteins. Its function is closely regulated by numerous control mechanisms, in order to prevent lysis of the organism's own cells.

The activation of complement can be triggered in two ways, known as the classical and alternative pathways. The classical pathway is usually stimulated by the presence of immune complexes on the surface of the particle, which implies that a specific antibody has already been produced. On the other hand, the alternative pathway is aspecific, and is therefore more often implicated in response to polymeric particles, for which specific antibodies do not exist. The two pathways converge on the cleavage of the protein C3, which is therefore a key component of the system. The presence of C3 fragments on a particle or surface triggers a common, terminal sequence.

In the classical pathway, the first component, C1, is activated by the presence of immune complexes formed by IgG or IgM and antigen. In the presence of calcium ions, this protein can bind to the F_c portion of IgG; however, the complement cascade is only triggered if the binding is at least divalent. In this case, C1 is able to cleave the proteins C2 and C4, yielding fragments C2a and C4b which combine to form the classical C3 convertase. This enzyme generates two fragments of C3: the small C3a and the larger C3b. The reaction is limited by the fact that the convertase has a half-life of 10 min at 37°C. However, if C3b is adsorbed or bound onto an activating surface, the activation can be amplified by the alternative pathway, as described below. C3a is an anaphylatoxin which can recruit phagocytes by chemotaxis.

The alternative pathway of complement activation depends on a small proportion of the protein C3 being in an extremely reactive conformation, known as "C3b-like C3". In this conformation, an unstable thioester group which is normally buried within a hydrophobic pocket becomes available and is able to form a covalent bond with nucleophilic groups on the target surface (Sim et al., 1993). A similar reaction occurs with true C3b generated by the convertase and with C4b. Furthermore, it has been shown that C3b-like C3 adsorbed onto hydrophobic surfaces such as polypropylene and polystyrene, without covalent interaction, can also trigger the activation cascade (Nilsson et al., 1993). The subsequent reactions depend on the local environment: in the presence of divalent magnesium ions, protein B and protein D, the alternative pathway C3 convertase (C3bBb) is formed. Native C3 can then be cleaved to C3a and true C3b and an amplifying cycle occurs, if C3b continues to bind to the surface. However, other, inhibitory, proteins may also bind C3b or C3b-like C3. Protein H has a differing affinity for these proteins depending on the nature of the surface to which they are bound. Once H has bound, it can be followed by factor I, converting C3b to C3bi, which has no C3-convertase activity. Therefore, surfaces which activate complement are those on which C3b has a low affinity for H. In this case, protein B, whose affinity for C3b is constant, binds preferentially and the surface acquires convertase activity. On the other hand, if the surface binds C3b in a way which favors H binding, convertase activity will be inhibited and the reaction limited.

The presence of C3b on a surface in the vicinity of a C3 convertase site also catalyses the cleavage of protein C5 to C5a and C5b, and thus triggers the terminal sequence of the complement cascade. C5a is an anaphylatoxin, even more efficient than C3a, which can attract

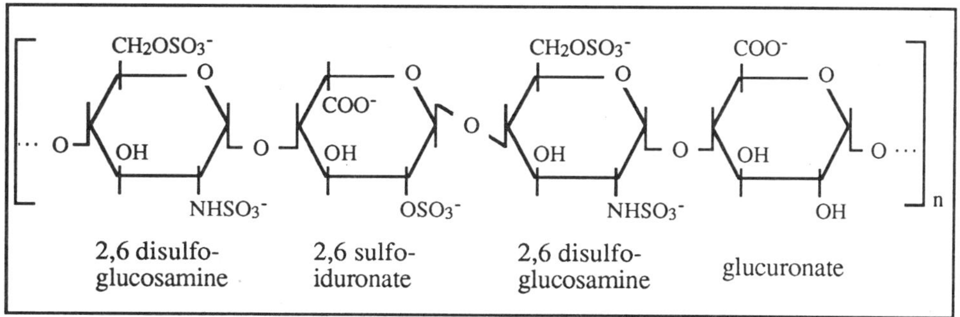

Figure 1. Structure of heparin

and activate phagocytes. These cells have surface receptors for C3b and its degradation products and will thus take up particles opsonized with these proteins, or bind to surfaces expressing them. C5b interacts with proteins C6, C7, C8 and C9 to form complexes capable of lysing bacterial and cell membranes. The production of these complexes around artificial surfaces which cannot be phagocytosed or lysed can lead to inflammatory reactions and cell damage, hence the importance of developing non activating surfaces for implants, dialysis membranes etc.

Interactions of Heparin with the Complement System

Heparin is a naturally occuring polysaccharide, made up principally of sulfated glucosamine and iduronic acid (Figure 1). It is present in polynuclear basophiles and tissue mastocytes, from which it is released during inflammatory reactions, suggesting that it plays a role in limiting inflammation. Large quantities of heparin for therapeutic uses can be extracted from tissues such as porcine or bovine intestinal mucosa or bovine lung. Depending on the extraction procedure, the molecular weight of the polysaccharide chain varies from 5000 to 30,000 Daltons. The best known biological activity of heparin is its anticoagulant property, but it has a number of other effects, including regulation of angiogenesis, stimulation of cell growth and the inhibition of complement activation.

Heparin has been shown to interfere with many different stages of the complement cascade: in activation by both the classical and the alternative pathway and in the formation of the membrane attack complex (Weiler and Linhardt, 1989, indicated on Figure 2). Sahu and Pangburn (1993) studied the interaction of heparin with 22 proteins of the complement system and found that 13 of them were able to bind heparin (C1q, C2, C4, C4bp, C1INH, B, D, H, P, C6, C8, C9 and vitronectin).

In solution, heparin inhibits the interactions of C1, C4 and C2 during formation of the classical C3b convertase. Heparin also prevents the formation of the alternative pathway convertase (Weiler et al., 1978; Linhardt et al., 1988; Kazatchkine et al., 1981) by blocking the binding site for B on C3b (Maillet et al., 1983). Furthermore, it increases the binding of the factor H to adsorbed C3b (Kazatchkine et al., 1979), by direct binding to the protein (Pangburn et al., 1991). These two latter activites will synergize to prevent amplification by the alternative pathway. Finally, heparin can also interfere with formation of the terminal sequence (Baker et al., 1975). Evidence that these reactions can occur *in vivo* as well as *in vitro* comes from experiments of Weiler et al., (1992), who pretreated animals with heparin before administering cobra venom factor. This toxin depletes C3 by a massive activation of complement. The loss of C3 activity was less rapid in animals preteated with heparin, suggesting that complement activation was inhibited.

Several studies have attempted to establish structure-activity relationships for the action of heparin on the complement pathway. It appears that the presence of O-sulfated groups and an amino-sugar is necessary but that this latter need not be sulfated (Pangburn et al., 1991; Kazatchkine et al., 1981). The minimum structure necessary seems to be a tetrasaccharide with 3 to 6 sulfated groups. The inhibitory capacity increases with size up to 14 saccharide units, while larger fragments (MW>4700) have an activity similar to that of native heparin (Maillet et al., 1988).

The anticoagulant and complement-inhibiting activities of heparin do not seem to reside in the same sites (Maillet et al., 1988; Linhard et al., 1988), and fractions with high and low affinity for antithrombin III have been shown to inhibit complement to a similar extent (Kazatchkine et al., 1981). Thus it should be possible to design systems which will avoid complement activation without interfering with coagulation.

Of interest for the present work is the fact that heparin conserves its anti-complement activity when present on the surface of biomaterials used for artificial organs (Kazatchkine et

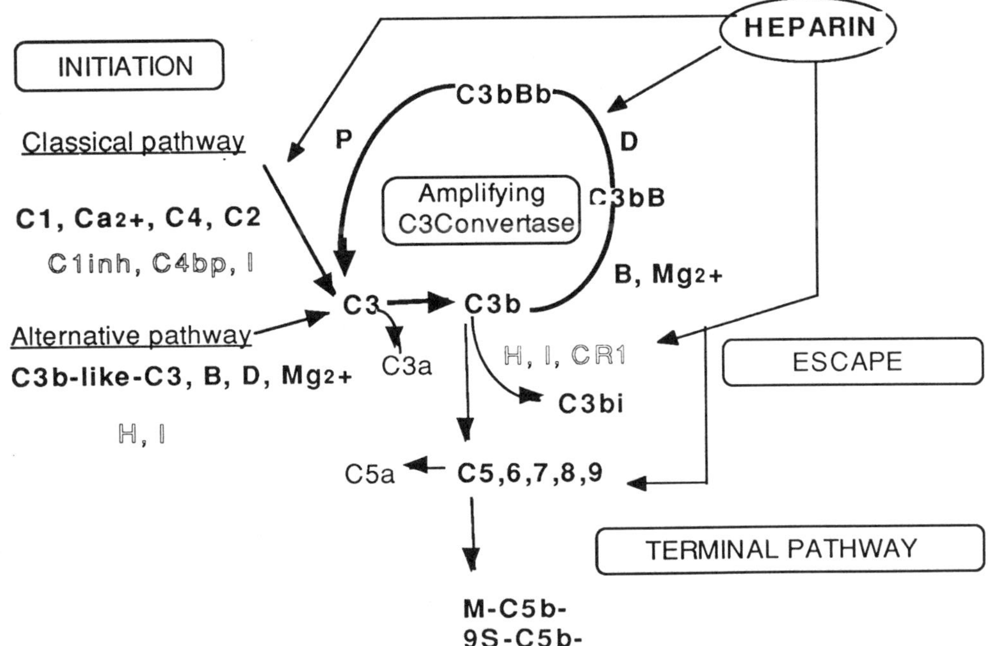

Figure 2. The influence of heparin on complement activation

al., 1979; Maillet et al., 1983). Heparin has been coupled to polymers used in such systems to improve their hemocompatibility (Riesenfeld et al., 1995; Marconi et al., 1997). For example, the biocompatibility of intraocular implants prepared from PMMA-heparin copolymers is improved, reducing the infiltration of macrophages and fibroblasts, the activation of granulocytes and platelet adhesion (Larssen et al., 1989). Complement activation by these surfaces is greatly reduced (Pekna et al., 1993) and local coagulation reactions are controlled by means of an increase in inhibitory proteins (Labarre, 1990).

Despite these observations, to our knowledge, this is the first time that heparin-grafted copolymers have been suggested for drug delivery applications.

Interactions of Dextran with the Complement System

Dextran is a neutral polysaccharide composed of α (1→6) linked glucose units. Depending on its molecular weight, it can activate the alternative pathway of complement by an unknown mechanism. However, anti-dextran antibodies have been detected in the serum of some individuals (Maillet and Kazatchkine, 1991). Although the alternative pathway can function in the absence of antibody, this reaction can be amplified by the presence of antibody on the surface. Cross-linked dextran beads (Sephadex®) have been shown to be a powerful activator of complement (Carreno et al., 1988a and b). PMMA-dextran copolymers were therefore chosen as a control for this study.

METHODOLOGY

Preparation of Nanoparticles

Amphiphilic copolymers were synthesized by graft polymerization of heparin (heparin sodium salt, from porcine intestinal mucosa, Sigma) or dextran (produced by *Leuconostoc mesenteroides* strain B-512, average molecular weight 70,000, Sigma) and methyl methacrylate monomers (Fluka). Polymerization was initiated by polysaccharide radicals produced in the presence of cerium ions. Within a defined range of monomer and cerium ion concentrations, the copolymers spontaneously formed submicronic particles, which were stable in the absence of added surfactant. Particle suspensions were dialysed to remove cerium ions and unreacted monomer.

Fluorescent nanoparticles for *in-vivo* studies were prepared by including N-vinyl-carbazole monomers (Fluka) in the copolymerization reaction medium at 5% (w/v) with respect to MMA.

Mean particle diameters were determined by quasi-elastic light scattering using a Nanosizer N4MD (Coultronics, Margency, France). The zeta potential of particles dispersed in 20 mM phosphate buffer, pH 7.0, was measured using a Zetasizer 4 (Malvern Instruments, Orsay, France).

Assessment of Interactions with Complement *in-vitro*

Two well established methods were used to assess the interactions with complement: the hemolytic CH50 test and two-dimensional electrophoresis of C3 and its fragments.

The CH50 test (Mayer, 1961) is based on the lysis of sensitized sheep erythrocytes (by the classical pathway) in the presence of human serum and divalent cations. To test the complement-activating capacity of the different formulations, serum was preincubated with a known amount of particles for 1h at 37°C. The volume of serum necessary to produce 50% hemolysis was then determined in a standard assay and compared with that obtained with control serum. Thus, if the particles possess an activating surface, C3b will be adsorbed during the preincubation step and C3 will be consumed, so that a larger volume of serum will be required to produce 50% lysis in the second incubation. In order to compare particles with different mean diameters, the quantities are expressed in terms of surface area.

Two-dimensional electrophoresis was performed according to the method of Laurell (1966). Again, particles were preincubated with human serum for 1h at 37°C. Serum was then subjected to iso-electric focusing on 1% agarose gels. The second-dimension electrophoresis was carried out in the presence of a goat antibody to human C3 (recognizing both C3 and C3b), followed by Coomassie blue staining. The relative sizes of the immunoprecipitation peaks give a semi-quantitative estimation of the proportions of native and cleaved C3.

Clearance *in-vivo*

Female CD1 mice, weighing 20-22g, were obtained from Charles River, France. Fluorescent nanoparticles suspensed in isotonic glucose solution were injected at doses of 50

Table 1. Characteristics of Nanoparticle Formulations

Name	Polymer Composition	Mean Diameter (nm)	Zeta Potential (mV)
Nanohep	Heparin-PMMA	80	-44
Nanodex	Dextran-PMMA	80	-6
Nanohep*	Heparin-PMMA-PNVC*	160	-51
Nanodex*	Dextran-PMMA-PNVC*	240	-8

* Poly(N-vinyl carbazole), fluorescent nanoparticles

or 150 mg/kg in a volume of 0.2 ml into the tail vein. Blood samples (120 μl) were drawn from the retro-orbital sinus at various time intervals and immediately diluted 15 times in normal saline. After centrifugation to remove cells, the fluorescence in the supernatant was measured with an excitation wavelength of 335 nm and an emission wavelength of 363 nm in a Spex spectrofluorimeter.

Preliminary experiments showed that the nanoparticle fluorescence could be determined in the diluted plasma samples without any extraction step. Addition of known quantities of nanoparticles to plasma from control mice showed that the fluorescence intensity was proportional to the quantity of copolymer.

Since the blood samples taken were small, each mouse could be used for a range of time points. The values shown on the graphs are the mean for 3-5 animals.

RESULTS AND DISCUSSION

Characterization of Nanoparticles

Table 1 shows the characteristics of the different formulations. The non fluorescent nanoparticles (Nanohep and Nanodex) were small, monodisperse and stable for several months in the absence of added surfactant. This fact, and the zeta potential results, suggest that the copolymers were organized with the polysaccharide moieties on the surface and the more hydrophobic PMMA in the core. Nanoparticles containing heparin, an anionic polysaccharide, had a strongly negative zeta potential, while those containing dextran were close to neutrality. The nanoparticles containing a fluorescent monomer (Nanohep* and Nanodex*) were slightly larger (about 200 nm), but remained within the size range of colloidal systems used for drug targeting purposes.

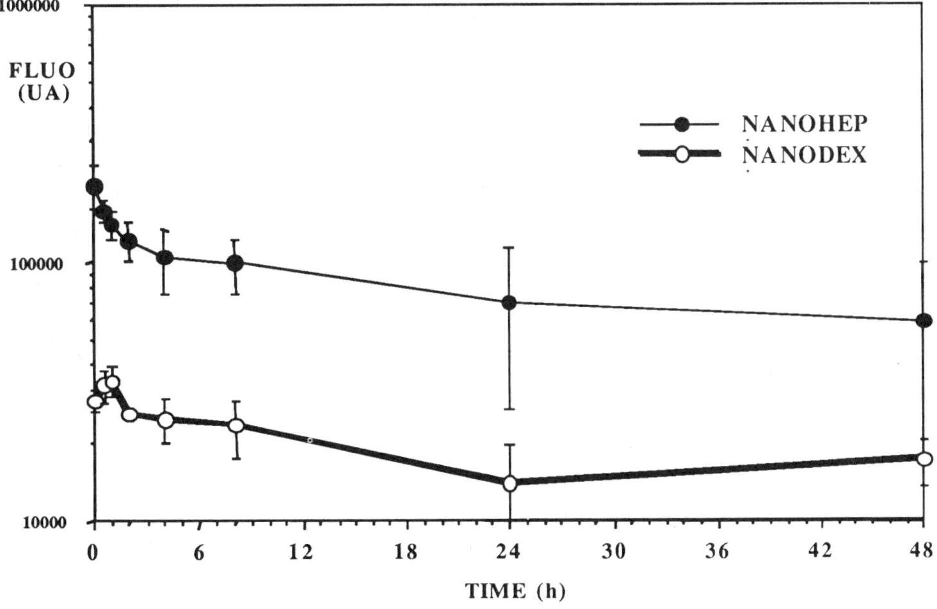

Figure 3. Clearance of Nanohep* and Nanodex* after administration at 150 mg/kg

Interactions with Complement *in-vitro*

CH50 Assay. The consumption of complement, as a function of particle surface area was determined for Nanodex, the "control" particles. These nanoparticles were found to activate complement, but to a lesser extent than Sephadex® beads at an equivalent surface area. This can probably be explained by the structural difference between the two forms: Sephadex® beads are crosslinked and have a mean diameter of 30 µm, whereas Nanodex have a mean diameter of 100 nm and probably carry the dextran chains as a brush-like layer on the surface.

In the case of Nanohep, the situation was more complicated. The results obtained in the CH50 test indicated increasing complement consumption as a function of surface area, suggesting an activating surface. However, the CH50 test is inadequate for substances which inhibit complement activation since, in order to obtain hemolysis, the classical pathway must function. As explained earlier, heparin can inhibit complement activation at a number of steps, including the classical pathway. In fact, the addition of soluble heparin to the assay system also gives the appearance of complement consumption. Therefore, this test was not able to determine the nature of the surface of Nanohep.

Two-dimensional Electrophoresis. Immunoelectrophoresis of control serum showed a large peak corresponding to intact C3 and only a small one corresponding to C3b. Very similar results were obtained with serum preincubated with Nanohep. In contrast, preincubation with Nanodex yielded a large peak of C3b, indicating activation (results not shown).

The results of this more specific test of complement activation thus validate the hypothesis that grafting of heparin produces a non activating surface, whereas similar particles with dextran are activators. It therefore seemed interesting to study the clearance of these two types of particle *in vivo*.

Clearance *in-vivo*

Figure 3 shows the plasma concentration-time prolifes of Nanohep* and Nanodex* after intravenous injection into mice at a dose of 150 mg/kg. In both cases, after an initial period in which the plasma fluorescence declined rapidly, both types of particle showed a slow clearance, and could still be detected in the blood 48h after injection.

To determine whether this slow clearance could be due to the saturation of a cellular uptake compartment, Nanohep* were also injected at a dose of 50 mg/kg. In order to take into account the different amounts of fluorescent material injected, in Figure 4 the fluorescence of the samples has been normalized by taking the fluorescence measured at 4h as 100. It can be seen that during this initial 4-h period, the clearance of the two doses was identical.

In the case of 50 mg/kg, the same clearance rate was maintained for 24h and linear regression analysis allowed us to calculate a half-life of 5h. This is much longer than the half-lives reported in the literature for unmodified polymeric nanoparticles, for example polyalkylcyanoacrylate (Couvreur et al., 1986) or polystyrene (Illum et al., 1982) and very similar to that observed for nanoparticles prepared from PLA-PEG diblock copolymers (Verrecchia et al., 1995). This suggests that the biomimetic strategy can be as effective as the steric barrier approach for the preparation of long-circulating drug delivery systems.

When a larger dose of Nanohep* was given, the inital uptake compartment seemed to be saturated and the remaining particles are cleared extremely slowly. We do not yet know the nature of the compartment responsible for this clearance. The half-life of 5h suggests that it is not the MPS. It is known that vascular endothelial cells express receptors for heparin (Patton et al., 1995), so it is tempting to hypothesize that these cells could bind, and perhaps internalize Nanohep*, but this hypothesis remains to be investigated experimentally.

The slow clearance of Nanodex*, observed in Figure 3, is somewhat surprising. *In-vitro* experiments indicated that these particles were activators of complement and would therefore

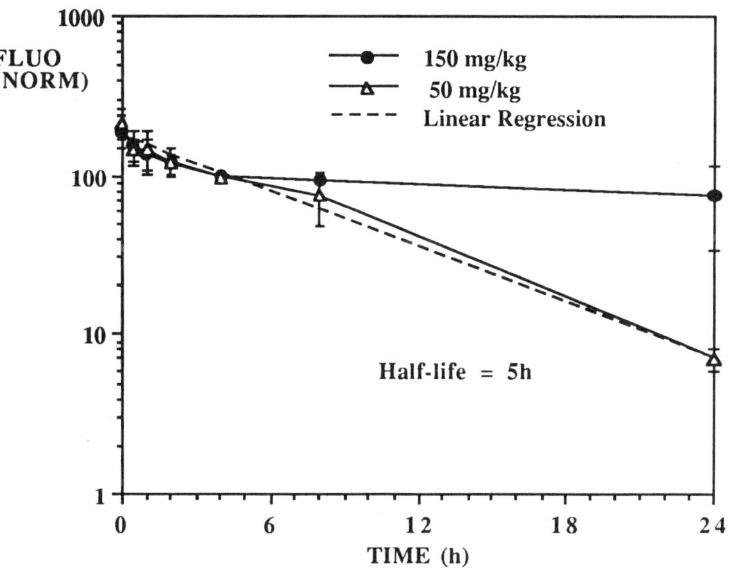

Figure 4. Clearance of Nanohep*: effect of injected dose

be expected to disappear from the circulation rapidly. However, the observations that Nanodex* were much poorer activators of complement than Sephadex® microparticles at equivalent surface area and that Nanodex, despite having a surface charge close to neutrality, remained in suspension without aggregation for several months, suggested a brush-like organization of the polysaccharide chains on the surface, which perhaps provided a steric barrier analogous to that of PEG.

The ideal control formulation for Nanohep would of course be plain PMMA particles. However, it is impossible to prepare nanoparticles from this hydrophobic polymer without adding surfactant and, as demonstrated by Tröster and Kreuter (1992), surfactant adsorbed onto the surface of nanoparticles can also influence their distribution.

CONCLUSION

In this work, we have investigated the biomimetic approach to the development of colloidal drug carriers which escape capture by the mononuclear phagocyte system. By coupling an entity which inhibits complement activation, heparin, to the surface, we have succeeded in preparing nanoparticles with long circulating half-lives. The use of amphiphilic copolymers which self-assemble in aqueous medium has allowed us to prepare submicronic particles without the use of surfactants. Obviously, much work remains to be done: transposition to a biodegradable polymer, incorporation of an active ingredient, as well as more fundamental studies of the interaction of these particles with different cell populations. Nevertheless, these preliminary results confirm the interest of the biomimetic approach to drug carrier formulation.

Acknowledgements

This work was supported in part by the Centre National de la Recherche Scientifique (France). Catherine Passirani was the recipent of a grant from the French Ministry of Education and Research. The authors wish to thank Prof. M.D. Kazatchkine and Drs. M.P. Carreno and F. Maillet (Hôpital Broussais, Paris) for their help in this study.

REFERENCES

Allen, T.M. and Chonn, A., 1987, Large unilamellar liposomes with low uptake into the reticulo-endothelial system, *FEBS Lett.* 223:42.

Allen, T.M., Hansen, C., Martin, F., Redemann, C. and Yau-Young, A., 1991, Liposomes containing synthetic lipids derivatives of poly(ethylene glycol) show prolonged circulation half-lives *in vivo*, *Biochim. Biophys. Acta* 1066:29.

Baker, P., Lint, T., McLeod, B., Behrends, C. and Gewurz, H., 1975, Studies on the inhibition of C5-6-induced lysis (reactive lysis) VI. Modulation of C5-6-induced lysis by polyanions and polycations, *J. Immunol.* 114:554.

Couvreur, P., Grislain, L., Lenaerts, V., Brasseur, F., Guiot, P., and Biornacki, A., 1986, Biodegradable polymeric nanoparticles as drug carriers for antitumor agents, in: *Polymeric Nanoparticles and Microspheres*, P. Guiot and P. Couvreur, eds., CRC Press, Boca Raton.

Carreno, M.P., Labarre, D., Jozefonvicz, M. and Kazatchkine, M.D. 1988a, The ability of Sephadex to activate human complement is suppressed in specifically substituted functional Sephadex derivatives, *Molec. Immunol.* 25:165.

Carreno, M.P., Maillet, F., Labarre, D., Jozefonvicz, M. and Kazatchkine, M.D. 1988b, Specific antibodies enhance Sephadex-induced activation of the alternative complement pathway in human serum, *Biomaterials* 9:514.

Ekre, H.P., Naparstek, Y., Lider, O., Hydén, P., Hägermark, Ö, Nilsson, T., Vlodavsky, I. and Cohen, I., 1992, Anti-inflammatory effects of heparin and its derivatives inhibition of complement and of lymphocyte migration, in: *Heparin and Related Polysaccharides*. D.A. Lane, ed., Plenum Press, New York.

Gref, R., Minamitake, Y., Peracchia, M.T., Trubetskoy, V., Torchilin, V. and Langer, R., 1994, Biodegradable long-circulating polymeric nanospheres, *Science* 263:1600.

Gregoriadis, G., McCormack, B., Wang, Z. and Lifely, R., 1993, Polysialic acids: potential in drug delivery, *FEBS Lett.* 315:271.

Huve P., 1994, Comprendre et éviter la capture des nanoparticules de poly (acide-lactique) par le système des phagocytes mononucléaires, Ph.D. Thesis, Université Paris-Sud, Sci. Pharm.

Illum, L., Davis, S.S., Wilson, C.G., Thomas, N.W., Firer, M. And Hardy, J.G., 1982, Blood clearance and organ disposition of intravenously administered colloidal particles: the effects of particle size, nature and shape, *Int. J. Pharm.* 12:135.

Jeon, S.I., Lee, J.H., Andrade, J.D. and De Gennes, P.G., 1991, Protein-surface interactions in the presence of polyethylene oxide; 1. Simplified theory, *J. Colloid Interf. Sci.* 142:149.

Kazatchkine, M.D., Fearon, D., Silbert, J. and Austen, K., 1979, Surface-associated heparin inhibits zymosan-induced activation of the human alternative pathway by augmenting the regulatory action of the control proteins on particle-bound C3b, *J. Exp. Med.* 150:1202.

Kazatchkine, M.D., Fearon, D., Metcalfe, D., Rosenberg, R. and Austen, K., 1981, Structural determinants of the capacity of heparin to inhibit the formation of the amplification C3 convertase. *J. Clin. Invest.* 67:223.

Labarre, D.J., 1990, Heparin-like polymer surfaces: control of coagulation and complement activation by insoluble functionalized polymers, *Int. J. Artif. Organs* 13:651.

Larsson, R., Selén, G., Björklund, H. and Fagerholm, P., 1989, Intraocular PMMA lenses modified with surface-immobilized heparin : evaluation of biocompatibility *in vitro* and *in vivo*. *Biomaterials* 10:511.

Laurell, C.B., 1966, Quantitative estimation of proteins by electrophoresis in agarose gels containing antibodies, *Anal. Biochem.* 15:45.

Linhardt, R., Rice, K., Kim, Y., Engelken, J. and Weiler, J., 1988, Homogeneous, structurally defined heparin-oligosaccharides with low anticoagulant activity inhibit the generation of the amplification pathway C3 convertase *in vitro*, *J. Biol. Chem.* 263:13090.

Maillet, F., Kazatchkine, M.D., Glotz, D., Fischer, E. and Rowe, M., 1983, Heparin prevents formation of the human C3 amplification convertase by inhibiting the binding site for B on C3b, *Molec. Immunol.* 20: 1401.

Maillet, F., Petitou, M., Choay, J. and Kazatchkine, M.D. 1988, Structure-function relationships in the inhibitory effect of heparin on complement activation: independency of the anti-coagulant and anti-complementary sites on the heparin molecule, *Molec. Immunol.* 25:917.

Maillet, F. and Kazatchkine, M.D. (1991) Specific antibodies enhance alternative complement pathway

activation by cuprophane, *Nephrol. Dial. Transplant.* 6 :193.
Marconi, W., Benvenuti, F. and Piozzi, A., 1997, Covalent bonding of heparin to a vinyl copolymer for biomedical applications, *Biomaterials* 18:885.
Mayer, M.M., 1961, Complement and complement fixation, in: *Experimental Immunochemistry*, EA. Kabat and M.M. Mayer, eds., 2nd Edn, Thomas, Springfield.
Nilsson, U.R., Storm, K.E., Elwing, H. and Nilsson, B., 1993, Conformational epitopes of C3 reflecting its mode of binding to an artificial polymer surface, *Mol. Immunol.* 30:211.
Olivier J.C., Vauthier, C., Tarverna, M., Puisieux, F., Ferrier, D. and Couvreur, P., 1996, Stability of orosomucoid-coated polyisobutylcyanoacrylate nanoparticles in the presence of serum, *J. Control. Rel.* 40:157.
Pangburn, M., Atkinson, M. and Meri, S., 1991, Localization of the heparin-binding site on complement factor H, *J. Biol. Chem.* 266:16847.
Patton, W.A. II, Granzow, C.A., Getts, L.A., Thomas, S.C., Zotter, L.M., Gunzel, K.A. and Lowe-Krentz, L.J., 1995, Identification of a heparin-binding protein using monoclonal antibodies that block heparin binding to porcine aortic endothelial cells, *Biochem. J.* 311:461.
Pekna, M., Larsson, R., Formgren, B., Nilsson, U.R. and Nilsson, B., 1993, Complement activation by polymethylmethacrylate minimized by end-point heparin attachment, *Biomaterials* 14:189.
Puisieux, F., Barratt, G., Couarraze, G., Couvreur, P., Devisssaguet, J.P., Dubernet, C., Fattal, E., Fessi, H., Vauthier, C. and Benita, S., 1994, Polymeric micro- and nanoparticles as drug carriers, in: *Polymeric Biomaterials*, S. Dumitriu, ed., M. Dekker, New York.
Riesenfeld, J, Olsson, P., Sanchez, J. and Mollnes, E. Surface modification with functionally active heparin. *Med. Dev. Tech.* March 1995:24.
Sahu, A. and Pangburn, M. K., 1993, Identification of multiple sites of interaction between heparin and the complement system. *Molec. Immunol.* 30:679.
Sim, E., Parker, K.E. and Jones, A., 1993, Nucleophilic compounds acting on C3 and C4 in: *Activators and Inhibitors of Complement*, Sim R.B. Ed., Kluwer Academic Publishers, Dordrecht.
Tröster, S.D. and Kreuter, J., 1992, Influence of the surface properties of low contact angle surfactants on the body distribution of ^{14}C-poly(methyl methacrylate) nanoparticles, *J. Microencap.* 9:19.
Verrecchia, T., Spenlehauer G., Bazile, D.V., Murry-Brelier, A., Archimbaud, Y. and Veillard, M., 1995, Non-stealth (poly(lactic acid / albumin)) and stealth (poly (lactic acid-polyethylene glycol)) nanoparticles as injectable drug carriers, *J. Control. Rel.* 36 :49.
Vittaz, M., Bazile, D., Spenlehauer, G., Verracchia, T., Veillard, M., Puisieux, F. and Labarre, D., 1996, Effect of PEO surface density on long-circulating PLA-PEO nanoparticles which are very low complement activators, *Biomaterials* 17:1575.
Weiler, J. and Linhardt, R., 1989, Comparison of the activity of polyanions and polycations on the classical and alternative pathways of complement. *Immunopharmacology* 17 :65.
Weiler, J., Yurt, R., Fearon, D. and Austen, K., 1978, Modulation of the formation of the amplification convertase of complement C3bBb, by native and commercial heparin, *J. Exp. Med.* 147:409.
Weiler, J., Edens, R., Linhardt, R. and Kapelanski, D., 1992, Heparin and modified heparin inhibit complement activation *in vivo*, *J. Immunol.* 148:3210.
Woodle, M.C. and Lasic, D.D., 1992, Sterically stabilized liposomes, *Biochim. Biophys. Acta* 1113:171.

COATING OF NANOPARTICLES WITH SURFACTANTS:

TARGETING VERSUS PROLONGED CIRCULATION

Jörg Kreuter

Institut für Pharmazeutische Technologie, Biozentrum,
J. W. Goethe-Universität, D-60439 Frankfurt, Germany

INTRODUCTION

Nanoparticles were introduced as drug delivery systems about 25 years ago. They share with liposomes a common problem: they are rapidly taken up by the macrophages of the so-called monocular phagocytic system (MPS) (also known as the reticuloendothelial system (RES)) and accumulate mainly in the liver, spleen, and lungs, whereas other targets in the body are much harder to reach. Great efforts have been undertaken to overcome this rapid uptake by the liver and other parts of the MPS, and the present symposium is mainly devoted to this problem. The most successful strategies to prevent the rapid RES uptake are coating of the particles with surfactants or covalent linkage of polyoxyethylene chains to their surfaces (see other chapters of this book for references). The properties of some of these systems to achieve prolonged blood circulation times are often refered to as "stealth" properties acknowledging their non-recognition by the MPS. However, keeping these colloidal drug carriers in the blood circulation for extended times does not necessarily mean that they reach their desired therapeutic target site. In order to reach this objective other strategies may have to be employed.

MACROPHAGE TARGETING FOR AIDS THERAPY

In the case of AIDS, for instance, the macrophages of the RES represent one of the most important therapeutic targets. In addition to the $CD4^+$ T lymphocytes, these cells play a decisive role as a reservoir for the human immunodeficiency virus (HIV). In tissues such as the lung and the brain, HIV is located primarily in macrophage-like cells (i.e. alveolar macrophages and microglia, respectively) (Levy, 1993; Weiss, 1993). In contrast to $CD4^+$ T cells, in which HIV replication is proliferation-dependent and finally leads to cell death, macrophages in a mature nonproliferating but immunologically active state can be productively infected with HIV type 1 (HIV-1) and HIV-2 (Gartner et al, 1986; Kühnel et al, 1989; Nicholson et al, 1986; von Briesen et al, 1990). Altered cellular functions in the macrophage population may contribute

Targeting of Drugs 6: Strategies for Stealth Therapeutic Systems
Edited by Gregoriadis and McCormack, Plenum Press, New York, 1998

to the development and clinical progression of AIDS. A large proportion of AIDS patients show severe HIV encephalitis with diffuse neuronal damage which is thought to be mediated by viral proteins and/or factors with neurotoxic activity (i.e. cytokines). This can occur through proinflammatory cytokines or by HIV-1-specific proteins secreted from cells of the mononuclear phagocyte system, including brain macrophages, microglia, and multinucleated giant cells, which have been shown to be productively infected with HIV (Bender et al, 1996; Esser et al, 1991; Gendelman et al, 1994).

There is evidence that, apart from having a function in the pathogenesis of the disease, cells of the macrophage lineage are vectors for the transmission of HIV. The placental macrophage is likely to be the primary cell type responsible for vertical (maternofetal) transmission of HIV (McGann et al, 1994). For mucosal transmission it was found that an important property of the transmitted HIV variant is its ability to infect macrophages (Milman and Sharma, 1994). The phenotypic characterization of virus populations which were transmitted sexually or vertically revealed a selection of variants with a predominant tropism for macrophages in the recipient (van't Wout et al, 1994). Because of the important role of cells of the monocyte/macrophage (Mo/Mac) lineage in the pathogenesis of HIV, fully effective anti-HIV therapy must reach Mo/Mac in addition to other target cells (Bender et al, 1996).

Consequently, nanoparticles without stealth properties may represent promising drug carriers for this disease. Therefore, the potential of nanoparticles to deliver antiviral drugs to HIV-infected and uninfected macrophages was investigated by us (Schäfer et al, 1991; 1992). Indeed, it was found in *in vitro* cell cultures of human monocytes/macrophages (Mo/Mac) that the uptake of nanoparticles by HIV-1-infected cells was about 30% higher after 7 days of culture and about 65% after 21 days than by uninfected cells. Similar results were observed with macrophages obtained from AIDS patients, depending on the state of disease. The uptake by these cells was also dependent on the surface coating of nanoparticles with different surfactants (Table 1). Interestingly, despite their possible stealth properties these surfactants did not reduce the *in vitro* uptake of the nanoparticles: poloxamer 338 had no influence and poloxamer 188 even increased the uptake (Schäfer et al, 1991; 1992).

The following investigation of the *in vitro* inhibition of the development of infection of the above human macrophage cultures by HIV-1 using the antiviral drugs azidothymidin (AZT) and dideoxycytidine (ddC), however, revealed no difference between free and nanoparticle-bound drugs (Bender et al, 1994): nanoparticle-bound AZT or ddC retained their activity but were not superior to free drug. The situation was totally different with the antiviral protease inhibitor saquinavir (Bender et al, 1996). Here a 10-fold increase in efficacy of nanoparticle-bound drug over free drug was observed. The difference in these results probably is due to the fact that AZT and ddC easily penetrate into cells whereas saquinavir does not, and therefore requires the mediation of cell uptake by the nanoparticles.

Table 1: Influence of surfactant on phagocytosis of PBCA nanoparticles by human MAC [reproduced from Schäfer et al., 1992]

	Phagocytosed nanoparticles (μg/ml ± SD)[a]		
	Uncoated	Pluronic F108	Pluronic F68
IFN-gamma[b]	8.6±2.0	9.5±1.8	13.0±2.9
Control	6.4±0.3	7.1±1.4	9.8±2.6

[a] Mean values of three parallel cultures are presented.

[b] The cells were stimulated with IFN-gamma for 24 hr (200 U/ml).

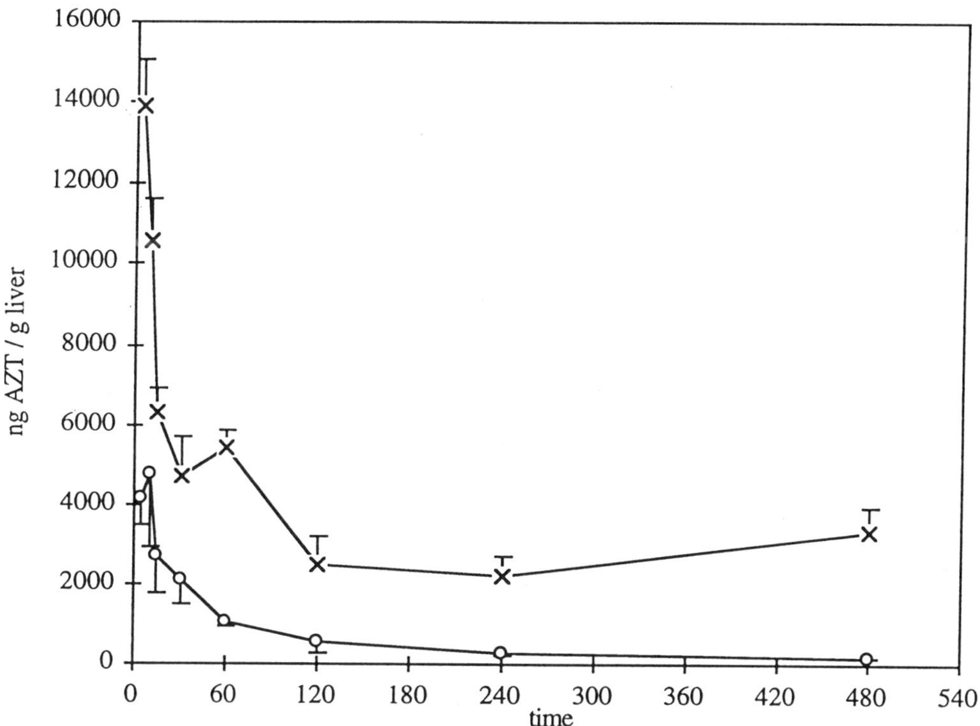

Figure 1. ^{14}C-AZT-label concentrations in the liver of rats after i.v. injection of an AZT solution o, or of AZT bound to poly(butyl cyanoacrylate) nanoparticles, x. (From Löbenberg and Kreuter, 1996; reproduced with kind permission of the copyright holder, Mary Ann Liebert, Inc. Lachmont, NY, USA).

However, it has to be considered that in tissue cultures a homologous cell population exists, which is totally different to the situation in the human or animal body: the body consists of a huge number of different cell types. Macrophages represent only a small percentage of the total number of cells. As a consequence, it was conceivable that even with AZT or ddC nanoparticles still may achieve a better targeting to these cells despite their inability to show a better antiviral efficacy of these drugs *in vitro* in the homologous cell cultures. Indeed that was observable: Löbenberg and Kreuter (1996) bound ^{14}C-labeled AZT to nanoparticles. After i.v. injection of nanoparticle-bound drug the concentrations of the AZT-label in the organs that were rich in macrophages, such as the liver (Fig. 1), were up to 18 times higher than with an AZT solution. Likewise, after oral administration the nanoparticle formulation delivered the AZT-label more efficiently to the RES than the aqueous solution. In addition, the blood concentration was significantly higher after oral administration of nanoparticles. Autoradiographic investigations using radioluminography supported the above results (Löbenberg et al, 1998): The radioactivity in the liver, lung, and spleen, organs with a high number of macrophages, was much higher than in other parts of the body with the nanoparticle formulation. The above organs showed a spotted appearance of the radioactivity that is typical for the accumulation in macrophages after i.v. as well as after oral administration of the drug bound to the particles. In contrast, after administration of the solution the radioactivity in the above organs was homogeneously distributed and much lower. These results show that the nanoparticles seemed to have reached their target and that non-stealth nanoparticles in some important cases are the delivery system of choice. Stealth properties in these situations would be counterproductive.

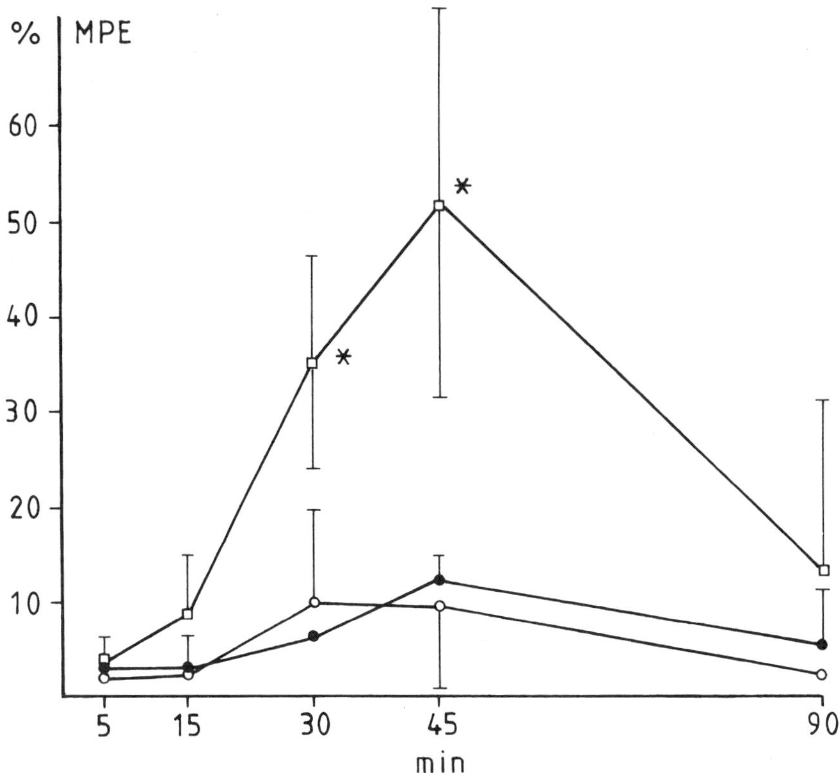

Figure 2. Maximal possible analgesic effect (MPE) determined by the tail-flick test after i.v. injection of one of the following preparations to mice: (○), 10 mg/kg dalargin in phosphate buffered saline, pH 7.2 (PBS); (●), a mixture of 10 mg/kg dalargin, poly(butyl cyanoacrylate) nanoparticles and polysorbate 80, mixed immediately before injection; (□), 7.5 mg/kg dalargin adsorbed to poly(butyl cyanoacrylate) nanoparticles overcoated with polysorbate 80. (From Kreuter et al., 1995; reproduced with kind permission of the copyright holder, Elsevier Science, Amsterdam, Netherlands).

BRAIN TARGETING

Another example is brain targeting. Tröster et al. (1990, 1992, a,b) already observed that besides the surfactants with stealth properties, such as poloxamer 338 and poloxamine 908 and 1508, other surfactants, for instance some of the polysorbates, may be as efficient or even more efficient in increasing the concentrations in organs that do not belong to the RES. Tröster et al. (1990) at that time assumed that the coating of the nanoparticles with the surfactants may have led to an accumulation of the nanoparticles in the brain blood vessels possibly by an increased adhesion to the blood vessel wall, but these authors did not believe that in the brain the particles were taken up by or transported across the endothelial cells lining these vessels. Later, however, Borchard et al, (1994) observed an increased uptake of surfactant-coated poly(methyl methacrylate) nanoparticles by bovine brain microvessel endothelial cell monolayers *in vitro*. The highest and fastest uptake (>300% compared to uncoated controls) was observed after coating with polysorbate 80. Poloxamer 407 also yielded a very high uptake that was, however, delayed by several hours. Poloxamers 184 and 188, polysorbate 80 and polyoxyethylen-23-laurylether showed an intermediate uptake enhancement (>100%) whereas poloxamer 338 and 908 only yielded a minimal and insignificant uptake enhancement.

These findings were later supported by *in vivo* experiments. These *in vivo* experiments with fluorescent polysorbate 80-coated fluorescence isothiocyanate dextrane 70000-labeled poly(butyl cyanoacrylate) nanoparticles in mice (Kreuter et al, 1995) indicated, using fluorescence and electron microscopy, that the particles were taken up by the endothelial cells lining the brain blood vessels and showed transport of fluorescence into the Punkinje cells of the brain 45 min after injection. No uptake occurred without polysorbate 80-coating. In addition, a strong analgesic effect exhibiting a maximum after 45 min was observed in mice by the tail-flick test following i.v. injection of the polysorbate 80-coated poly(butyl cyanoacrylate) nanoparticles that were previously loaded with dalargin (Fig. 2). Dalargin is a hexapeptide leu-enkephalin analogue with good blood stability that normally does not cross the blood-brain-barrier (BBB) after i.v. injection but possesses analgesic activity after direct intraventricular brain injection. Indeed, in the above experiments by Kreuter et al. (1995) and by Alyautdin et al. (1995) an intravenously injected dalargin solution as well as other controls showed no analgesia. The controls included empty nanoparticles, polysorbate 80 alone, dalargin plus polysorbate 80, dalargin bound to the nanoparticles but without polysorbate 80, and the simple mixture of all three components, dalargin, nanoparticles and polysorbate 80. In this latter mixture these three components were mixed immediately before injection, whereas in the analgetically active preparations dalargin was bound to the nanoparticles by incubation under stirring for 3 hours followed by the adsorption of polysorbate 80 by another 30 min. The analgesic effect achieved with this preparation was dose-dependent and it was totally blocked by pretreatment with naloxone, an opioid receptor antagonist, given i.v. 10 min before injection of the polysorbate 80 nanoparticle preparation. These experiments with the tail-flick test were later confirmed by Schröder and Sabel (1996) using the hot plate test.

The observed analgesia with dalargin bound to nanoparticles depended on the surfactant employed: only polysorbate 20, 40, 60, and 80 achieved an analgesic effect after overcoating of the dalargin nanoparticles and i.v. injection (Kreuter et al, 1997). Polysorbate 80 enabled the highest induction of analgesia at both doses of dalargin employed, i.e. 7 mg/kg and 10 mg/kg. All other surfactants including poloxamer 184, 188, 388, and 407, poloxamine 908, polyoxyethylen-23-laurylether, Cremophor®EL and RH 40, and polysorbate 81 and 85 exhibited no effect. This shows that the transport of dalargin cannot be due to a simple surfactant effect on the endothelial cells or the tight junctions since most of the surfactants were inactive. It might be argued that the polysorbates may interact much more strongly with the nanoparticles than the other surfactants. However, experiments in our laboratory have shown that this is not the case: On the contrary, poloxamine 908 interacts much more strongly with the nanoparticles and influences the body distribution at much lower concentrations than polysorbate 80 after intravenous injection. It has to be noted that none of the "stealth" surfactants, poloxamer 338 or poloxamine 908, exhibited any effect although both had achieved a high brain concentration of the nanoparticles in the study of Tröster et al. (1990). Also poloxamer 407 which was very active in the cell culture work of Borchard et al. (1994) showed no effect in the above-mentioned study of Kreuter et al. (1997). It is possible, that the high brain concentrations of the nanoparticles observed after coating with poloxamer 338 and poloxamine 908 by Tröster et al. (1990) are a reflection of their high blood concentrations but that these surfactants do not enable an interaction with the endothelial cells as already indicated by the *in vitro* results of Borchard et al, (1994). Poloxamer 407-coated particles may be able to interact with endothelial cells, but because of the very slow onset of this process with this surfactant poloxamer 407 may not have led to a relevant response *in vivo* or only after very extended times, i.e. after the observation interval.

The binding of other drugs to nanoparticles and overcoating with polysorbate 80 also led to pharmacologic responses. Loperamid too is an opioid that does not lead to central effects because it cannot cross the blood-brain-barrier and, therefore, is used perorally against diarrhea. This drug also became centrally active after binding to nanoparticles overcoated with polysorbate 80 (Alyautdin et al, 1997). In the same manner tubocurarine, a muscle relaxant,

Table 2: Doxorubicin concentrations [µg/g organ] in plasma after i.v. injection to rats as determined by HLPC

	10 min	1 h	2 h	4 h	6 h	8 h
DX	2.40±0.09	0.41±0.08	0.17±0.08	0.07±0.01	0.04±0.01	0.03±0.01
DX-80	2.70±0.12	0.37±0.04	0.15±0.02	0.05±0.01	0.05±0.01	0.03±0.00
DX-NP	1.68±0.07	0.21±0.03	0.10±0.04	0.08±0.01	0.09±0.01	0.07±0.00
DX-NP-80	3.15±0.42	0.89±0.10	0.21±0.07	0.09±0.01	0.09±0.01	0.06±0.08

DX = Doxorubicin [5 mg/kg] in saline
DX-80 = Doxorubicin [5 mg/kg] in 1% polysorbate 80/saline solution
DX-NP = Doxorubicin [5 mg/kg] bound to nanoparticles
DX-NP-80 = Doxorubicin [5 mg/kg] bound to nanoparticles and overcoated with 1% polysorbate 80

induced spikes in the EEC in brain perfusion studies (Alyautdin et al, 1998). Again the controls, uncoated tubocurarine-loaded nanoparticles or a mixture of the drug with polysorbate 80, were inactive in enabling a passage of this drug across the blood-brain-barrier and, consequently, the EEC remained normal.

Another example is doxorubicin. This drug was bound to poly(butyl cyanoacrylate) nanoparticles, and the blood plasma, liver, spleen, lung, heart, and kidney and brain concentrations were measured by the method of Alexandrova et al., (1986) after i.v. injection into rats at a dose of 5 mg doxorubicin/kg (unpublished data). No significant differences were found between a solution of doxorubicin in saline and a doxorubicin solution in saline containing 1% polysorbate 80. Coating of the doxorubicin-loaded nanoparticles with this surfactant enhanced the plasma concentrations significantly and led to prolonged circulation times (Table 2). Both nanoparticle preparations, i.e. doxorubicin bound to nanoparticles without or with coating with polysorbate 80, reduced the heart concentration very significantly (Table 3). This observation is in excellent agreement with earlier data by Couvreur and coworkers (Couvreur et al, 1982; 1986; Chiannilkulchai et al., 1990).

However, in the brain only the polysorbate 80-coated nanoparticles achieved detectable concentrations (Table 4). The brain concentrations appeared rather late, after about 2 to 4 hours, but they were very considerable (>6µg/g). This level is more than 60 times above the detection

Table 3: Doxorubicin concentrations [µg/g organ] in heart after i.v. injection to rats as determined by HPLC

	10 min	1 h	2 h	4 h	6 h	8 h
DX	0.74±0.06	4.27±0.20	6.57±0.55	4.00±0.37	3.30±0.24	1.10±0.05
DX-80	1.10±0.35	8.40±0.62	5.73±0.50	2.10±0.30	0.75±0.08	0.40±0.05
DX-NP	0.28±0.04	0.30±0.05	<0.1	<0.1	<0.1	<0.1
DX-NP-80	0.42±0.07	<0.1	<0.1	<0.1	<0.1	<0.1

DX = Doxorubicin [5 mg/kg] in saline
DX-80 = Doxorubicin [5 mg/kg] in 1% polysorbate 80/saline solution
DX-NP = Doxorubicin [5 mg/kg] bound to nanoparticles
DX-NP-80 = Doxorubicin [5 mg/kg] bound to nanoparticles and overcoated with 1% polysorbate 80

Table 4: Doxorubicin concentrations [µg/g organ] in brain after i.v. injection to rats as determined by HPLC

	10 min	1 h	2 h	4 h	6 h	8 h
DX	<0.1	<0.1	<0.1	<0.1	<0.1	<0.1
DX-80	<0.1	<0.1	<0.1	<0.1	<0.1	<0.1
DX-NP	<0.1	<0.1	<0.1	<0.1	<0.1	<0.1
DX-NP-80	<0.1	2.20±0.49	6.13±0.27	6.20±0.80	2.67±0.18	1.10±0.07

DX = Doxorubicin [5 mg/kg] in saline
DX-80 = Doxorubicin [5 mg/kg] in 1% polysorbate 80/saline solution
DX-NP = Doxorubicin [5 mg/kg] bound to nanoparticles
DX-NP-80 = Doxorubicin [5 mg/kg] bound to nanoparticles and overcoated with 1% polysorbate 80

limit whereas all other preparations were below the detection limit at all times. This result shows that this delivery system, polysorbate 80-coated nanoparticles, holds great promise for the delivery of antitumor drugs to the brain and may open up totally new possibilities in the treatment of brain cancers.

As already mentioned above, the mechanism of the delivery seems to be endocytotic uptake of the drug-loaded nanoparticles by the endothelial cells lining the blood vessels of the brain. Endocytosis seems to be mediated by certain substances that are adsorbed from the blood after injection of the surfactant-coated particles (Kreuter, 1983). Accordingly, the adsorption pattern of serum components after incubation of nanoparticles in serum was highly surfactant specific (Blunk et al, 1993 a, b). No increased uptake of inulin was observed when the inulin was injected shortly after injection of the nanoparticles (unpublished). This indicates that the so-called inulin spaces were not significantly increased and that the tight junctions of the brain blood vessel endothelium were not opened.

Nevertheless, besides endocytosis another possibility exists for the increased brain uptake of the above drugs, namely inactivation of the P-glycoprotein efflux pump. This glycoprotein is present in the brain endothelial cells (Cordon-Cardo et al, 1989) and is, for instance, also responsible for the multidrug resistance which represents a major obstacle to cancer chemotherapy (Woodcock et al, 1990, 1992; Zordan-Nudo et al, 1993). Surfactants including polysorbate 80 were shown to inhibit this efflux system and to reverse multidrug resistance (Nerurkar et al., 1996). However, as mentioned above, addition of polysorbate 80 to the drug solutions was totally inefficient. On the other hand, it is conceivable that this surfactant and, of course, the other active polysorbates, may be delivered more efficiently to the brain endothelial cells if they were adsorbed to the nanoparticle surface. This could explain why polysorbate-coated nanoparticles provided high brain concentrations in contrast to simple surfactant solutions. Nevertheless, we believe that the latter mechanism, i.e. blockage of the efflux system by polysorbate, may contribute to the higher brain/drug uptake mechanism but that induction of the endocytotic uptake may play an equal or much larger role. The reason for this assumption is the observation that in contrast to other organs significant brain concentrations were obtained only after some time. Such a delayed response is typical for a delivery mediated by the polysorbate-coated nanoparticles (Alyautdin et al, 1995; Kreuter et al, 1995) and seems to be a reflection of the time-consuming process of endocytosis.

CONCLUSIONS

As mentioned above, the stealth surfactants, poloxamer 338 and poloxamine 908, were not able to mediate the transport of dalargin across the blood-brain-barrier. They were also useless for the targeting to macrophages in AIDS therapy. Therefore it has to be kept in mind, that for each type of disease different surface properties of particulate carriers are required to obtain an optimal therapeutic result.

REFERENCES

Alexandrova, L.G., Rubasheva, A.M., Zbarsky, V.B., Salamova., E.I. and Brazhnikova, M.G., 1986, Quantitative assay of doxorubicin by HPLC technique, *Antibiot.Med.Biotechnol.* 11:851.

Alyautdin, R., Gothier, D., Petrov, V., Kharkevich, D. and Kreuter, J., 1995, Analgesic activity of the hexapeptide dalargin adsorbed on the surface of polysorbate 80-coated poly(butyl cyanoacrylate) nanoparticles, *Europ.J.Pharm.Biopharm.* 41:44.

Alyautdin, R. N., Petrov, V. E., Langer, K., Berthold, A., Kharkevich, D. A., and Kreuter, J., 1997, Delivery of loperamide across the blood-brain barrier with polysorbate 80-coated polybutylcyanoacrylate nanoparticles, *Pharm.Res.* 14:325.

Alyautdin, R. N., Tezikov, E. B., Ramge, P. Kharkevich, D. A., Begley, D. J., and Kreuter, J., 1998, Significant entry of tubocurarine into the brain of rats by adsorption to polysorbate 80-coated polybutyl-cyanoacrylate nanoparticles: an in situ brain perfusion study, *J.Microencapsul.* 15:67.

Bender, A.R., von Briesen, H., Kreuter, J., Duncan, I.B., and Rübsamen-Waigmann, H., 1996, Efficiency of nanoparticles as a carrier system for antiviral agents in human immunodeficiency virus-infected human monocytes/macrophages in vitro, *Antimicrob.Agents Chemother.* 40:1467.

Bender, A., Schäfer, V., Steffan, A. M., Royer, C., Kreuter, J., Rübsamen-Waigmann, H., and Briesen, H., 1994, Inhibition of HIV in vitro by antiviral drug-targeting using nanoparticles, *Res.Virol.* 145:215.

Blunk, D.F., Hochstrasser, Müller, B.W., and Müller, R.H., 1993a, Differential adsorption: Effect of plasma protein adsorption patterns on organ distribution of colloidal drug carriers, *Proceed.Intern. Symp. Control.Rel.Bioact.Mater.* 20:256.

Blunk, D.F., Hochstrasser, Sanchez, J.-C., Müller, B.W., and Müller, R.H., 1993 b, Colloidal carriers for intravenous drug targeting: Plasma protein adsorption patterns on surface-modified latex particles evaluated by two-dimensional polyacrylamide gel electrophoresis, *Electrophoresis.* 14:1382.

Borchard, G., Audus, K. L., Shi, F., and Kreuter, J., 1994, Uptake of surfactant-coated poly(methyl-methacrylate)-nanoparticles by bovine brain microvessel endothelial cell monolayers, *Int.J.Pharm.* 110:29.

Chiannilkulchai, N., Ammoury, N., Caillou, B., Devissaguet, J.P., and Couvreur, P., 1990, Hepatic tissue distribution of doxorubicin-loaded nanoparticles after i.v. administration in reticulosarcoma M 5076 metastases-bearing mice, *CancerChemother.Pharmacol.* 26:122.

Cordon-Cardo, C., O'Brien, J.P., Casals, D., Rittmann-Grauer, L., Biedler, J.L., Melamed, M.R., and Bertino, J.R., 1989, Multidrug resistance gene (P-glycoprotein) is expressed by endothelial cells at blood brain barrier sites, *Proc.Natl.Acad.Sci. USA.* 86:695.

Couvreur, P., Kante, B., Grislain, L., Roland, M., and Speiser, P., 1982, Toxicity of polyalkylcyanoacrylate nanoparticles II: Doxorubicin-loaded nanoparticles, *J.Pharm.Sci.* 71:790.

Couvreur, P., Grislain, L., Lenaerts, V., Brasseur, F., Guiot, P., and Biernacki, A., 1986, Biodegradable polymeric nanoparticles as drug carrier for antitumor agents; in: *Polymeric Nanoparticles and Microspheres,* P. Guiot and P. Couvreur eds., CRC Press, Boca Raton, pp. 27-93.

Esser, R., von Briesen, H., Brugger, W., Ceska, M., Glienke, W. Müller, S., Rehm, A., Rübsamen-Waigmann, H., and Andreesen, R., 1991, Secretory repertoire of HIV-infected human monocytes/macrophages, *Pathobiology* 59:219.

Gartner, S., Markovitis, P., Markovitz, D.M., Kaplan, M.H., Gallo, R.C., and Popovic, M., 1986, The role of mononuclear phagocytes in HTLV-III/LAV infection, *Science.* 223:215.

Gendelman, H. E., Lipton, S.A., Tardieu, M., Bukrinsky, M.I., and Nottet, H.S.L.M., 1994, The neuropathogenesis of HIV-1 infection, *J. Leukocyte Biol.* 56:389.

Kreuter, J., 1983, Evaluation of nanoparticles as drug-delivery systems III: Materials, stability, toxicity, possibilities of targeting, and use, *Pharm.Acta.Helv.* 58:242.

Kreuter, J., Alyautdin, R. N., Kharkevich, D. A., and Ivanov, A. A. , 1995, Passage of peptides through the blood-brain barrier with colloidal polymer particles (nanoparticles), *Brain Res.* 674:171.

Kreuter, J., Petrov, V. E., Kharkevich, D. A., and Alyautdin, R. N., 1997, Influence of the type of surfactant on the analgesic effects induced by the peptide dalargin after its delivery across the blood-brain

barrier using surfactant-coated nanoparticles, *J. Controlled Rel.* 49:81.
Kühnel, H., von Briesen, H., Dietrich, U., Adamski, M., Mix, D., Biesert, L., Kreutz, R., Immelmann, A., Henco, K., Meichsner, C., Andreesen, R., Gelderblom, H., and Rübsamen-Waigmann, H., 1989, Molecular cloning of two West African human immunodeficiency virus type 2 isolates that replicate well in macrophages: a Gambian isolate, from a patient with neurologic acquired immunodeficiency syndrome, and a highly divergent Ghanian isolate, *Proc.Natl.Acad.Sci. USA.* 86:2383
Levy, J.A., 1993, Pathogenesis of human immunodeficiency virus infection, *Microbiol.Rev.* 57:183.
Löbenberg, R., and Kreuter, J., 1996. Macrophage targeting of azidothymidine: A promising strategy for AIDS therapy, *AIDS Res. Human Retrovir.* 12:1709.
Löbenberg, R., Maas, J., and Kreuter, J., 1998, Improved body distribution of ^{14}C-labelled AZT bound to nanoparticles in rats determined by radioluminography, *J. Drug Target.* (in press).
McGann, K.A., Collmann, R., Kolson, D.L., Gonzalez-Scarano, F., Coukos, G., Coutifaris, C., Strauss, J.F., and Nathanson, N., 1994, Human immunodeficiency virus type 1 causes productive infection of macrophages in primary placental cell cultures. *J.Infect.Dis.* 169:746.
Milman, G., and Sharma, O., 1994, Mechanisms of HIV/SIV mucosal transmission, *AIDS Res. Hum. Retroviruses.* 10:1305.
Nerurkar, M.M., Burto, P.S., and Borchardt, R.T., 1996, The use of surfactants to enhance the permeability of peptides through Caco-2 cells by inhibition of an apically polarized efflux system, *Pharm.Res.* 13:528.
Nicholson, J.K., Cross, G.D., Callaway, C.S., and McDougal, J.S., 1986, *In vitro* infection of human T lymphotropic virus type III./Lymphadenophaty-associated virus (HTLV-III/LAC), *J.Immunol.*, 137:323.
Schäfer, V.M., von Briesen, H., Tröster, S.D., Andreesen, R., Kreuter, J., and Rübsamen-Waigmann, H., 1991, How to get antiviral drugs into HIV-infected cells? Phagocytosis of nanoparticles by human macrophages, *Microbiologist* 2:117.
Schäfer, V., von Briesen, H., Andreesen, R., Steffan, A.-M., Royer, C., Tröster, S., Kreuter, J., and Rübsamen-Waigmann, H., 1992, Phagocytosis of nanoparticles by human immunodeficiency virus (HIV)-infected macrophages: A possibility for antivral drug targeting, *Pharm.Res.* 9:541.
Schröder, U., and Sabel, B.A., 1996, Nanoparticles, a drug carrier system to pass the blood-brain-barrier, permit central analgesic effects of i.v. dalargin injections, *Brain Res.* 710:121.
Tröster, S. D., Müller, U., and Kreuter, J., 1990, Modification of the body distribution of poly(methyl-methacrylate) nanoparticles in rats by coating with surfactants, *Int.J.Pharm.* 61:85.
Tröster, S.D., and Kreuter, J., 1992, Influence of the surface properties of low contact angle surfactants on the body distribution of ^{14}C-poly(methyl methacrylate) nanoparticles, *J.Microencapsul.* 9:19.
Tröster, S.D., Wallis, K.H., Müller, R.H., and Kreuter, J., 1992, Correlation of the surface hydrophobicity of ^{14}C-poly(methyl methacrylate) nanoparticles to their body distribution, *J. Controlled Rel.* 20:247.
Van't-Wout, A.B., Kootstra, N.A., Mulder-Kampinga, G.A., Albrecht-van-Lent, N., Scherpbier, H.J., Veenstra, J., Boer, K., Coutinho, R.A., Miedema, F., and Schuitemaker, H., 1994, Macrophage-tropic variants initiate human immunodeficiency virus type 1 infection after sexual, parenteral, and vertical transmission, *J.Clin.Invest.* 94:2060.
Von Briesen, H., Andreesen, R., Esser, R., Brugger, W., Meichsner, C., Becker, K., and Rübsamen-Waigmann, H., 1990, Infection of monocytes/macrophages by HIV in vitro, *Rés. Virol.* 141:225.
Weiss, R.A., 1993, How does HIV cause AIDS?, *Science* 260:1273.
Woodcock, D.M., Jefferson, S., Linsenmeyer, M.E., Crowther, P.J., Chojnowski, G.M., Williams, B., and Bertoncello, I., 1990, Reversal of multidrug resistance phenotype with Cremophor EL, a common vehicle for water-insoluble vitamins and drugs, *Cancer Res.* 50: 4199.
Woodcock, D.M., Linsenmeyer, M.E., Chojnowski, G., Kriegler, A.B., Nink, V., Webster, L.K., and Sawyer, W.H., 1992, Reversal of multidrug resistance by surfactants, *Br. J. Cancer* 66:62.
Zordan-Nudo, T., Ling, V., Liv, Z., and Georges, E., 1993, Effects of nonionic detergents on P-glycoprotein drug binding and reversal multidrug resistance, *Cancer Res.* 53:5994.

RECENT DEVELOPMENTS AND LIMITATIONS OF POLOXAMINE-COATED LONG-CIRCULATING PARTICLES IN EXPERIMENTAL DRUG DELIVERY

S. M. Moghimi

Department of Pharmacy, University of Brighton,
Brighton BN2 4GJ, UK

INTRODUCTION

There has been growing interest in the engineering of colloidal carrier systems that upon intravenous administration avoid rapid recognition by scavenger cells of the reticuloendothelial system and adequately remain in the blood. There are several reasons as to why the search for such 'phagocyte-resistant' or long-circulating particles is so extensive. First, long-circulating drug carriers may act as a 'sustained release system' in the blood and slowly discharge their entrapped materials for better treatment and diagnosis. Secondly, long-circulating drug carriers may passively accumulate in pathological sites with affected vasculature (e.g., selected tumours, rheumatoid arthritis, infections) and improve or enhance treatment or diagnosis in those areas (Bakker-Woudenberg et al., 1993; Wu et al., 1993; Huang et al., 1994; Yuan et al., 1994). Thirdly, grafting or covalent attachment of homing molecules such as antibodies, peptides that mimic ligands for αvß3 and αvß5 integrins, oligosaccharides that mimic ligands for P-selectin (e.g., synthetic sialyl-Lewis X and small heparin chains containing four or more monosaccharide residues), folate, etc. to long-circulating carries can dramatically improve the targeting precision to sites beyond the reticuloendothelial system (e.g., vascular and tumour endothelial cells, circulating blood cells, circulating toxic materials) (Mulligan et al., 1993; Nelson et al., 1993; Lee and Low, 1994; Bloemen et al., 1995; Gabizon, 1995; Moghimi, 1995a,c; Torchilin, 1995; Lee and Huang, 1996; Huang et al., 1997; Pasqualini et al., 1997).

During the late sixties, in a pioneering experiment Geyer (1967) demonstrated that intravenously injected lipid emulsions prepared with high molecular weight members of polyoxyethylene (POE)/polyoxypropylene (POP) polymer nonionic surfactants (e.g., poloxamer 238 and 338) as emulsifiers remained in the blood for relatively long periods. Later, it was suggested that these surfactants in some way prevent lipid particles from sticking to the blood vessel endothelium or recognition by macrophages (Jeppsson and Rossner, 1975). Others also suggested that such polymers could interfere with the lipoprotein-lipase activity (Hart and

Payne, 1971; Jeppsson and Rossner 1975). It has now been realized that POP/POE surfactants can adsorb on to the surface of hydrophobic particles, a process which increases particle stability in the biological environment and affects the mode of blood protein interaction with particle surface (Troster et al., 1990; Moghimi , 1995a). Based on these observations, the concept of steric-stabilization (Naper, 1983; Needham et al., 1992; Torchilin et al., 1994) has now been used routinely to engineer an array of drug carrier systems which remain relatively "invisible" to resident macrophages of the reticuloendothelial system, particularly the Kupffer cells, and exhibit prolonged circulation time.

COATING OF NANOBEADS WITH POE/POP COPOLYMERS

The POE/POP block polymer nonionic surfactants are surface active agents prepared by the sequential addition of two or more alkylene oxides to a low molecular weight water-soluble organic compound containing one or more active hydrogen atoms (Schmolka, 1977). Two series of commercially available block polymer surface-active agents -the poloxamers and the poloxamines- are of particluar interest. The poloxamers are triblock copolymers and consist of a segment of hydrophobic POP flanked by two polymers of hydrophilic POE. A slightly different structure is exhibited by the poloxamines, which are prepared from the tetra-functional initiator ethylenediamine. They resemble the poloxamers in having the same sequential order of addition of alkylene oxides but they have four alkylene oxide chains rather than two. By varying the overall size of the molecule and the ratios of POE to POP, a large number of poloxamer and poloxamine compounds have been produced that span a broad range of surface-active properties.

Block polymers of the poloxamer and the poloxamine types adsorb strongly to the surface of hydrophobic particles via their hydrophobic POP center-block (Moghimi et al., 1993d; Moghimi 1995a). This mode of adsorption leaves the hydrophilic POE side-arms in a mobile state as they extend outward from the particle surface and provide stability to the particle suspension by a repulsion effect through a steric mechanism of stabilization involving both enthalpic and entropic contribution (Moghimi, 1995a). Generally, the strength of polymer adsorption and the resultant polymer conformation is dependent on the proportion and the size of the POP and POE segments of the co-polymers as well as other forces, such as the initial bead surface charge, hydrogen bonding between the polyoxyethylene ether groups and bead surface hydroxyl or sulfonic acid groups, hydrophobic forces among the polymer chains, polymer-solvent interaction, etc. Indeed, with the help of field-flow fractionation, electron spin resonance techniques, as well as conventional labelling techniques, Caldwell and her associates have accomplished the detailed analytical characterization of the adsorption complexes formed between such block polymers and polystyrene beads of different sizes (Li and Caldwell, 1994). These studies further indicated the importance of the surface curvature of beads on polymer chain mobility and conformation.

Among various members of poloxamer and poloxamine series tested, poloxamine-908 appears to be the material of choice for biological evaluation. This polymer has a molecular weight of 22.5 kDa and contains four chains each of POE and POP, each POE chain consisting of 122 ethylene oxide units and each POP chain consisting of 17 propylene oxide units. In our hands, poloxamine-908 effectively adsorbs onto the surface of 60 nm polystyrene beads (Moghimi *et al.,* 1993d). For example, we have measured an electrophoretic mobility value of $-2.70 \pm 0.11\ \mu ms^{-1}/Vcm^{-1}$ for uncoated beads as compared to $-0.64 \pm 0.05\ \mu ms^{-1}/Vcm^{-1}$ for poloxamine-coated particles. Furthermore, the stability of the poloxamine coating was also confirmed by the electrolyte flocculation test.

THE BIOLOGICAL FATE OF POLOXAMINE-COATED NANOBEADS

Following intravenous administration, poloxamine-908 coated radiolabelled beads (60 nm) are not accumulated in the liver but stay in circulation for prolonged periods (Moghimi et al., 1991,1993a,d). Similarly, subcutaneously administered poloxamine-908 coated beads are rapidly drained across tissue lymph interface in dermal lymphatic capillaries, escape clearance by both type I and type II macrophages in regional nodes, reach the systemic circulation and remain in the blood (Moghimi et al., 1994; Moghimi and Rajabi-Siahboomi, 1996). The long-circulation time of polymer engineered beads is believed to be due to the surface steric-barrier imposed by the mobile POE segments of poloxamine molecules. For instance, the POE chains may minimize blood protein adsorption and most importantly it may suppress the natural blood opsonization processes that are necessary for the recognition of particles by tissue macrophages (e.g., Kupffer cells). A number of investigators have tried to study these possibilities following incubation of naked and polymer-coated particles with serum or plasma. For instance Illum and co-workers (Norman et al., 1992,1993) have illustrated that when polystyrene particles are exposed to serum or plasma, they rapidly acquire a coating of proteinaceous molecules. This protein binding, which was demonstrated by separating beads from serum or plasma incubations and then analyzing the bead-protein complexes by sodium dodecylsulfate-polyacrylamide gel electrophoresis (SDS-PAGE), differed considerably in amount and in pattern depending on the polymer coating. These studies, however, may be misleading. First, it is possible that other proteinaceous molecules of functional significance bind loosely to the particle surface and are lost in the routine separation procedures (ultracentrifugation and repeated washes, size exclusion chromatography, etc.) prior to SDS-PAGE analysis. Secondly, the procedures used to desorb proteins from the bead surface fails to release all the adsorbed proteinaceous materials; in some cases not more than 20-30% of total adsorbed proteins are released. With the aid of SDS-PAGE analysis these investigators (Norman et al., 1993) also tried to explain why certain sterically stabilized particles (poloxamer-407 coated latex beads) avoid Kupffer cells but are cleared exclusively by bone marrow sinusoidal endothelial cells in rabbit. Despite the fact that the clearance of poloxamer-407 coated particles by bone marrow

Table 1. D-Galactose-specific, lectin-mediated agglutination of gold particles

Procedures	Agglutination result
Rat pFn-Au-17	
+ Antiserum (1:100)	Positive
+ RCA120 (200 μg/ml)	Positive
+ RCA120 + D-Galactose (0.1M)	Negative
Poloxamine 908-Au-17	
+ Antiserum	Negative
+ RCA	Negative

RCA: Ricinus communis agglutinin A (a D-galactose-specific plant lectin)
Some agglutination did occur when poloxamine-904 was used in place of poloxamine-908 to stabilize gold particles; poloxamine-904 is approximately a 3200 Da co-polymer that contains four chains each of POE and POP, each POE chain consisting of 15 ethylene oxide units and each POP chain consisting of 17 propylene oxide units.

Table 2. Uptake of gold particles by freshly isolated rat Kupffer cells

Particle choice	Added component	Uptakeresults
Lactosylated BSA-Au-17		+++
Plasma coated-Au-17		+++
pFn-Au-17		+++
pFn-Au-17	Lactosylated BSA 1μM	+
Plasma coated-Au-17	Lactosylated BSA 1μM	+
Poloxamine 908-Au-17		No uptake
Poloxamine 908-Au-17	Rat plasma	No uptake

is exclusive to this species (Moghimi, 1995c), Norman et al. (1993) exposed poloxamer-coated particles to human serum, but not to rabbit serum, prior to SDS-PAGE analysis in order to identify this putative "rabbit bone marrow endothelial cell-specific" opsonin! A number of other studies have repeatedly emphasized that, in mice, certain "Stealth" or long-circulating particles avoid recognition by the liver macrophages as they inhibit complement activation. This suggestion is rather confusing as resident mouse Kupffer cells do not express the complement receptor (CR) 3 (Lepay et al., 1985; Lee et al., 1986; Gordon et al., 1992; Moghimi and Patel, 1998). In addition, mouse Kupffer cells do not express other complement receptors equivalent to human CR1 and CR4. However, CR3 is expressed by mouse spleen marginal zone macrophages.

Therefore, to test the validity of the steric barrier hypothesis in suppressing or inhibiting particle opsonization in serum or plasma, colloidal gold particles of 17 nm in diameter were prepared as the phagocytic substrate for freshly isolated rat Kupffer cells. Pre-opsonization of gold particles in autologous rat plasma results in immobilization of fibronectin on their surface, a process which stimulates the uptake of gold particles by isolated Kupffer cells via a peripheral membrane-bound C-reactive protein (recognizing the glycan portion of fibronectin) and an integrin (recognizing the tri-peptide sequence arginine-glycine-aspartate on the carboxyl-terminal cell adhesion domain of fibronectin) (Kolb-Bachofen and Abel, 1991). Gold particles can be successfully stabilized with poloxamine-908 (Moghimi et al., 1993d). Coating of gold particles with poloxamine-908 prior to plasma opsonization prevented the adsorption of fibronectin onto their surface as demonstrated by the agglutination test using antifibronectin antibodies and the lectin *Ricinus communis* agglutinin (a D-galactose-specific plant lectin) (Table 1). Simultaneously, Kupffer cells failed to recognize poloxamine-908 coated particles before and prior to plasma opsonization (Table 2) (Moghimi et al., 1993d). Therefore, in addition to preventing or suppressing opsonization (at least with respect to fibronectin), the steric POE barriers can also suppress non-specific interaction of particles with the macrophage cell surface receptors. Another plausible hypothesis is that the presence of a suppressive heat-stable 100kD proteinaceous component in the blood (a dysopsonin) exerts some regulatory role in the clearance of poloxamine-coated particles by Kupffer cells (Moghimi et al., 1993d). For instance, the surface-adsorbed poloxamines may thwart the immune system by binding with dysopsonin, thereby suppressing opsonization or reducing the amount of particle-bound opsonins.

In relation to the above mechanisms, it should also be mentioned that the concept of surface coating with hydrophilic polymers is not restricted to the design and engineering of long-circulating and target-specific drug carriers and diagnostic agents; other applications include coating of cells and implants. For instance coating of red blood cells with polyethyleneglycol (PEG) may enable patients to receive the "incorrect" type of blood

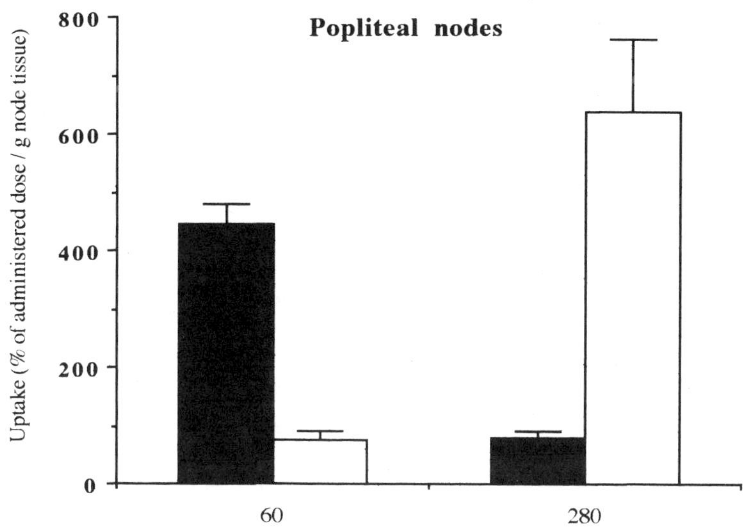

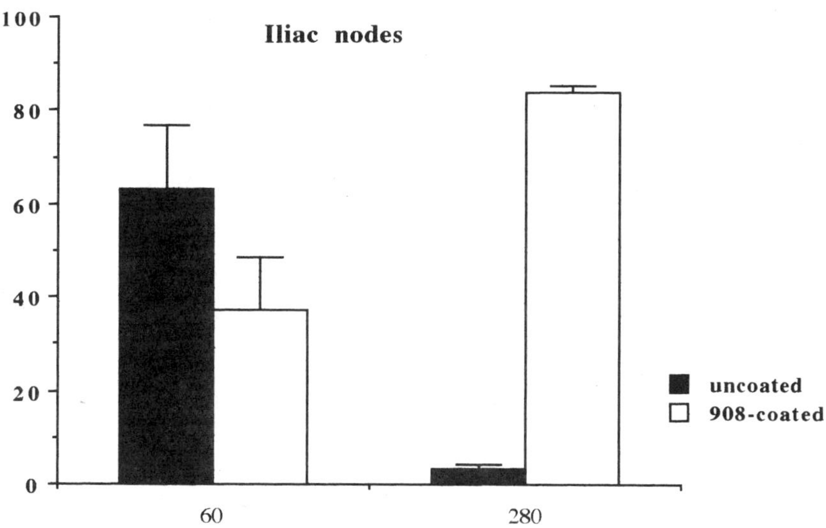

Fig. 1 - Lymphatic distribution of uncoated and poloxamine-908 coated radiolabelled polystyrene beads 2h after interstitial administration into rat footpads. Results are presented as a percentage of the administered dose ± SEM of 3 rats.

transfusion when the appropriate blood is in short supply as PEG may stop antibodies recognising and rejecting erythrocytes (Scott et al., 1997). Such manipulations could even further increase the life span of drug-loaded erythrocytes (resealed erythrocytes).

THE EFFECT OF PARTICLE SIZE ON THE BIOLOGICAL FATE OF POLOXAMINE-COATED BEADS

When the size of the poloxamine-908 coated polystyrene beads is increased to 150 nm or greater, hepatic extraction slightly increases but the splenic clearance becomes dominant (Moghimi et al., 1991, 1993b,c, 1995b). For example, the rat spleen can remove approximately 50% of the dose of poloxamine-908 coated spheres of 220 nm within a few hours after intravenous administration. The mechanism for the splenic clearance of poloxamine-coated beads is initially mechanical filtration and appears to be at the level of interendothelial cell slits (venous sinuses). This process, which is a function of both particle characteristics (deformability, shape, size) and the splenic blood flow pathways, is discussed in detail elsewhere (Moghimi et al., 1993c,1995b).

The filtered poloxamine-908 coated beads (220 nm) are eventually phagocytosed by macrophages of the red-pulp despite the presence of the initial steric POE segments on the particle surface (Moghimi et al., 1993c). Therefore, the steric barrier of the POE on the surface of larger beads cannot be as efficient as the corresponding chains on the surface of smaller particles because they are removed more efficiently by macrophages. This statement is further supported by the observation that macrophages of regional lymph nodes can also phagocytose drained large poloxamine-coated particles following interstitial administration. Therefore, the size-dependent interaction of poloxamine-coated beads with macrophages may be the result of differences in surface characteristics of nanobeads, such as the surface conformational clouds of POP/POE chains, the flexibility and hydration of the surface hydrophilic POE segment in biological fluids, etc. It has also been suggested that adsorption of POP/POE copolymers on larger particles is relatively weaker than that on smaller particles and, hence, the adsorbed polymer can be partially displaced, thus decreasing the steric shielding at particular patches on the surface of large particles (Lee et al., 1989; Li and Caldwell, 1994). This process may occur efficiently while particles are adhered to the reticular meshwork of the red-pulp or trapped at the sites of interendothelial cell slits in the rat spleen. Studies with dually labeled substrates (radiolabelled beads and radiolabelled poloxamine) are necessary to resolve this matter.

CONVERSION OF 'PHAGOCYTE-PRONE' NANOBEADS TO LONG-CIRCULATING PARTICLES IN VIVO

Recently, I demonstrated that intravenously injected uncoated small (60 nm) and large (250 nm) size model polystyrene particles (which are cleared rapidly from the blood by macrophages of the reticuloendothelial system) can be converted to long-circulating and splenotropic particles *in vivo*, respectively, if such particles are injected shortly (up to 3 h) after an appropriate dose (20 mg / 150 g body weight in rats) of the block polymer non-ionic surfactant, poloxamine-908 (Moghimi, 1997). The altered biodistribution profile of beads was apparently independent of the 'hepatic-blockade' concept which could have been caused by the administered poloxamine-908. Detailed experimental evidence indicated that small and large size polystyrene beads have aquired a coating of poloxamine 908 and/or poloxamine-protein complexes in vivo; this event could explain their altered body distribution (Moghimi, 1997). These unusual findings have raised questions as to the mode of interaction of poloxamine molecules with plasma proteins *in vivo*.

PROBLEMS WITH REPEATED ADMINISTRATION OF POLOXAMINE-COATED NANOBEADS

From experimental and clinical points of view it is often necessary to administer repeated doses of poloxamine engineered vehicles over a period of days. Accordingly, experiments were conducted to study the pharmacokinetics and tissue distribution of such particles after repeated intravenous dosing.

It was found that a single intravenous administration of poloxamine-908 coated particles (60 nm) into rats can dramatically affect the circulation half-life and body distribution of a second subsequent dose (Moghimi and Gray, 1997). The degree of this alteration was dependent on the interval between the two doses. For example, when the interval between the two doses was 1 or 2 days, poloxamine-coated beads behaved as usual and remained in the blood. However, when the second dose of poloxamine-coated particles was given 3 days after the first dose, particles were cleared rapidly from the blood by Kupffer cells and splenic red-pulp macrophages. On the other hand, when the second dose was given at day 14, poloxamine-coated beads again remained "invisible" to macrophages of liver and spleen. The failure of liver and spleen to clear poloxamine-coated particles at day 14 was not associated with macrophage depletion or defects in phagocytosis. Subsequently, the coating polymer (poloxamine-908) was found to trigger bead clearance by liver and spleen, since a single intravenous dose of an endotoxin-free solution of poloxamine (levels as low as 3.5 mg polymer per kg body weight) 3 days before administration of long-circulatory particles induced similar effects. Again, when the interval between the two injections was 2 weeks, poloxamine-coated particles exhibited long circulation half-life. This cycle could be repeated following intravenous administration of a second poloxamine dose 2 weeks after the first poloxamine injection. The mode of particle recognition (poloxamine-coated beads) by resident tissue macrophages was also found to be independent of blood opsonization processes.

The surfactants of POP/POE series are known to exhibit a wide spectrum of immunological activities (Hunter et al., 1981; Howerton et al., 1990; Ingram et al., 1992). For example, a report has emphasized the ability of such surfactants to activate the phagocytic activity of neutrophils (Ingram et al., 1992). Therefore, activated neutrophils, upon homing to the liver and spleen, may be likely candidates to clear poloxamine-coated beads from the blood. In a series of experiments we pre-labelled the resident macrophage population in the liver and spleen prior to the administration or poloxamine-908 and long-circulating beads (poloxamine-coated beads) with fluorescent latex beads in order to discriminate between the resident and recruited cells from distant locations. These studies together with immunohistochemical observations indicated that in poloxamine-treated animals only $ED2^{+}$ resident periportal and mid-zonal Kupffer cells and certain resident sub-populations of spleen macrophages could clear poloxamine-coated beads from the blood (Moghimi and Gray, 1997). Elsewhere, we demonstrated that zymosan-stimulated Kupffer cells can effectively phagocytose long-circulating beads by a process independent of blood opsonization (Moghimi et al., 1993a). This effect of zymosan was attributed to changes in macrophage cell surface hydrophobicity as well as changes in the mobility of certain plasmalemma receptors. Interestingly, poloxamine-908 is known to interact with macrophage plasma membrane via its POP segments (Watrous-Peltier et al., 1992). This interaction could, perhaps, enhance the fluidity of cell membrane and as a result expose cryptic receptors or enhance the mobility of certain plasmalemma receptors leading to the recognition of some structural determinants of long-circulating beads *in vivo*. Alternatively, the interaction of poloxamine with other cells in the vicinity of tissue macrophages may lead to the release of a lymphokine that upon acting on macrophages, can trigger phagocytosis (Vogel and Rosenstreich, 1979; Griffin and Mullinax, 1981). We also noted that in poloxamine-treated animals, the blood half-life of phagocyte-prone particles

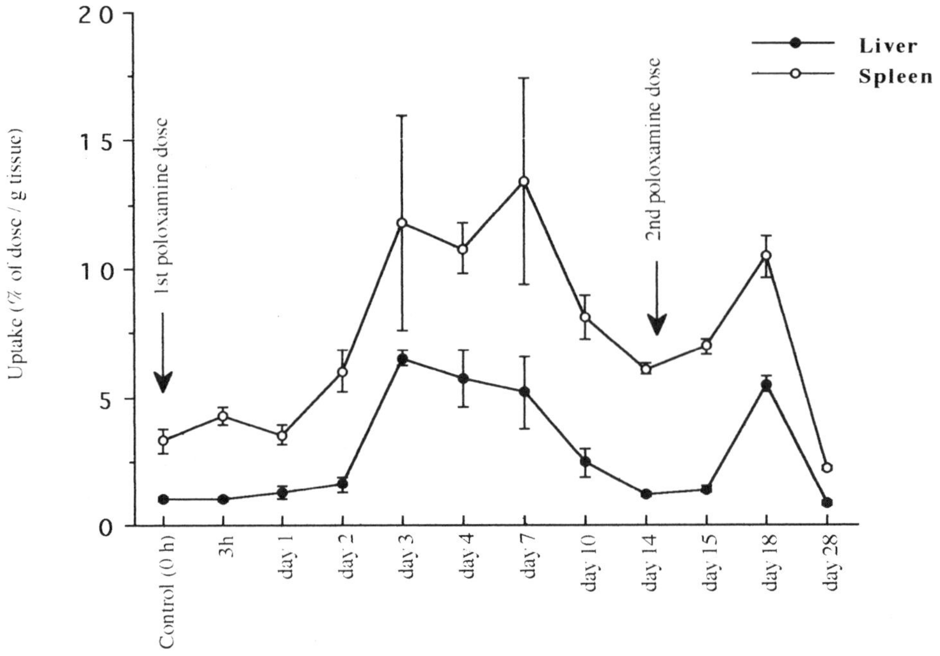

Figure 2. Kinetics of organ distribution of poloxamine-908 coated beads (60 nm) following intravenous administration of free poloxamine-908 (7.0 mg poloxamine per kg body weight) into male Wistar rats. Poloxamine was injected at days 0 and 14.

(uncoated latex beads) was also reduced dramatically (Moghimi and Gray, 1997). Detailed biochemical and immunological mechanisms responsible for these observed effects are currently under investigation.

Experiments were also performed to assess the capacity of splenic uptake mechanism(s) of splenotropic particles from the blood in the rat model. Here, rats received daily intravenous administration of splenotropic beads (220 nm) for 4 days. Control animals were injected with saline. On the fifth day all animals were injected with either radiolabelled splenotropic or long-circulating particles. Predosing dramatically decreased the splenic uptake of the test splenotropic beads but had no effect on the uptake of long-circulating nanobeads. The failure of the spleen to take up the test splenotropic particles was not associated with an increased circulatory level of beads. Surprisingly, both small (long-circulating) and large (splenotropic) beads were sequestered by Kuppfer cells. Here, recognition of both particles by Kupffer cells was the result of opsonization by an unknown serum component (Moghimi et al., 1993c).

These studies could have important implications in biomedical applications, design and engineering of poloxamine-based long-circulating or splenotropic drug carriers for repeated intravenous administration.

PRIOR INTRAVENOUS ADMINISTRATION OF RELATED POLYMERS CAN ALSO AFFECT THE BODY DISTRIBUTION OF LONG-CIRCULATING BEADS

Prior intravenous administration of an endotoxin-free solution of poloxamer-188 also stimulated the hepatic and splenic clearance of poloxamine-908 coated beads from the blood in both rats and C3H/HeJ mice (Moghimi and Murray, 1996). Similarly to the previous case, the interval between the two injections was 3 days. In addition to this effect, phagocytic activation (superoxide anion production and degranulation) of human neutrophils by poloxamer-188 has also been reported (Ingram et al., 1992). Poloxamer-188 is a linear copolymer of 8400 molecular weight and consists of 30 POP units surrounded by two hydrophilic chains, each comprising of 52 POE units (Schmolka, 1977). This polymer has well-known effects on blood, reducing viscosity and improving circulation, particularly in ischemic tissue (Colbassani et al., 1989; Mayer et al., 1994). Silk and Sigman (1972) reported that intravenous administration of poloxamer-188 following inoculation of Walker 256 tumour cells

Table 3. Organ distribution of poloxamine-908 coated beads (60 nm) in rats pre-treated with designated polymers 3 to 4 days before intravenous administration of the particles

Treatment	Uptake (% of administered dose)			Total radioactivity recovered (% of administered dose)
	Liver	Spleen	Blood	
Saline	8.5 ± 0.5	2.9 ± 0.1	57.6 ± 3.1	71.9 ± 2.9
Poloxamine	34.6 ± 0.1	10.5 ± 0.8	19.4 ± 0.8	70.9 ± 2.6
Tween-80	31.9 ± 3.6	8.9 ± 1.2	18.7 ± 2.0	65.6 ± 3.6
PEG 1000	9.1 ± 2.5	3.1 ± 0.9	54.5 ± 1.0	66.7 ± 1.0
PEG 8000	12.8 ± 1.3	4.5 ± 0.3	50.4 ± 1.2	69.7± 2.9
PEG 20000	42.8 ± 1.1	7.8 ± 0.3	7.0 ± 1.5	69.7 ± 6.6

All polymers were free of endotoxins. Each group of animals (Wistar rats, n=3) was injected intravenously with designated polymers at 18.5 mg / kg body weight. Animals were sacrificed 3h after intravenous administration of poloxamine-coated beads.

decreased the incidence of pulmonary metastases; this effect was assumed to be secondary to the effects on blood coagulation. Therefore, in light of the above observed effects on macrophage functions by poloxamer-188, we believe that it may be appropriate to re-evaluate the use of POP/POE copolymers in the context of an adjuvant therapy to prime or stimulate macrophage function to destroy tumour cells. We are currently examining cytokine secretion, NO-synthase activity, oxygen radical production, and tumourcidal capacity of tissue macrophages following treatment with certain members of POP/POE surfactants.

Another interesting observation was the ability of nonionic surfactants which contain polyethoxyl hydrophilic side chains attached to hydrophobic moieties (e.g., Tween-80) and also PEGs to stimulate the clearance of long-circulating beads by macrophages of the liver and spleen. Earlier, Woodcock et al. (1992) demonstrated that Tween-80 can reverse drug exclusion by cells that exhibit the multidrug resistance phenotype. The blockage of drug efflux was

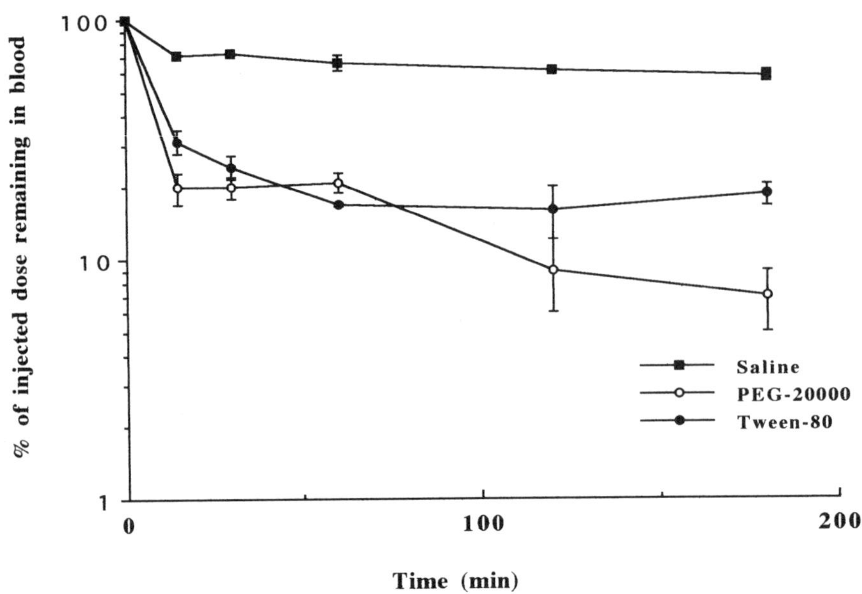

Figure 3. Blood clearance of poloxamine-908 coated nanospheres (60 nm) in control and surfactant-treated (Tween-80 and PEG-20000) rats. Results are presented as percentage of the administered dose ± SEM of 3 rats. Removal of any endotoxins associated with polymers was achieved by chromatography using polymyxin-B-agarose and confirmed by the Limulus lysate assay. Animals (male Wistar rats, 150 ± 10 g) were injected intravenously via the tail vein with 18.5 mg/kg body weight of the appropriate surfactants. Four days later, animals received intravenously poloxamine-coated radiolabelled latex beads (3.5 mg polystyrene/kg body weight).

apparently mediated by a substantial alteration in the fluidity of cell membrane induced by the surfactant. Fluorescence polarisation measurements are planned in order to determine alterations in the fluidity of macrophage plasma membrane and its effect on phagocytosis following treatment with Tween-80. Among various PEG molecules tested only those with high molecular weights (e.g., PEG-20000) were able to stimulate particle recognition by tissue macrophages. Following intravenous administration, PEG-20000 exhibits longer blood half-life in comparison to PEGs with low molecular weight and as a result it may have a better chance to exert an effect on cells of the immune system. Nevertheless, the above observations raises the issue of specificity which is difficult to address *in vivo*. Detailed *in vitro* investigations are necessary to unravel whether PEGs and POP/POE-based nonionic surfactants act on tissue macrophages themselves or their effect is secondary via other cell types (e.g., regional endothelial cells, hepatocytes).

REFERENCES

Bakker-Woudenberg, I.A.J.M., Lokerse, A.F., Ten Kate, M., Mouton, J.W., Woodle, M.C., and Storm, G., 1993, Liposomes with prolonged blood circulation and selective localization in *Klebsiella pneumoniae* infected lung tissue, *J. Infect Dis* 168: 164

Bloemen, P.G.M., Henricks, P.A.J., van Bloois, L., van den Tweel, M.C., Bloem, A.C., Nijkamp, F.P., Crommelin, D.J., and Storm, G., 1995, Adhesion molecules: a new target for immuno liposome-mediated drug delivery, *FEBS Lett* 357: 140

Colbassani H.J., Barrow D.L., Sweeny, K.M., Bakay, R.A., Check, I.J., and Hunter, R.L., 1989, Modification of acute focal ischemia in rabbits by poloxamer 188, *Stroke* 20: 1241

Gabizon, A.A., 1995, Liposome circulation time and tumor targeting: implications for cancer chemotherapy, *Adv Drug Deliv Rev* 16: 285

Geyer, R.P., 1967, Studies on the metabolism of intravenous fat emulsions, *Fette Med* 6: 59

Gordon, S., Lawson, L., Rabinowitz, S., Crocker, P.R., Morris, L., and Perry, V.H., 1992, Antigen markers of macrophage differentiation in murine tissues, *Curr Topic Microbiol Immunol 181*: 1

Griffin, F.M., Jr., and Mullinax, P.J., 1981, Augmentation of macrophage complement receptor function *in vitro*. II. C3b receptors that promote phagocytosis migrate within the plane of the macrophage plasma membrane, *J Exp Med* 154: 291

Hart, P.D., and Payne, S.N., 1971, Effects of non-ionic surfactants that modify experimental tuberculosis on lipase activity of macrophages, *Br J Pharmacol* 43: 190

Howerton, D.A., Hunter, R.L., Ziegler, H.K., and Check, I.J., 1990, Induction of macrophage Ia expression *in vivo* by a synthetic block copolymer, L81, *J Immunol* 144: 1578

Huang, X., Molema, G., King, S., Watkins, L., Edgington, T.S., and Thorpe, P.E., 1997, Tumor infarction in mice by antibody-directed targeting of tissue factor to tumor vasculature, *Science* 275: 547

Huang, S.K., Stauffer, P. R., Hong, K., Guo, J.W.H., Phillips, T.L., Huang, A., and Papahadjopoulos, D., 1994, Liposome hyperthermia in mice: increased tumor uptake and therapeutic efficacy of doxorubicin of sterically stabilized liposomes, *Cancer Res* 54: 2186

Hunter, R., Strickland, F., and Kézdy, F., 1981, The adjuvant activity of nonionic block polymer surfactants. I. The role of hydrophile-lipophile balance, *J Immunol 127:1244.*

Ingram, D.A., Forman, M., and Murray, J.J., 1992, Phagocytic stimulation of human neutrophils by the detergent component of Fluosol, *Am J Pathol* 140: 1081.

Jeppsson, R., and Rossner, S., 1975, The influence of emulsifying agents and of lipid soluble drugs on the fractional removal rate of lipid emulsions from the blood stream of rabbit, *Acta Pharmacol Toxicol* 37: 134

Kolb-Bachofen, V., and Abel, F., 1991, Participation of D-galactose-specific receptors of liver macrophages in recognition of fibronectin-opsonized particles, *Carbohydrate Res* 213: 201

Lee, S.H., Crocker, P., and Gordon, S., 1986, Macrophage plasma membrane and secretory properties in murine malaria. Effects of *Plasmodium yoelii* blood-stage infection on macrophages in liver, spleen, and blood, *J Exp Med* 163: 54

Lee, R.J., and Huang, L., 1996, Folate-targeted, anionic liposome-entrapped polylysine-condensed DNA for tumor cell-specific gene transfer, *J Biol Chem* 271: 8481

Lee, R.J., and Low, P.S., 1994, Delivery of liposomes into cultured KB cells via folate receptor-mediated endocytosis, *J Biol Chem* 269: 3198

Lee, J., Martic, P.A., and Tan, J.S., 1989, Protein adsorption on pluronic co-polymers coated polystyrene particles, *J Colloid Interface Sci* 131: 252

Lepay, D.A., Steinman, R.M., Nathan, C.F., Murray, H.W., and Cohen, Z.A., 1985, Liver macrophages in murine listeriosis. Cell-mediated immunity is correlated with an influx of macrophages capable of generating reactive oxygen intermediates, *J Exp Med* 161: 1503

Li, J.T., and Caldwell, K.D., 1994, Rapoport, N. Surface properties of pluronic-coated polymeric colloids, *Langmuir* 10: 4475

Mayer D.C., Strada S.J., Hoff, C., Hunter, R.L., and Artmann, M., 1994, Effects of poloxamer 188 in a rabbit model of hemorrhagic shock, *Annal Clin Lab Sci* 24: 302

Moghimi, S.M., 1995a, Mechanisms regulating body distribution of nanospheres conditioned with pluronic and tetronic block co-polymers, *Adv Drug Deliv Rev* 16: 183

Moghimi, S.M., 1995b, Mechanisms of splenic clearance of blood cells and particles: towards development of new splenotropic agents, *Adv Drug Deliv Rev* 17: 103

Moghimi, S.M., 1995c, Exploiting bone marrow microvascular structure for drug delivery and future therapies, *Adv Drug Deliv Rev* 17: 61

Moghimi, S.M., 1997, Prolonging the circulation time and modifying the body distribution of intravenously injected polystyrene nanospheres by prior intravenous administration of poloxamine-908. A reticuloendothelial blockade event or manipulation of nanosphere surface *in vivo*?, *Biochim Biophys Acta* 1336:1

Moghimi, S.M., and Gray, T.A., 1997, A single dose of intravenously injected poloxamine-coated long-circulating particles triggers macrophage clearance of subsequent doses in rats, *Clin Science* 93:371

Moghimi, S.M., Hawley, A.E., Christy, N.M., and Gray, T., 1994, Surface engineered nanospheres with enhanced drainage into lymphatics and uptake by macrophages of the regional lymph nodes, *FEBS Lett* 344, 25

Moghimi, S.M., Hedeman, H., Christy, N.M., Illum, L., and Davis, S.S., 1993a, Enhanced hepatic clearance of intravenously administered sterically stabilized microspheres in zymosan-stimulated rats, *J Leukoc Biol* 54: 513

Moghimi, S.M., Hedeman, H., Illum, L., and Davis, S.S. 1993b, Effect of splenic congestion associated with haemolytic anaemia on filtration of 'spleen-homing' microspheres, *Clin Science* 84: 605

Moghimi, S.M., Hedeman, H., and Muir, I.S., 1993c, An investigation of the filtration capacity and the fate of large filtered sterically-stabilized microspheres in rat spleen, *Biochim Biophys Acta* 1157: 233

Moghimi, S.M., Muir, I.S., Illum, L., Davis, S.S., and Kolb-Bachofen, V., 1993d, Coating particles with a block co-polymer (poloxamine-908) suppresses opsonization but permits the activity of dysopsonins in the serum, *Biochim Biophys Acta* 1179: 157

Moghimi, S.M., and Murray, J.C., 1996, Poloxamer-188 revisited: a potential valuable immune modulator?, *J Natl Cancer Inst* 88: 766

Moghimi, S.M., and Patel, H.M., 1998, Serum-mediated recognition of liposomes by phagocytes of the reticuloendothelial system - The concept of tissue specificity, *Adv Drug Deliv Rev* (in press)

Moghimi, S.M., Porter, C.J.H., Muir, I.S., Illum, L., and Davis, S.S., 1991, Non-phagocytic uptake of intravenously injected microspheres in rat spleen: influence of particle size and hydrophilic coating, *Biochem Biophys Res Commun* 177: 861

Moghimi, S.M., and Rajabi-Siahboomi, A.R., 1996, Advanced colloid based systems for efficient delivery of drugs and diagnostic agents to the lymphatic tissues, *Prog Biophys Mol Biol* 65: 221

Mulligan, M.S., Paulson, J.C., Frees, S.D., Zheng, Z., Lowe, J.B., and Ward, P.A., 1993, Protective effects of oligosaccharides in P-selectin-dependent lung injury, *Nature* 364: 149

Naper, D.H., 1983, Polymeric Stabilization of Colloidal Dispersions, Academic Press, New York

Needham, D., McIntosh, T.J., and Lasic, D.D., 1992, Repulsive interactions and mechanical stability of polymer-grafted lipid membranes, *Biochim Biophys Acta* 1108: 40

Nelson, R.M., Cecconi, O., Roberts, W.G., Aruffo, A., Linhardt, R.J., and Bevilacqua, M.P., 1993, Heparin oligosaccharides bind L- and P-selectin and inhibit acute inflammation, *Blood* 82: 3253

Norman, M.E., Williams, P., and Illum, L., 1992, Human serum albumin as a probe for protein adsorption (opsonisation) to block co-polymer coated microspheres, *Biomaterials* 13: 841

Norman, M.E., Williams, P., and Illum, L., 1992, Influence of block copolymers on the adsorption of plasma proteins to microspheres, *Biomaterials* 14: 193

Pasqualini, R., Koivunen, E., and Ruoslahti, E., 1997, αv Integrins as receptors for tumor targeting by circulating ligands, *Nature Biotech* 15:542

Schmolka, I.R., 1977, A review of block polymer surfactants, *J Am Chem Soc* 54: 110

Scott, M.D., Murad, K.L., Koumpouras, F., Talbot, M., and Eaton, J.W., 1997, Chemical camouflage of antigenic determinants: Stealth erythrocytes, *Proc.Natl.Acad.Sci.USA*, 94:7566

Silk, M., and Sigman, E., 1972, Effect of Pluronic-F68 on the development of tumor metastasis, *Cancer* 29:171

Torchilin, V.P., 1995, Targeting of drugs and drug carriers within the cardiovascular system, *Adv Drug Deliv Rev* 17: 75

Torchilin, V.P., Omelyanenko, V.G., Papisov, I.M., Bogdanov, A.A., Jr, Trubetskoy, V.S., Herron, J.N., and Gentry, C.A., 1994, Poly(ethylene glycol) on the liposome surface: on the mechanism of polymer-coated liposomes, *Biochim Biophys Acta* 1195: 11

Troster, S.D., Müller, U., and Kreuter, J., 1990, Modification of the body distribution of poly(methyl methacrylate) nanoparticles in rats by coating with surfactants, *Int J.Pharm* 61: 85

Vogel, S.N., and Rosenstreich, D.L., 1979, Defective Fc receptor-mediated phagocytosis in C3H / HeJ macrophages. I. Correction by lymphokine-induced stimulation, J Immunol 123: 2842

Watrous-Peltier, N., Uhl, J., Steel, V., Brophy, L., and Merisko-Liversidge, E., 1992, Direct suppression of phagocytosis by amphipathic polymeric surfactants, *Pharm Res* 9: 1177

Woodcock, D. M., Linsenmeyer, M. E., Chojnowski, G., Kriegler, A.B., Nink, V., Webster, L.K., and Sawyer, W.H., 1992, Reversal of multidrug resistance by surfactants, *Br J Cancer* 66: 62

Wu, N. Z., Da, D., Rudoll, T.L., Needham, D., Whorton, A.R., and Dewhirst, M.W., 1993, Increased microvascular permeability contributes to preferential accumulation of stealth liposomes in tumor tissue, *Cancer Res* 53:3765

Yuan, F., Leunig, M., Huang, S.K., Berk, D.A., Papahadjopoulos, D., and Jain, R., 1994, Microvascular permeability and interstitial penetration of sterically stabilized (Stealth) liposomes in a human tumor xenograft, *Cancer Res* 54: 3352

PEG-COATED NANOSPHERES : SURFACE OPTIMIZATION AND THERAPEUTIC APPLICATIONS

R. Gref[1], P. Quellec[1], M.J. Alonso[2], M. Tobio[2], M. Lück[3], R.H. Müller[3], E. Dellacherie[1]

[1]Laboratoire de Chimie Physique Macromoléculaire, ENSIC, 1, Rue Grandville bp 451, Nancy 54001, France
[2]Universidad de Santiago, Laboratorio de Farmacia Galénica, 15706 Santiago de Compostella, Spain
[3]Department of Pharmacy, Biopharmacy and Biotechnology, Freie University of Berlin, 12169 Berlin, Germany

INTRODUCTION

Biodegradable polyethylene glycol (PEG)-coated nanospheres have important therapeutic applications as injectable blood persistent particulate carriers for the controlled release of drugs, site-specific drug delivery, or medical imaging. They are obtained by emulsion/solvent evaporation (Gref et al., 1994), solvent displacement (Bazile et al., 1995), or block copolymer adsorption on preformed particles (Stolnick et al., 1994). The hydrophilic PEG coating reduces the surface charge, avoids plasma protein adsorption, minimizes the interaction with phagocytic cells and therefore increases the blood circulation time. The blood half-life of such particles is increased to up to six hours in rats (Bazile et al., 1995), compared to uncoated ones, which are removed within minutes, essentially by macrophages located in the liver and spleen. These recent nanospheres have essential advantages over other long-circulating systems, such as increased stability, resistance against plasma protein insertion in the matrix, and the possibility of obtaining controlled release, on the basis of an appropriate choice of the core composition.

However, mechanistic studies of PEG-coated nanoparticle formation using block copolymers, by emulsion (o/w) followed by solvent evaporation, and optimization of the coating in terms of coat density and thickness to minimize interactions with proteins and phagocytic cells, are challenging topics still to be explored. Our first aim was to study the surface properties of nanospheres coated with PEG with an average molecular weight (MW) of 2 to 20 Kg/mole, at different surface densities, with special attention to competitive plasma protein adsorption, analyzed by two-dimensional polyacrylamide gel electrophoresis (2D-PAGE). The particles with optimal coatings were further tested for drug encapsulation.

So far, only the encapsulation of hydrophobic drugs was reported in the case of PEG-coated nanospheres. Various hydrophobic drugs, among them lidocaine (Peracchia et al., 1997) and cyclosporin A (Quellec et al., 1996), were successfully entrapped in the biodegradable hydrophobic cores, with high efficiencies and high loadings.

Protein delivery is a promising area of research, due to the necessity to improve the therapeutic index of newly developed macromolecular drugs and antigens. Therefore, it was one of our main goals to explore the possibility of entraping albumin, a model protein, in PEG-coated nanospheres of less than 200 nm in diameter. Finally, we study the feasibility of encapsulating tetanus toxoid within PEG-coated nanospheres and releasing active antigen.

NANOSPHERE SURFACE OPTIMIZATION

PEG-coated nanospheres were prepared from diblock PEG-R (Gref et al., 1994; Bazile et al., 1995) or multibloc PEG_n-R (Peracchia et al., 1997) copolymers, where R is hydrophobic and biodegradable (eg. poly(lactic acid) (PLA), poly(lactic-co-glycolic acid) (PLGA), poly ε–caprolactone (PCL)), and PEG which is hydrophilic and can be eliminated from the body (MW less than 20K). These blocks have a tendency to phase-separate during nanosphere preparation, due to their mutual incompatibility and different affinity with water and organic solvent, leading to the formation of a hydrophobic core of R and a hydrophilic PEG brush orientated towards the dispersing aqueous phase. Evidence of this type of structure was given by ^{1}H NMR studies (Hrkach et al., 1997), complement activation studies (Vittaz et al, 1996) and PEG controlled detachment by alkaline hydrolysis (Bazile et al, 1995).

It is generally assumed that the PEG coating prevents opsonization by plasma proteins and subsequent recognition by the macrophages of the reticulo-endothelial system (Tan et al., 1993). The growing perception that the *in vivo* fate of the intravenously injected particles may be decisively influenced by the adsorption of plasma proteins initiated investigation on the type and nature of these proteins. Mathematical modeling of PEG chains attached in a brush configuration on hydrophobic surfaces, has shown that a long PEG chain length (i.e. high MW) and primarily a high PEG surface density are optimal conditions for protein resistance (Jeon et al., 1991).

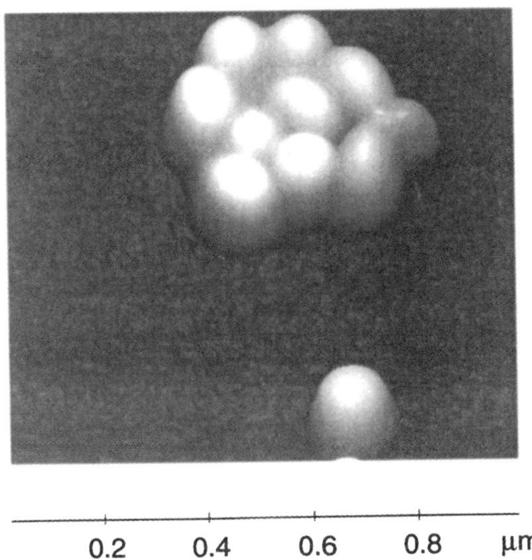

Figure 1. Atomic force microscopy (tapping mode) picture of PEG5K-PLA45K nanospheres

In a previous study, PEG-coated nanospheres prepared from various PEG-PLGA polymers showed a distinctly reduced plasma protein adsorption as compared to PLGA polymers alone (Gref et al., 1995). The PEG-PLGA polymers differed in MW of both PEG and PLGA blocks. To investigate the effect of PEG chain length and density in the coating brush, nanospheres were produced from a copolymer series of the type PEG-PLA45K (45K= MW 45000 g/mole) with different MW of PEG (2, 5, 10, 15 and 20K). This resulted in the particle types PEG2K-PLA45K, PEG5K-PLA45K, etc. Moreover, nanospheres were prepared from blends of PEG-PLA and PLA in order to vary the PEG surface density. In all cases, the nanospheres appear spherical and with a smooth surface, as can be seen on typical pictures (Fig. 1) taken by atomic force microscopy, a non-destructive technique.

Fig. 2 presents the size of nanospheres prepared from blends of PEG-(2 to 20K)PLA45K and PLA90K, from dichloromethane-in-water emulsions stabilized with sodium cholate (SC), an anionic surfactant of low MW. We chose this surfactant because of the possibility of removing it from the nanosphere surface by extensive washing, as shown by surface analysis studies (SIMS). Emulsion stabilization is obtained by steric repulsion (due to the PEG chains on the surface) and electrostatic repulsion (due to the presence of SC and COO^- endgroups of PLA). In the case of nanosphere formation using the homopolymer PLA, the ionic surfactant SC prevents droplet coalescence by electrostatic repulsion between the electrical double layers and steric repulsion between the hydrophilic heads of the surfactant.

The size of the nanospheres formed using PEG(2 or 5K)-PLA45K : PLA blends is practically constant, whatever the PEG content in the blend. In contrast, the size of nanospheres prepared from blends with longer PEG chains (10 and 20K) drastically increases up to a maximum at about 2% wt PEG, then decreases with increasing PEG content. So, the addition of PEG up to 2% wt destabilizes the emulsion, probably because of a competition between SC and PEG at the interface, resulting in a displacement of SC and a consequent decrease in the electrostatic repulsion between the droplets. However, by increasing the PEG content above 2% wt, the PEG surface density becomes high enough to prevent droplet coalescence by

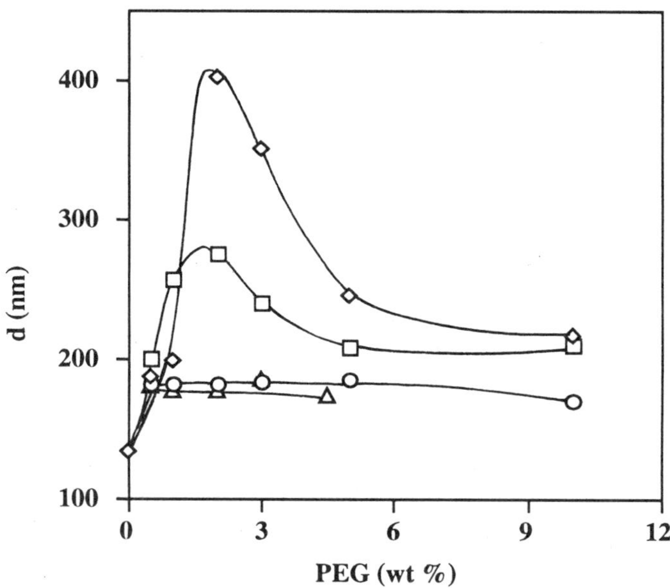

Figure 2. Influence of the PEG wt% in the matrix on the size of nanospheres prepared from PEG-PLA45K : PLA blends, where the molecular weight of PEG is 2K (△), 5K (○), 10K (□) and 20K (◇).

efficient steric stabilization. It is interesting to point out that the steric repulsion effect diminishes with the decrease in the PEG MW as predicted by the theories of steric stabilization (Jeon et al., 1991).

Plasma protein adsorption on PEG-coated particles was analyzed by two dimensional polyacrylamide gel electrophoresis (2D-PAGE). 2D-PAGE provides the unique possibility of simultaneously detecting hundreds of proteins from complex mixtures such as human plasma. The nanospheres were incubated in human plasma, then separated from excess plasma by centrifugation. Subsequently, the adsorbed plasma proteins were desorbed from the surface and applied to the 2D-PAGE analysis (Blunk et al., 1993; Blunk, 1994). 2D-PAGE was performed essentially as described by Hochstrasser et al. (1988) for standard plasma samples. In the first dimension, isoelectric focussing (IEF), the proteins were separated in a pH gradient according to their isoelectric point. In the second dimension, SDS-PAGE, separation was performed solely on the basis of the MW of the proteins. Routinely, silver-stained protein spots appearing on the 2D-PAGE patterns of adsorbed plasma proteins were identified by matching the obtained gels with the master map of human proteins (Golaz et al., 1993). Proteins of unknown identity were identified by means of N-terminal microsequencing.

Fig. 3 shows the effect of PEG chain length in the PEG-PLA45K polymers on the total amount of protein adsorbed on the nanospheres. The plasma protein adsorption on the PEG-coated nanospheres was drastically reduced as compared to the reference PLA50K nanospheres. The main protein spots detected in the adsorption patterns on PEG-PLA nanospheres were albumin, immunoglobulin (Ig) IgG, Ig light chains, and the apolipoproteins apoA-IV and apoE. The amount of protein detected on PEG2K-PLA45K was reduced to 57% of the amount adsorbed on PLA ones. A steep decrease was observed on increasing the PEG MW from 2 to 5K; the total protein amount was reduced by more than 50% as compared to PEG2K-PLA45K. However, on increasing the PEG MW to above 5K, only small changes in plasma protein adsorption were obtained. These results are consistent with previous studies

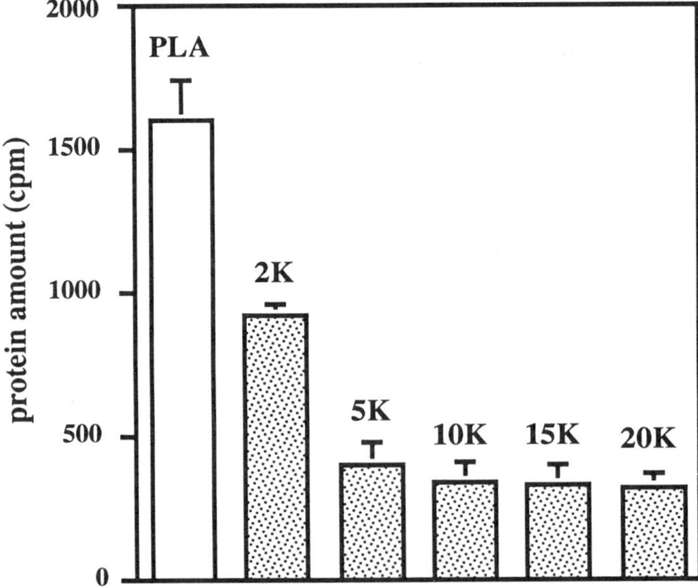

Figure 3. Total amounts of adsorbed proteins on PLA50K and PEG-PLA45K nanospheres with different PEG chain length (2; 5; 10; 15 and 20K). The protein amount is expressed in arbitrary units (cpm) (Appel et al., 1991). The values are the means of two experiments

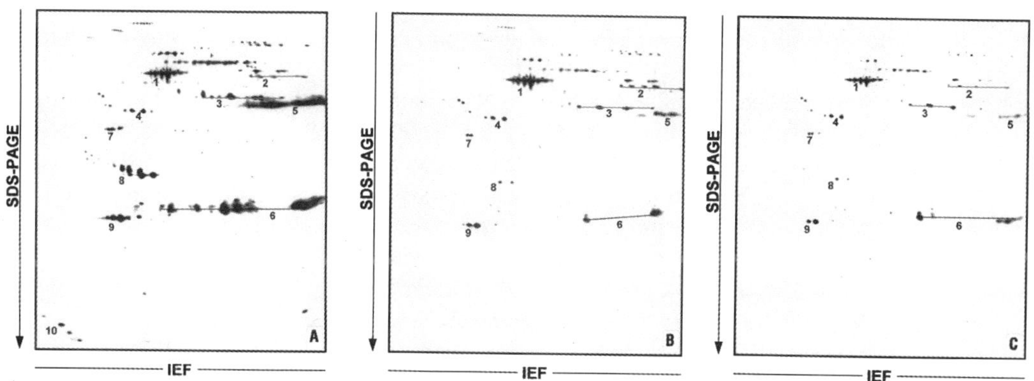

Figure 4. 2D-PAGE gels of adsorbed plasma proteins on nanospheres (of approx. 200 nm) prepared using PEG5K-PLA20K : PLA50K blends at 2 % (A), 5 % (B) and 8 % (C) weight % of PEG in the matrix. The whole gels are shown, pI 4.0 to 8.0 (from left to right, not linear), MW 6-250Kg/mole (from bottom to top, not linear).
(1) albumin, (2) fibrinogen a, (3) fibrinogen b, (4) fibrinogen g, (5) IgG g, (6) Ig light chains, (7) apoA-IV, (8) apoJ, (9) apoE, (10) apoA-I, (11) apoC-III.

(Gombotz et al., 1991), suggesting a PEG MW of approximately 3.5K as a critical value for the reduction of the adsorption of model proteins (albumin and fibrinogen) onto poly(ethylene terephtalate) hydrophobic substrates.

As a PEG MW of 5K appears in our study as a threshold for protein reduction in terms of coating thickness, it is further interesting to determine a threshold in terms of PEG5K surface density. For this, we studied protein adsorption on a series of nanospheres made from

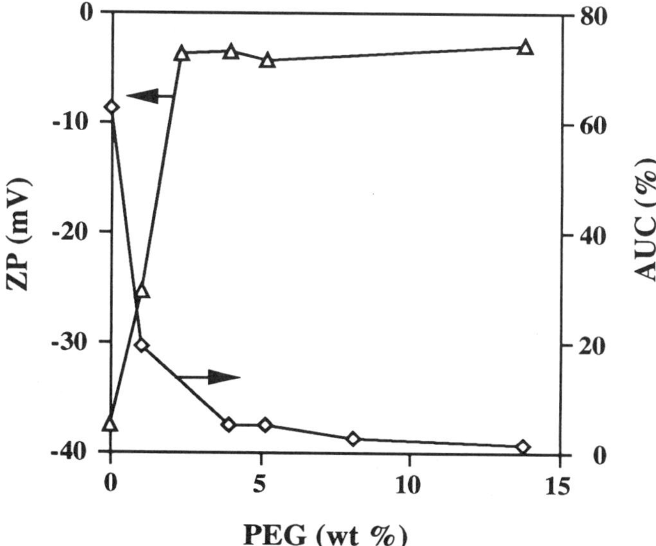

Figure 5. Influence of the PEG5K wt% on the Zeta potential (ZP) (Δ) and uptake (area-under-the-curve) (AUC) (◇) of nanospheres prepared from blends of PEG5K-PLA20K and PLA50K. ZP is measured in the same medium (PBS/glucose) where the uptake studies are performed using the chemiluminescence technique. AUC is expressed as % with regard to the reference (Zymosan).

PEG5K-PLA20K : PLA50K blends, containing 0. 5 to 20% wt PEG. A distinct reduction in plasma protein adsorption was detected for the PEG-coated particles, with PEG contents as low as 0.5% wt, compared to the PLA nanospheres, enhancing the hypothesis that the PEG chains migrate to the surface. Fig. 4 shows typical 2D-PAGE gels obtained at three different PEG surface densities, clearly showing a drastic protein reduction (by a factor of three) when the PEG content is increased from 2 to 5% wt. The most significant reduction was found for apoE. Also the amount of IgG and Ig light chains was clearly diminished. However, the total amount of proteins adsorbed could not be further reduced by increasing the PEG content above 5% wt and up to 20% wt; the plasma protein adsorption patterns of the particles containing 5% or more PEG were almost identical.

Assuming that all the PEG in the blends migrates to the surface to form the PEG brush, and taking into account particle size and density, we can calculate a surface density threshold for protein reduction corresponding to a distance of about 2.2 nm between two terminally attached PEG chains. We further checked if the PEG coating was effective in preventing nanosphere uptake by phagocytic cells. For this, we put into contact the above-studied nanospheres (with different PEG 5K surface densities) with human polymorphonuclear (PN) cells isolated from donor blood. The oxidative burst of PN cells following phagocytosis was quantified by chemiluminescence (CL) and the uptake of opsonized zymosan was used as a reference for all studies. The area under the curve (AUC), in arbitrary units, was calculated as the time integral of CL emission accumulated over 40 min.

Fig. 5 clearly shows the drastic reduction of the uptake of the nanospheres which contain PEG in amounts as small as 1% wt. The recognition is practically negligible above 4% wt. There is a good correlation between the zeta potential of the nanospheres and their interaction with the cells; the lower the surface charge, the lower the particle uptake. These observations support the hypothesis that, even at low concentrations, the PEG chains migrate preferentially to the nanosphere surface, sterically stabilize it, therefore masking the negative surface charge, reduce the zeta potential and ensure efficient protection against uptake. The threshold of the PEG5K surface density corresponds to the one previously determined for protein adsorption, by the 2D-PAGE studies (i.e. 4-5 wt%).

The PEG5K-coated nanospheres were further used for encapsulation studies. The polymer PEG5K-PLA45K (10% wt PEG) and its blends with PLA primarily attracted our attention.

THERAPEUTIC APPLICATIONS

Albumin entrapment

Lipophilic drugs, such as lidocain (Peracchia et al., 1997) and cyclosporin A (Quellec et al., 1996), were successfuly entrapped and released from PEG-coated nanospheres. The hydrophobic biodegradable cores of these particles were able to entrap these compounds with high efficiencies. The release was controlled by the physico-chemical properties of the polymer (Peracchia et al., 1997). In contrast, lipophilic drugs were difficult to entrap in liposomes, due to the possibility that these compounds, at high loading, destabilize the liposome bilayer membrane, whereas water-soluble drugs were more easily entrapped in the aqueous liposome inner core (Delattre et al., 1993). *A priori*, PEG-coated nanospheres, with hydrophobic dense cores, seem inappropriate for the encapsulation of water-soluble compounds using a water-in oil-in water ($w_1/o/w_2$) emulsion method, because of their high specific surface area, leading to a high probability of diffusion towards the outer aqueous phase (w_2) of the inner aqueous droplets (w_1) containing the drug. Therefore, PEG-coated nanospheres and liposomes appear as complementary with regard to drug encapsulation.

It was our goal to explore the possibility of entrapping highly water-soluble compounds, such as proteins, into PEG-coated nanospheres of a size of about 200 nm, with high enough loading and entrapment efficiencies, and to study the main parameters which govern drug release. For this, we used in a first step, a model protein, serum albumin. This protein is often chosen as a model for encapsulation studies (Youxin et al., 1994; Park et al., 1995; Sharif and O'Hagan, 1995; Cohen et al., 1991; Hora et al., 1990), due to its well established physico-chemical properties, low cost and the existence of previous litterature.

The nanosphere preparation method, using a w/o/w emulsion method previously described (Blanco and Alonso, 1997), was adapted from the microsphere preparation. Briefly, a w/o emulsion was prepared by sonicating a human serum albumin (HSA) solution (w_1) with a polymer solution in methylene chloride, then a $w_1/o/w_2$ emulsion was formed by sonication, with this first emulsion and an aqueous solution (w_2) containing a surfactant (SC). After solvent evaporation, the nanospheres were extensively washed to remove adsorbed surfactants and HSA. *In vitro* release studies were performed in phosphate buffer solution (PBS), pH 7.4, at 37°C.

Table 1 presents the influence of preparation conditions on the size of the resulting nanospheres. A size of about 200 nm was obtained at low polymer concentration (column c). Under identical preparation conditions and at low c (25 mg/ml), the sizes of the nanospheres prepared using PLA50K or PEG5K-PLA45K were very similar. The size is independent of the theoretical loading (L_t) (from 5 to 20 wt%), defined as the weight ratio of HSA and HSA and polymer used for the nanospheres preparation. The entrapment yield (Y), defined as the weight ratio of HSA entrapped in the polymer and HSA used for the nanosphere preparation, was calculated in assaying (Peterson, 1977) HSA remaining in the supernatant following nanosphere recovery by centrifugation. The real loading (L_r), or the weight of HSA effectively entrapped in a well defined mass of lyophilized nanospheres, was determined by destroying

Table 1. Influences of the type of polymer (PEG5K-PLA45K or PLA50K, or 3:1 blends), its concentration (c) in the organic phase and the theoretical loading (L_t) on the diameter of the nanospheres (d), the encapsulation yield (Y) and the real loading (L_r) of HSA.

Polymer	c (mg/ml)	L_t (wt%)	d (nm) ±5nm	Y (wt%) ±0.5	L_r (wt%) ±0.3
PEG-PLA	25	5	186	56.3	3.1
		10	209	57.1	6.6
		20	186	35.9	7.8
	100	5	452	81.7	4.6
		10	456	88.4	8.9
PLA	25	10	203	56.9	6.7
	100	5	367	96.4	7.2
		10	410	95.9	12.6
PEG-PLA : PLA 3 : 1	25	10	195	50.5	8.2
	100	5	321	90.5	5.1
		10	443	92.3	9.8

Table 1 shows that it is possible to entrap up to 12.6% wt HSA into PLA nanospheres. The loading is lower for PEG-PLA nanospheres, but still satisfactory; an upper limit is reached (about 8% wt) and it appears that increasing L_t above 10% does not increase L_r. The entrapment yield Y was high (82-96 %) for large nanospheres, and about 57% wt for 200 nm particles. Possibly, during the preparation of smaller particles, there is an enhanced diffusion of the inner aqueous phase (w_1) containing HSA towards the outer aqueous phase (w_2), leading to a loss of HSA. These results clearly show that it is possible to entrap high amounts of albumin into PEG-coated nanospheres. Moreover, surface analysis techniques (ESCA and SSIMS) enabled us to show that HSA was effectively entrapped inside the matrix and was not present in the upper layers of PEG-PLA nanospheres.

The cumulative HSA release from these particles is presented in Fig. 6. About 12 wt % HSA is released from PLA nanospheres in the first five hours, then the amount of HSA in the supernatant diminishes and stabilizes at about 7% wt. This can be explained by a readsorption of the released HSA on the hydrophobic surface. This hypothesis was checked by independent experiments where empty PLA nanospheres were put into contact with a HSA solution (results not shown): significant HSA amounts were adsorbed, whereas adsorption was almost undetectable on PEG-PLA particles.

In contrast to PLA nanospheres, HSA release from PEG-PLA nanospheres continuously increased, reaching about 10% wt in the first day, then progressing and reaching a plateau after ten days, at about 20% wt. Moreover, it was possible to further increase the release rate by using PEG-PLA : PLA blends; the most effective blend (3:1) enabled us to obtain 16% wt release in the first day (Fig. 6), and after ten days the release still continued and reached 30% wt. These nanospheres could also entrap HSA more efficiently (as shown in Table 1, for L_t equal to 10%, L_r was as high as 8.2% wt, compared to 6.7 % in the case of PLA or PEG-PLA).

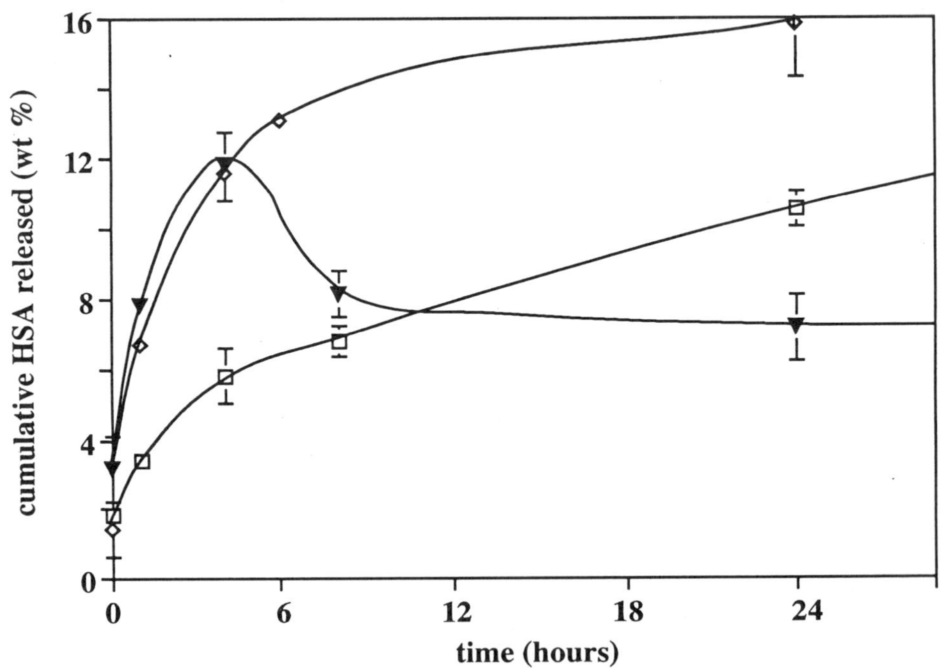

Figure 6. Release kinetics of HSA from nanospheres made of PLA45K (▽) (203 nm, L_r 6.7 %), PEG5K-PLA45K (□) (209 nm, L_r 6.6 %) and a blend PEG5K-PLA45K : PLA50K 3:1 (◇) (194 nm, L_r 8.2 %).

The different entrapment and release properties of PEG-PLA and PEG-PLA : PLA nanospheres could be explained in terms of differences in their internal structure, in particular the organization of the PEG chains during the emulsification and solvent evaporation steps.

During two weeks of release, no noticeable particle size or polymer molecular weight changes were detected. The PEG surface density of PEG-PLA and PEG-PLA : PLA 3:1 nanospheres was high enough to prevent their uptake by polymorphonuclear cells, as determined by chemiluminescence studies. Possibly, most of the PEG chains migrate to the particle surface to ensure this efficient protection. Moreover, the *in vitro* PEG detachment was relatively low (less than 10% detachment in one day) making these particles interesting for the long-term intravenous administration of proteins.

However, it is possible that the efficient HSA encapsulation is mostly due to hydrophobic interactions between the hydrophobic parts of the protein and the PLA chains. This hypothesis is supported by the fact that other macromolecular compounds such as dextran, with no hydrophobic regions, could not be entrapped in these nanospheres.

Vaccine entrappment

The above presented studies showed that it is possible to entrap albumin, a model protein into small size PEG-coated nanospheres and that the release was controlled by fabrication parameters and core composition. It is important now to determine if a therapeutically active protein can be entrapped and released in an active form. We investigated the feasability of encapsulating tetanus toxoid (TT) within nanospheres made of PEG5K-PLA45K and PLA50K. The previous work on the microencapsulation of tetanus vaccine into PLGA microspheres showed an interaction between TT and the polymer which resulted in the inactivation of the encapsulated vaccine (Swendeman et al., 1997).

It was our hypothesis that the presence of hydrophilic PEG chains in the matrix could reduce the interaction of the vaccine with PLA and its inactivation. In addition, taking into account the excellent properties of gelatin for the stabilization of TT (Swendeman et al., 1997), the co-encapsulation of gelatin in PEG-PLA nanospheres was also investigated. Gelatin was incorporated into the inner aqueous phase at a concentration of 2% (w/w in respect to polymer).

Table 2. Physicochemical properties of PLA50K and PEG5K-PLA45K tetanus toxoid-loaded nanospheres.

Polymer	**Stabilizer**	**Particle size (nm)**	**Zeta Potential (mV)**	**EE (wt%)**
PLA	----	153,1±3,1	-47,9±1,5	36,7±0,3
PLA-PEG	----	142,8±1,5	-30,1±4,2	31,1±0,5
PLA-PEG	Gelatin	136,8±1,1	-31,7±0,8	35,3±0,6

To prepare the nanospheres, ethyl acetate was used as a solvent and a w/o/w emulsion technique (Blanco and Alonso, 1997) was chosen in order to facilitate the orientation of the hydrophilic PEG chains towards the aqueous medium in which the protein is incorporated. The theoretical TT loading of the nanospheres was 1% (w/w in respect to polymer). For the determination of the real TT loading in the nanospheres, TT was radiolabelled with ^{125}I and the radioactivity was measured in a gamma counter. The vaccine encapsulation efficiency (EE) was calculated as the ratio of the real and the theoretical loadings. *In vitro* release studies of TT from PLA and PLA-PEG nanospheres were performed by incubating (37°C) the nanospheres in PBS (pH 7.4) containing 0.02% Tween® 80. At time intervals, samples were centrifuged and the supernatant removed for analysis. The determination of the antigenically active TT released was performed using an ELISA test.

The physicochemical properties of the nanospheres are presented in Table 2. These results show the feasibility of encapsulating significant amounts of TT (more than 30%) within nanospheres of a size less than 200 nm. In addition, it was found that nanosphere size and encapsulation efficiency were not affected by the presence of the PEG in the matrix or by the co-encapsulation of the stabilizer gelatin.

The *in vitro* release of antigenically active TT from the nanospheres is shown in Table 3. The presence of the hydrophilic PEG blocks led to an increase of the active antigen released over the first week, however, it had no effect on the long term release of antigenically active TT. Nevertheless, the incorporation of the stabilizer gelatin in the PEG-PLA nanospheres enhanced remarkably the stability of the encapsulated toxoid. In fact, these novel nanospheres released significant amounts of active antigen for at least 28 days. Consequently, TT-loaded PEG-PLA nanospheres can be considered as potential TT carriers for nasal administration.

Table 3. Distributive percentages of antigenically active TT relased *in vitro* after 1 to 28 days from PLA50K and PEG5K-PLA45K nanospheres containing or not gelatin as a stabilizer.

Polymer	**Stabilizer**	**Antigenically active TT released (wt %)**						
		1 d	**4 d**	**8 d**	**12 d**	**16 d**	**21 d**	**28 d**
PLA	----	7,39 ±1,56	1,50 ±0,04	0,35 ±0,04	0,60 ±0,13	0,32 ±0,06	n.d†.	n.d.
PLA-PEG	----	6,22 ±0,84	6,42 ±0,01	1,69 ±0,56	0,48 ±0,01	0,29 ±0,01	n.d.	n.d.
PLA-PEG	Gelatin	18,21 ±2,50	1,06 ±0,03	0,61 ±0,02	1,33 ±0,49	0,62 ±0,16	0,27 ±0,01	1,61 ±0,02

†not detected.

CONCLUSION

Nanospheres were prepared from o/w emulsions, with a PEG coating brush of various PEG density and thickness. Surface optimization studies carried out by 2D-PAGE showed that a PEG MW of 5K is required to reduce the total amount of plasma proteins adsorbed. Further, a minimal surface density corresponding to about 2.2 nm between the PEG 5K chains is needed, to reduce simultaneously the amount of proteins adsorbed and nanosphere zeta potential and uptake by the polymorphonuclear cells.

We studied the possibility to encapsulate proteins in PEG5K-coated nanospheres. Albumin, a model protein was entrapped up to 9% wt in PEG-PLA nanospheres, and was released continualy over several days. Blends of PEG-PLA and PLA enabled us to increase both the release rate and the entrapment efficiency.

Further, a vaccine, tetanus toxoid, was entrapped with high efficiency (more than 30% wt). The presence of PEG chains increased the amount of active antigen released over the first week. However, the use of gelatin was needed to obtain significant amounts of vaccine for at least 28 days. Preliminary data indicate that these nanospheres are able to cross some mucosal surfaces, such as nasal mucosa, to a higher extent than the PLA ones. With further studies, we believe these types of nanospheres will find many potential applications for intravenous macromolecular compound administration, as well as for other routes of administration. Further research will have to be carried out to check the feasability of using the PEG-coated nanoparticles as drug and vaccine carriers through the mucosa.

REFERENCES

Appel, R.D., Hochstrasser, D.F., Funk, M., Vargas, J.R., Pellegrini, C., Muller, A.F. and Scherrer, J.R., 1991, The MELANIE project: From a biopsy to automatic protein map interpretation by computer, *Electrophoresis* 12:722.

Bazile, D., Prud'homme, C., Bassoullet, M.T., Marlard, M., Spenlehauer, G. and Veillard, M., 1995, Stealth MePEG-PLA nanoparticles avoid uptake by the mononuclear phagocytes system, *J. Pharm. Sci.* 84:493.

Blanco, M.D. and Alonso, M.J., Development and characterization of protein-loaded poly(lactide-co-glycolide) nanospheres, *Eur. J. Pharm. Biopharm.*, in press

Blunk, T., Hochstrasser, D.F., Sanchez, J.C., Müller B.W., and Müller, R.H., 1993, Colloidal carriers for intravenous drug targeting: Plasma protein adsorption patterns on surface-modified latex particles evaluated by two-dimensional polyacrylamide gel electrophoresis, *Electrophoresis*, 14:1382.

Blunk, T., 1994, Ph. D. thesis, University of Kiel, Germany.

Cohen, S., Yoshioka, T., Lucarelli, M., Hwang, L.H. and Langer, R., 1991, Controlled delivery systems for proteins based on poly(lactic/glycolic acid) microspheres, *Pharm. Res.*, 8:713.

Delattre, J., Couvreur, P., Puisieux, F. and Philippot, J.-R., 1993, Les liposomes, aspects technologiques, biologiques et pharmacologiques, Ed. InanospheresERM, Tec &Doc Lavoisier, Paris.

Golaz, O., Hughes, G.J., Frutiger, S., Paquet, N., Bairoch, A., Pasquali, C., Sanchez, J.-C., Tissot, J.-D., Appel, R.D., Walzer, C., Balant, L., and Hochstrasser, DF., 1993, Plasma and red blood cell protein maps: Update 1993, *Electrophoresis* 14:1223.

Gombotz, W.R., Guanghui, W., Horbett, T.A., and Hoffmann, A.S., 1991, Protein adsorption to poly (ethylene oxide) surfaces, *J. Biomed. Mater. Res.*, 25:1547.

Gref, R., Minamitake, Y., Peracchia, M.T., Trubetskoy, V., Torchilin, V., and Langer, R., 1994, Biodegradable long-circulating nanospheres, *Science* 263:1600.

Gref, R., Domb, A., Quellec, P., Blunk, T., Müller, R.H., Verbavatz, J.M., and Langer, R., 1995, The controlled intravenous administration of drugs using PEG-coated sterically stabilized nanospheres, *Adv. Drug Deliv. Rev.* 16:215.

Hochstrasser, D.F., Harrington, M.G., Hochstrasser, A.-C., Miller, M.J., and Merril, C.R., 1988, Methods for increasing the resolution of two-dimensional protein electrophoresis, *Anal. Biochem.*, 173:424.

Hora, M.S., Rana, R.K., Nunberg, J.H., Tice, T.R., Gilley, R.M. and Hudson, M.E., 1990, Release of human serum albumin from poly(lactide-co-glycolide) microspheres, *Pharm. Res.*, 7:1190.

Hrkach, J.S., Peracchia M.T., Domb, A., Lotan N., and R. Langer, Nanotechnology for biomaterials engineering: structural characterization of amphiphilic polymeric nanospheres using 1H-NMR

spectroscopy, *Biomaterials*, in press
Jeon, S.I., Lee, J.H., Andrade, J.D., and De Gennes, P.G., 1991, Protein-surface interactions in the presence of polyethylene oxide, I. Simplified theory, *J. Colloid Interf. Sci.*, 142:149.
Park, T.G., Lu, W. and Crotts, G., 1995, Importance of the in vitro experimental conditions on protein release kinetics, stability and polymer degradation in protein encapsulated P(DL-LA-co-GA) MS, *J. Controlled Rel.* 33:211.
Peracchia, M.T., Gref, R., Minamitake, Y., Domb, A., Lotan, N., and Langer, R., 1997, PEG-coated nanospheres from amphiphilic diblock and multiblock copolymers: investigation of their drug encapsulation and release characteristics, *J. Controlled Rel.* 46:223.
Peterson, G.L., 1977, A simplification of the protein assay method of Lowry and al. which is more generally applicable, *Anal. Biochem.*, 83:346.
Quellec, P., Gref, R., Calvo, P., Alonso, M.J., and Dellacherie, E., 1996, Encapsulation of a model protein and of a hydrophobic drug into long-circulating biodegradable nanospheres, *Procced. Intern. Symp. Control. Rel. Bioact. Mater.*, 23:815.
Sharif, S. and O'Hagan, DS., 1995, A comparison of alternative methods for the determination of the levels of proteins entrapped in PLGA microparticles, *Int. J. Pharm.* 115:259.
Stolnick, S., Dunn, S.E., Garnett, M.C., Davies, M.C., Coombes, A.G.A., Taylor, D.C., Irving, M.P., Purkiss, S.C., Tadros, T.F., Davis, S. and Illum, L., 1994, Surface modification of poly(lactide-co-glycolide) nanospheres by biodegradable poly(lactide)-poly(ethylene glycol) copolymers, *Pharm. Res.* 11:1800.
Swendeman, S.P., Tobio, M., Joworowicz, M., Alonso, M.J. and Langer, R., New strategies for the micro-encapsulation of tetanus vaccine, *J. Microencapsulation*, in press.
Tan, J., Butterfield, D., Voycheck, C., Caldwell, C. and Li, J., 1993, Surface modification of nanoparticles by PEO/PPO block copolymers to minimize the interactions with blood components and prolong blood circulation in rats, *Biomaterials*, 14:823.
Vittaz, M., Bazile, D., Spenlehauer, G., Verrecchia, T., Veillard, M., Puisieux, F., and Labarre, D., 1996, Effect of PEO surface density on long-circulating PLA-PEO nanoparticles which are very low complement activators, *Biomaterials*, 17:1575.
Youxin, L., Volland, C. and Kissel, T., 1994, *In vitro* degradation and BSA release of P(L-LA)-POE-P(L-LA) and PLGA-POE-PLGA, *J. Controlled Rel.*, 32:121.

Participants of the NATO Advanced Studies Institute "Targeting of Drugs: Strategies for Stealth Therapeutic Systems" held at Cape Sounion Beach, Greece, during 24 June - 5 July 1997. The Organizing Committee included Gregory Gregoriadis (ASI Director), Patrick Couvreur, Daan Crommelin, Brenda McCormack and Demetrios Papahadjopoulos.

CONTRIBUTORS

M.J. Alonso, Universidad de Santiago, Laboratorio de Farmacia Galénica, 15706 Santiago de Compostella, Spain

T.M. Allen, Department of Pharmacology, University of Alberta, Edmonton, Canada T6G 2H7

J. d'Angelo, URA CNRS 1843, Faculty of Pharmacy, University of Paris-Sud XI, France

I.A.J.M. Bakker-Woudenberg, Department of Clinical Microbiology, Erasmus University, Rotterdam, The Netherlands;

Y. Barenholz, Department of Biochemistry, The Hebrew University-Hadassah Medical School, P.O.B. 12272, Jerusalem 91120, Israel

G.Barratt, URA CNRS 1218, Centre d'Etudes Pharmaceutiques, Université Paris-Sud, Chatenay-Malabry, France

C.C. Benz, Division of Hematology and Oncology, Cancer Research Institute, Department of Cellular and Molecular Pharmacology, Liposome Research Laboratory, CPMCRI, San Francisco, CA, USA

O.C. Boerman, Department of Nuclear Medicine, University Hospital, Nijmegen, The Netherlands

K.D. Caldwell, Center for Biopolymers at Interfaces, Department of Bioengineering, University of Utah, Salt Lake City, UT 84112, USA

F.H.M. Corstens, Department of Nuclear Medicine, University Hospital, Nijmegen, The Netherlands

P. Couvreur,URA CNRS 1218, Faculty of Pharmacy, University of Paris-Sud XI, France

D.J.A. Crommelin, Utrecht Institute for Pharmaceutical Sciences, Utrecht University, P.O.Box 80.082, 3508 TB Utrecht, The Netherlands

T. Daemen, Department of Physiological Chemistry, Groningen Institute for Drug Studies,University of Groningen, Medical Sciences, Antonius Deusinglaan 1, 9713 AV Groningen, The Netherlands

E. Dellacherie, Laboratoire de Chimie Physique Macromoléculaire, ENSIC, 1, Rue Grandville bp 451, Nancy 54001, France

D. Desmaële, URA CNRS 1843, Faculty of Pharmacy, University of Paris-Sud XI, France

J.-P. Devissaguet, URA CNRS 1218, Centre d'Etudes Pharmaceutiques, Université Paris-Sud, Chatenay-Malabry, France

R. Duncan, Centre for Polymer Therapeutics, The School of Pharmacy, University of London, 29-39 Brunswick Square, London WC1N 1AX, UK

E. Estey, Department of Hematology, The University of Texas M.D. Anderson Cancer Center, Houston, TX 77030, USA

E. Fattal, URA CNRS 1218, Faculty of Pharmacy, University of Paris-Sud XI, France

A. Fernandes, Centro de Tecnologia Farmaceutica, Inst Superior de Ciencias da Saude - Sul, Quinta da Granja, Travessa da Granja, 2825 Monte da Caparica, Portugal

P. Ferruti, Dipartimento di Chimica Organica e Industriale, Università di Milano, Via Venezian 21, 20133 Milano, Italy

A. Gabizon, Department of Oncology, Hadassah Hebrew University, Hospital, Jerusalem, Israel

J. Gittelman, SEQUUS Pharmaceuticals, Inc., 960 Hamilton Court, Menlo Park, California 94025, USA

S. Gordon, Sir William Dunn School of Pathology, University of Oxfor, South Parks Road, Oxford OX1 3RE, UK

D. Goren, Department of Oncology, Hadassah Hebrew University, Hospital, Jerusalem, Israel

R. Gref, Laboratoire de Chimie Physique Macromoléculaire, ENSIC, 1, Rue Grandville bp 451, Nancy 54001, France

G. Gregoriadis, Centre for Drug Delivery Research, The School of Pharmacy, University of London, 29-39 Brunswick Square, London WC1N 1AX, UK

C.B. Hansen, Department of Pharmacology, University of Alberta, Edmonton, Canada T6G 2H7

J.A. Harding, SEQUUS Pharmaceuticals, Inc., 960 Hamilton Court, Menlo Park, California 94025, USA

D. Hirsch-Lerner, Department of Biochemistry, The Hebrew University–Hadassah Medical School, PO Box 12272, Jerusalem, Israel;

K. Hong, Department of Cellular and Molecular Pharmacology, Liposome Research Laboratory, CPMCRI, San Francisco, CA, USA

A.T. Horowitz, Department of Oncology, Hadassah Hebrew University, Hospital, Jerusalem, Israel

J.A.A.M. Kamps, Department of Physiological Chemistry, Groningen Institute for Drug Studies,University of Groningen, Medical Sciences, Antonius Deusinglaan 1, 9713 AV Groningen, The Netherlands

D. Kirpotin, Department of Cellular and Molecular Pharmacology, Liposome Research Laboratory, CPMCRI, San Francisco, CA, USA

G.A. Koning, Department of Physiological Chemistry, Groningen Institute for Drug Studies,University of Groningen, Medical Sciences, Antonius Deusinglaan 1, 9713 AV Groningen, The Netherlands

K. Kostarelos, Farmeco Co., 32 Ag. Glykerias Str., Galatsi 11146, Athens, Greece

J. Kreuter, Institut für Pharmazeutische Technologie, Biozentrum, J. W. Goethe-Universität, D-60439 Frankfurt, Germany

R. Krishna, Department of Advanced Therapeutics, BC Cancer Agency, 600 West 10 Avenue, Vancouver, BC, V5Z 4E6, Canada

D. Labarre, URA CNRS 1218, Centre d'Etudes Pharmaceutiques, Université Paris-Sud, Chatenay-Malabry, France

D. Lopes de Menezes, Department of Pharmacology, University of Alberta, Edmonton, AB, Canada T6G 2H7

G. Lopez-Berestein, Immunobiology and Drug Carriers Section, Department of Bioimmunotherapy and Department of Hematology, The University of Texas M.D. Anderson Cancer Center, Houston, TX 77030, USA

M. Lück, Department of Pharmacy, Biopharmacy and Biotechnology, Freie University of Berlin, 12169 Berlin, Germany

M. Mital, Centre for Drug Delivery Research, The School of Pharmacy, University of London, 29-39 Brunswick Square, London WC1N 1AX, UK

L.D. Mayer, Department of Advanced Therapeutics, BC Cancer Agency, 600 West 10 Avenue, Vancouver, BC, V5Z 4E6, Canada

B. McCormack, Centre for Drug Delivery Research, The School of Pharmacy, University of London, 29-39 Brunswick Square, London WC1N 1AX, UK

K. Mehta, Immunobiology and Drug Carriers Section, Department of Bioimmunotherapy and Department of Hematology, The University of Texas M.D. Anderson Cancer Center, Houston, TX 77030, USA

O. Meyer, Department of Cellular and Molecular Pharmacology, Liposome Research Laboratory, CPMCRI, San Francisco, CA, USA

E.H. Moase, Department of Pharmacology, University of Alberta, Edmonton, Canada T6G 2H7

S.M. Moghimi, Department of Pharmacy, University of Brighton, Brighton BN2 4GJ, UK

N. Mullah, SEQUUS Pharmaceuticals, Inc., 960 Hamilton Court, Menlo Park, California 94025, USA

R.H. Müller, Department of Pharmacy, Biopharmacy and Biotechnology, Freie University of Berlin, 12169 Berlin, Germany

W.J.G. Oyen, Dept of Nuclear Medicine, University Hospital, Nijmegen, The Netherlands

D. Papahadjopoulos, Department of Cellular and Molecular Pharmacology, Liposome Research Laboratory, CPMCRI, San Francisco, CA, USA

J.W. Park, Division of Hematology/Oncology, Cancer Research Inst, Department of Cellular and Molecular Pharmacology, Liposome Research Laboratory, CPMCRI, San Francisco, CA, USA

C. Passirani, URA CNRS 1218, Centre d'Etudes Pharmaceutiques, Université Paris-Sud, Chatenay-Malabry, France

M.T. Peracchia,URA CNRS 1218, Faculty of Pharmacy, University of Paris-Sud XI, France

W.T. Phillips, Department of Radiology, University of Texas Health Science Center at San Antonio, San Antonio, Texas 78284, USA

A. Priev, Department of Biochemistry, The Hebrew University-Hadassah Medical School, P.O.B. 12272, Jerusalem 91120, Israel

M.M. Qazen, SEQUUS Pharmaceuticals, Inc., 960 Hamilton Court, Menlo Park, California 94025, USA

P. Quellec, Laboratoire de Chimie Physique Macromoléculaire, ENSIC, 1, Rue Grandville bp 451, Nancy 54001, France

S. Richardson, Centre for Polymer Therapeutics, The School of Pharmacy, University of London, 29-39 Brunswick Square, London WC1N 1AX, UK

A.M. Samuni, Department of Biochemistry, The Hebrew University-Hadassah Medical School, P.O.B. 12272, Jerusalem 91120, Israel

G.L. Scherphof, Department of Physiological Chemistry, Groningen Institute for Drug Studies,University of Groningen, Medical Sciences, Antonius Deusinglaan 1, 9713 AV Groningen, The Netherlands

R.M. Schiffelers, Department of Pharmaceutics, Utrecht Institute for Pharmaceutical Sciences, Utrecht University, Utrecht, The Netherlands

Y. Shao, Department of Cellular and Molecular Pharmacology, Liposome Research Laboratory, CPMCRI, San Francisco, CA, USA

B. Sternberg, Research Institute and California Pacific Medical Center, 2340 Clay Street, San Francisco, CA 94115, USA

G. Storm, Utrecht Institute for Pharmaceutical Sciences, Utrecht University, P.O.Box 80.082, 3508 TB Utrecht, The Netherlands

O. Tirosh, Department of Biochemistry, The Hebrew University-Hadassah Medical School, P.O.B. 12272, Jerusalem 91120, Israel

M. Tobio, Universidad de Santiago, Laboratorio de Farmacia Galénica, 15706 Santiago de Compostella, Spain

C. Vauthier, URA CNRS 1218, Faculty of Pharmacy, University of Paris-Sud XI, France

S. Zalipsky, SEQUUS Pharmaceuticals, Inc., 960 Hamilton Court, Menlo Park, California 94025, USA

X. Zhang, Centre for Drug Delivery Research, The School of Pharmacy, University of London, 29-39 Brunswick Square, London WC1N 1AX, UK

W. Zheng, Department of Cellular and Molecular Pharmacology, Liposome Research Laboratory, CPMCRI, San Francisco, CA, USA

N.J. Zuidam, Department of Pharmaceutics, Utrecht University, Utrecht, The Netherlands

INDEX